Lecture Notes in Economics and Mathematical Systems

466

Springer
Berlin
Heidelberg
New York
Barcelona
Budapest
Hong Kong
London
Milan
Paris
Singapore
Tokyo

Abdolkarim Sadrieh

The Alternating Double Auction Market

A Game Theoretic
and Experimental Investigation

 Springer

381.17
S/2a

Author

Abdolkarim Sadrieh
University of Bonn
Laboratorium für experimentelle Wirtschaftsforschung
Adenauerallee 24 – 42
D-53113 Bonn, Germany

Library of Congress Cataloging-in-Publication Data

Sadrieh, Abdolkarim, 1965-
   The alternating double auction market : a game theoretic and
experimental investigation / Abdolkarim Sadrieh.
      p.   cm. -- (Lecture notes in economics and mathematical
systems, ISSN 0075-8442 ; 466)
   Includes bibliographical references (p.   ).
   ISBN 3-540-64895-X (softcover : alk. paper)
   1. Auctions--Mathematical models.   2. Economics, Mathematical.
3. Game theory.   I. Title.   II. Series.
   HF5476.S23 1998
   381'.17--dc21   98-37922
                                                                  CIP

ISSN 0075-8442
ISBN 3-540-64895-X Springer-Verlag Berlin Heidelberg New York

© Springer-Verlag Berlin Heidelberg 1998
Printed in Germany

The use of general descriptive names, registered names, trademarks, etc. in this publication does not imply, even in the absence of a specific statement, that such names are exempt from the relevant protective laws and regulations and therefore free for general use.

Typesetting: Camera ready by author
SPIN: 10649660   42/3143-543210 - Printed on acid-free paper

Acknowledgements

I dedicate this work to Reinhard Selten,
who is not only a true pioneer of scientific thought,
but also an outstanding teacher, advisor, co-researcher, and friend.
I am greatly indebted to him, for all his help and patience.

Many very special thanks to Klaus Abbink and Bettina Kuon for their important comments and helpful support.

Many thanks to Jan Vleugels for his extensive and helpful e-mail comments.

I also thank all my other colleagues at the *Laboratorium für experimentelle Wirtschaftsforschung* of the University of Bonn.

I deeply thank my mother for her seemingly endless patience and care.
And thanks to all my other family members for encouraging me so often.

I thank Dan Friedman for his helpful advice and Urs Schweizer for his support.

Many thanks also to Werner Hildenbrand and especially to Heike Hennig-Schmidt for the support throughout the years.

Finally, I very gratefully acknowledge the financial support by the Deutsche Forschungsgemeinschaft through the Sonderforschungbereich 303.

Table of Contents

I.   Introduction   1

II. The Alternating Double Auction Market Game   6

III. The Simplified Alternating Double Auction Market
    - A Game Theoretic Analysis   14

    A. Introduction   14

    B. Definitions and Lemmas   19

    C. 1-Trade Markets   51

       1. 1-Trade Markets Without Extra-Marginal Traders   52

       2. 1-Trade Markets With Extra-Marginal Traders on Only One Market Side   56

       3. 1-Trade Markets With Extra-Marginal Traders on Both Market Sides   80

    D. 2-Trade Markets Without Extra-Marginal Traders   95

    E. M-Trade Markets   118

    F. Résumé and Discussion   166

IV. An Experimental Investigation of the Alternating Double Auction Market   177

    A. Introduction   177

    B. The Experimental Setup   178

    C. The Results of the Experimental Investigation   187

       1. Benchmarks for Evaluating the Experimental Results   188

          a. The Walrasian Competitive Equilibrium   188

          b. Game Theoretic Approaches   189

          c. The Zero-Intelligence Traders Approach   192

       2. Market Efficiency   198

       3. Convergence of Prices to the Market Clearing Price   199

       4. Direction of Price Convergence   207

       5. Price Path Repetition   211

       6. Distribution of Contract Proposals   218

       7. The Proposer's Curse Phenomenon   221

       8. Bounded Rationality Approaches   235

    D. The Anchor Price Hypothesis - A Concluding Summary and Evaluation   249

V.   Concluding Remarks   258

VI. References   261

VII. Appendix   265

    A. Experimental Handouts and Screen Shots   265

    B. Experimental Data Panels and Figures   273

    C. Zero-Intelligence Trader Simulation Data Panels   301

I. Introduction

Double auction markets are fascinating institutions. They establish a simple set of rules that organize the interaction of a large number of individual traders. In spite of the diversity of choice and the complexity of interaction, the markets bring on efficient allocations. This is most probably true in the double auction markets in business and finance, but certainly true in experimental double auction markets.

Although the volume of research on double auction markets is quite impressive, little is known about the forces that drive the markets towards their fascinating outcomes. Most of the investigations treat the internal micro-economics of the double auction market as a black-box, simply placing trust in the predictions of the Walrasian competitive equilibrium. This trust has expanded ever since SMITH (1962) reported the seminal results of his first double auction market experiment, namely that experimental double auction markets exhibit an extremely high efficiency and a remarkable convergence to the market clearing price.

But, ignoring the interaction and the learning that takes place in the course of trading can be very costly. Any trader in a *real world* commodities market and any subject of a double auction market experiment will agree: high overall efficiency and a long term convergence of prices to the market clearing price are no comfort at all, after an important trade was missed or an exorbitant price was paid. In other words, *getting to equilibrium* is economically often of much more interest than *being in equilibrium*. This is the reason why, the investigation presented in the following is mainly about the path to the equilibrium of a double auction market.

Before we can start investigating the path to equilibrium, however, we first have to know where we are heading to. This is another largely ignored problem with the double auction market institution: there is no definite answer to the question what features the equilibrium of the game actually has. There are two good reasons why the game theoretic equilibrium of the double auction market game has not been scrutinized more intensively yet: the game is extremely complex and the game is not properly specified.

The paper by WILSON (1987) is the only investigation we know of, in which the attempt is made to overcome both difficulties.[1] WILSON presents "certain necessary conditions" (p. 376) for a symmetric Markov perfect sequential equilibrium of a double auction market game with incomplete information. The key idea is that the traders signal their values by choosing appropriate delays before submitting a *serious* offer. The existence of competing traders on the own market side determines the general size of delays. One main feature of the proposed equilibrium is that buyers and sellers are matched, one after the other, exactly in the order of their valuations: First, the buyer with the highest redemption value trades with the seller with the lowest unit cost. Next, the two second ranked traders trade, and so forth. Another feature is that, as the number of traders is increased, the efficiency in equilibrium rises and the prices come closer to the market clearing price.

Unfortunately, WILSON only specifies the *induction step* for the construction of the proposed equilibrium. The starting point needed for the recursive construction (the solution to the *endgame*, in which a single buyer and a single seller remain in the market) is left unspecified. Although he provides suggestions concerning the problem, the construction of the proposed equilibrium - in its presented form - for concrete market parameters does not seem possible. (Additionally, WILSON's model is described by a set of partial differential equations, about which CASON and FRIEDMAN (1994, p. 7) write: "No numerical algorithms presently are available to solve the equations, even for simple value distributions and auxiliary assumptions.")

---

1) FRIEDMAN (1984) also derives very interesting results for the double auction market institution by examining the game under special conditions, the most important being the *no-congestion equilibrium*. His main result is that, with discrete values under the conditions examined, the market outcome will be *almost* Pareto optimal and the last trade *close* to the market clearing price. Unlike the analysis in WILSON (1987) and the analysis presented here, however, that paper is not an attempt to present a fully characterized game theoretic solution of a double auction market game.

Apart from the difficulties mentioned above, we believe that, for the purposes of this investigation, a complete information game approach to the double auction market game is of greater interest than the incomplete information game approach. The main reason for being interested in the equilibria of the game with complete information is that most double auction market experiments allow for static repetitions of the game, typically called *market periods*. Market periods with identical demand and supply curves are run in order to *allow the prices to converge*. This - on the level of individual behavior - means that it is expected that, in the course of the repetitions, the subjects learn to adjust their offers in a way, that lets the contract prices converge to the market clearing price. The way they adjust their offers is precisely one of the main themes of this investigation. Thus, the equilibrium that is of greatest interest to our examination is the equilibrium to which subjects' *learning* drives behavior. This equilibrium, we believe, is more probably the equilibrium of the complete information game than that of the incomplete information game.

There are a number of clues hinting that subjects in experiments, in which a basic incomplete information game is played repeatedly, have a tendency to move towards the equilibrium of the corresponding complete information game. (For an extensive discussion, see FRIEDMAN 1993.) For example, JORDAN (1991) considers players who repeatedly play the Baysian equilibrium of a normal-form game with incomplete information. He proves that, if the players update their beliefs according to Bayes' Rule after each play, such a process converges to an equilibrium of the complete information game. For the complete information game equilibrium to be reached, (given certain conditions) it is not even necessary that the players are actually completely informed on the parameters of the game at any point in the process. Moreover, COX, SHACHAT, and WALKER (1997) have presented some experimental evidence for JORDAN's result. Even though, the results are still preliminary, the conjecture that the equilibrium of the complete information game is more likely to be learned than that of the incomplete information game, seems - at least tentatively - justified.

Given the above discussion, we have chosen the following path of research: First, - in chapter II - we introduce a new double auction market institution called the *alternating double auction market*. Next, - in chapter III - a game theoretic analysis of the complete information game follows. Following that, - in chapter IV - we report an experimental investigation of the alternating double auction market institution. Finally, following some concluding remarks - in chapter V - and the references - chapter VI - the experimental data and the zero-intelligence trader markets simulation data are contained in the Appendix (chapter VII),.

The alternating double auction market is introduced as an alternative to the traditional (oral or computerized) continuous-time double auction market. Because the new institution, unlike the traditional one, is a properly specified game in extensive form, it allows for a game theoretic analysis without the problem of ambiguity with respect to the *parameterization of time*. This is one focus of the presented work.

The game theoretic analysis of the alternating double auction market game is mainly concerned with the description of equilibria of the complete information game (more precisely: of the game with *almost perfect* information). The analysis leads to the specification of the unique outcome of all *impatience equilibria* of the game with any number of traders. Roughly speaking, impatience equilibria, which are subgame perfect equilibria consisting of strategies that are pure in most parts, fulfill the condition that no player can switch to another best reply strategy, which specifies submitting a better offer or submitting an equal offer earlier than specified by his equilibrium strategy. (We postpone the exact formulations to the corresponding chapter on the game theoretic analysis.)

Another focus of the study is to confirm that experimental outcomes of the alternating double auction market exhibit the typical efficiency and convergence properties that are well-established for the continuous-time double auction markets. The results of ten experimental sessions with inexperienced subjects and eight sessions with experienced subjects are reported. We find strong evidence for the efficiency and convergence properties. Thus, we will argue, that the results obtained from the examinations of this institution can be assumed to have some general validity.

Furthermore, we report on experimental findings regarding institutional and behavioral regularities, especially concerning the process of price formation in double auction markets. The experimental results are compared to the results of the game theoretic analysis as well as to the results of simulations with zero-intelligence trader markets, as introduced by GODE and SUNDER (1993). Additionally, a number of boundedly rational theories and hypotheses that relate to the process of price formation are compared to the collected data. Among these are the theories presented by EASLEY and LEDYARD (1983 and 1993), by FRIEDMAN (1991), by GJERSTAD and DICKHAUT (1997), and by SELTEN and BUCHTA (1998).

Finally, the two main results of the experimental analysis concerning the process of price formation are reported. First, we find, that the convergence of prices to the market clearing price is driven by a simple retrospective success control mechanism, as postulated by the *learning direction theory* (SELTEN and STOECKER 1986, SELTEN and BUCHTA 1998). Second, we conjecture, that the structure of price paths and the direction of convergence can be best predicted by the *anchor price hypothesis*, introduced in section IV.D of this study.

II. The Alternating Double Auction Market Game

The structure of the alternating double auction market is basically the same as in most experimental double auction markets. The market is an institution used for trading units of a hypothetical, homogenous, and indivisible good. It contains a set of players T with n members, divided distinctly into two subsets: buyers[2] B with n_B members and sellers S with n_S members, so that $T = B \cup S$ and $B \cap S = \{\}$ and $n = n_B + n_S$. There are no market entries and exits, so that the number of players in the market remains constant.

Without loss of generality, the players $T_m \in T$, with m = 1 ... n, are indexed in such a way that $T_m = B_m \in B$, for m = 1 ... n_B, and $T_m = S_{m-nB} \in S$, for m = (n_B + 1) ... n. Each player T_m, with m = 1 ... n, has a maximum number q of allowed trading actions.

Each buyer $B_i \in B$ is assigned q *redemption values* v_{ik}, with i = 1 ... n_B and k = 1 ... q, from the set of redemption values V, i.e. $v_{ik} \in V$. Each v_{ik} specifies the gross marginal value of the k-th purchase of a unit. The redemption values can also be seen as fictitious resale prices, i.e. each purchased unit is fictitiously resold to some outsider at this price after the market period. Not purchasing leaves the buyer's wealth unchanged. Without loss of generality, we will always assume that V is ordered from highest to lowest redemption value. The ordered set V will sometimes be called the *market demand*.

Each seller $S_j \in S$ is assigned q *unit costs* c_{jh}, with j = 1 ... n_S and h = 1 ... q, from the set of unit costs C, i.e. $c_{jh} \in C$. Each c_{jh} specifies the marginal cost of the h-th sale of a unit. The unit cost is not a sunk cost of production, i.e. only sold units will incur the unit cost onto the seller. Not selling leaves the seller's wealth unchanged. Without loss of generality, we will always assume that C is ordered

---

2) Without loss of equity, we will use the pronoun *she* when referring to a seller and the pronoun *he* when referring to a buyer or to a player (trader) in general.

from lowest to highest unit cost. The ordered set C will sometimes be called the *market supply*.

We will use the generic term *player valuation* w for any redemption value v_{ik} or unit cost c_{jh}, i.e. $w \in W = \{V \cup C\}$. All player valuations are assumed to be non-negative integers from a given range $[p, \bar{p}]$. This range is known to all players, but the player valuations are considered to be private information of the traders. The only information each trader has about the distribution of the valuations of the other players in the market is that they all lie in the same given range.

Each buyer B_i, who has already purchased k-1 units, with $k \leq q$, can submit a bid non-negative integer b^t_{ik} for buying the k-th unit at the time t. (The dynamic structure of the market will be specified more precisely later.) The bid would read: *I am willing to buy a unit at the price* b^t_{ik}. Each seller S_j, who has already sold h-1 units, with $h \leq q$, can submit a non-negative integer ask a^t_{jh} for selling the h-th unit. The ask would read: *I am willing to sell a unit at the price* a^t_{jh}. To submit or to refrain from submitting an offer are the only allowed actions of the traders in the market. The resulting trade contracts are the only valid contracts in the market. All bids and asks, and thus also all contracts, concern only single unit trades. Assuming the buyer B_i and seller S_j have made a contract at the trading price p, the profit of B_i from buying the k-th unit is

$$\pi^{B_i}_k = v_{ik} - p$$

and the profit of the S_j from selling the h-th unit is

$$\pi^{S_j}_h = p - c_{jh} .$$

The major difference between a typical continuous-time double auction and the alternating double auction lies in the dynamic structure of the trading process. Unlike the typical continuous-time double auction the alternating double auction is not organized as an *open outcry* market, in which traders can submit their offers at any given moment, in any given sequence. Instead, a strict and well-defined sequence of actions is installed by dividing the market period into offer cycles. Each offer cycle consists of exactly one bidding round, in which the buyers are active, and one asking round, in which the sellers are active. A market period begins with a random, equal probabilities draw that determines which market side will be the *opening* market side, i.e. first active in the market period. Thus, a market period consists of a number of offer cycles from the first opening offers to the market period termination. The opening market side is the first active market side in every offer cycle to the end of trading at market period termination. After market period termination a new market period with the same player valuations can begin (this is often referred to as a new *trading day*). Alternatively, a new market period with different player valuations can begin (this is often referred to as a trading day of a new *trading week*).

The traders on the active market side are allowed to submit sealed offers for their next trade, whereas the traders on the passive market side simply have to wait. After all traders on the active market side made their decisions, the right to make offers switches to the other market side. Thus, the activity on the market constantly alternates between the two market sides, where each cycle starts with a decision stage of the opening market side and ends after the decision stage of the *second moving* market side.

The market period ends when a given number $z \in \mathbb{N}$ of offer cycles have passed without any new offers submitted. If $z = 1$, for example, the market period will end after the first cycle in which, neither the opening market side, nor the second moving market side submit new offers. For values of $z > 1$ the market period ends, after the first adjacent z cycles in which neither market side submits new offers. Since any player can at any time choose not to submit an offer, the market period will end whenever all players decide to refrain form further trade. But, since valid offers are restricted by the *no-loss*, the *no-crossing*, and *ask-bid-spread-reduction* rules - as described below -, the period will also end when none of the traders is able to submit any further valid offers. Thus, the alternating double auction, when combined with the two mentioned rules, represents a finite game.

The discrete-time property of the alternating double auction, which is achieved by dividing the market period into offer cycles, could be represented in common continuous-time double auctions, if it is assumed that the traders in continuous-time double auctions need a certain time span $\bar{t} \gg 0$ for their reactions, which is constant and equal for all traders. Given this reaction time, the market period of such an auction can also be divided into *offer rounds*, each of the time length $\bar{t}$. Similar to the alternating double auction, traders can decide to submit or not to submit an offer in each offer round. However, the strict alternation between bidding and asking as well as the period ending rule in the alternating double auction would still present key differences to the modified continuous-time double auction.

In the common continuous-time auction, by definition, no two offers can arrive at exactly the same time. To put it differently: usually, there is no well defined rule, stating what will happen, if two offers arrive simultaneously. In some *real world* auctions and experimental setups, there may be rules for breaking the ties in time measurement, e.g. *floor offers go before telephone offers*, etc. These rules, however, only provide support for the fact that simultaneous offers are not accounted for, even though they are possible, because time - like all theoretically continuous entities - is discrete from a positive point of view. Due to the limits of exactness in perception and measurement (no matter where they may lie), empirically, time is always divided into countable portions, allowing for the simultaneity of occurrences within one smallest time unit.

In the alternating double auction simultaneity of offers is well defined and accounted for. All buyers' bids in a bidding round and all sellers' asks in an asking round are considered as having been made simultaneously. Thus, all players on the active side, who submit an offer, submit it without knowledge of the current action of the others on their own market side. In this sense, the offers of each bidding or asking round are *sealed*. Of all submitted offers of one such round, only the *best* offers are placed *on the market*. The best bids are the highest bids that were submitted. The best asks are the lowest asks that were submitted. Offers on the market are publicly made known to all players and remain *standing*, i.e. can be accepted by one or more players from the other market side. Offers already on the market expire only if better offers have been submitted, one or more trading contracts have been made, or the market period ends. Thus, the number of offers on the market will be zero at the beginning of each market period and after any trade. It will remain zero, if all players, decide not to submit new offers. Note that the number of offers on either side of the market can and will be larger than one, whenever, in a given round, more than one of the players submit best offers, i.e. equal offers that are better than all other submitted offers. The number of offers on the market is made known publicly. Thus, size and number of offers currently standing on each side of market are public information.

The inferior offers in each offering round, i.e. those offers that were not best offers, are neither made known publicly, nor are they recorded in a rank queue (*specialist's book*). They are simply rejected and do not affect the rest of the trading process. Each of the players, who had submitted an offer, is informed whether the offer was inferior, i.e. rejected, or was best, i.e. placed on the market, after the bidding or asking round.

Not every technically possible offer is valid. Valid offers must satisfy three conditions:

1) *No-Loss* - An offer is only valid, if it guarantees a non-negative payoff, in case it is accepted: a buyer's bid for purchasing the k-th unit must be smaller or equal to the buyer's k-th redemption value and a seller's ask for the sale of the h-th unit must be greater or equal to the seller's h-th unit cost. (If a trading bonus is paid, it is not included in the payoff calculation, i.e. a potential loss cannot be balanced with the bonus.) This condition is always binding.

2) *No-Crossing* - An offer is only valid, if it guarantees a potential payoff greater or equal to the payoff that is achievable by accepting the other side's offer standing on the market: a buyer's bid must be smaller or equal to the ask currently standing on the market and a seller's ask must be greater or equal to the bid currently standing on the market. This condition guarantees that the current market bid and ask do not cross, i.e. the best bid is always smaller or equal to the best ask. This condition is only binding, if an offer from the other market side is currently standing on the market.

3) *Ask-Bid-Spread-Reduction* - An offer is only valid, if it reduces the ask-bid spread on the market: a buyer's bid must be greater than the bid currently standing on the market and seller's ask must be smaller than the ask currently standing on the market. This condition is only binding, if an offer from the same market side is currently standing on the market.

Taken together (and remembering that there is a technical price range $[\underline{p},\bar{p}]$), these rules lead to the following conditions:

Let $\tilde{b}$ denote the bid and $\tilde{a}$ the ask currently standing on the market. The bid b_{ik} of buyer B_i for the k-th unit with the redemption value v_{ik} is valid under the condition

$$\left.\begin{array}{ll} \text{if } \exists \, \tilde{b} & \tilde{b} \\ \text{otherwise} & \underline{p}-1 \end{array}\right\} < b_{ik} \leq \left\{\begin{array}{ll} \min(v_{ik},\tilde{a}) & \text{if } \exists \, \tilde{a} \\ v_{ik} & \text{otherwise} \end{array}\right. , \text{ with } i=1...n_R \text{ and } k=1...q.$$

Thus, the range of valid bids is bounded from below by either the current market bid or the lower bound of the technical price range. If a current market bid $\tilde{b}$ exists, the bid b_{ik} is only valid, if it is greater than $\tilde{b}$, due to the *ask-bid-spread-reduction* rule. If no current market bid $\tilde{b}$ exists (e.g. at market opening or after a trade), b_{ik} can be any value greater or equals to the lower bound of the technical price range $\underline{p}$. From above, the range of valid bids is bounded by either the current market ask or the redemption value of the buyer submitting that bid. That means that, if a current market ask $\tilde{a}$ exists, the bid b_{ik} can only be valid, if it is smaller or equal to $\tilde{a}$, due to the *no-crossing* rule. In all cases, however, (i.e. no matter whether a current market ask $\tilde{a}$ exists or does not exist) b_{ik} must be smaller or equal to the redemption value v_{ik} of the buyer B_i for his k-th unit, due to the *no-loss* rule.

The ask a_{jh} of seller S_j for the h-th unit with the unit cost c_{jh} is valid under the condition

$$\left. \begin{array}{ll} \text{if } \exists \, \tilde{b} & \max(c_{jh}, \tilde{b}) \\ \text{otherwise} & c_{jh} \end{array} \right\} \leq a_{jh} < \left\{ \begin{array}{ll} \tilde{a} & \text{if } \exists \, \tilde{a} \\ \bar{p}+1 & \text{otherwise} \end{array} \right. \text{, with } j=1\ldots n_S \text{ and } h=1\ldots q.$$

The range of valid asks is bounded from above by either the current market ask or the upper bound of the technical price range. If a current market ask $\tilde{a}$ exists, the ask a_{jh} is only valid, if it is smaller than $\tilde{a}$, due to the *ask-bid-spread-reduction* rule. If no current market ask $\tilde{a}$ exists (e.g. at market opening or after a trade), a_{jh} can be any value smaller or equals to the upper bound of the technical price range $\bar{p}$. From below, the range of valid asks is bounded by either the current market bid or the unit cost of the seller submitting that ask. That means that, if a current market bid $\tilde{b}$ exists, the ask a_{jh} can only be valid, if it is greater or equal to $\tilde{b}$, due to the *no-crossing* rule. In all cases, however, (i.e. no matter whether a current market bid $\tilde{b}$ exists or does not exist) a_{jh} must be greater or equal to the unit cost c_{jh} of the seller S_j for her h-th unit, due to the *no-loss* rule.

In course of the market period, players might not be able to submit further valid offers, either because they have already made all possible trades, or because any new offer, with which they could reduce the current ask-bid spread, would lead to a potential loss. The former are forced to stay passive for the rest of the market period. The latter are barred from submitting further offers until at least one contract is made. All players who cannot submit valid offers are technically treated as if they had decided not to submit an offer. Thus, if a situation arises, in which none of the players can submit new valid offers, the market period will end. This will always be the case, when the market surplus, the sum of buyers' and sellers' rents, has been completely exploited.

A contract in the alternating double auction market is made, whenever at least one of the players of the active market side submits an offer equal to the other side's offer currently on the market. Thus, accepting and submitting an offer equal to the other side's current market offer are equivalent actions. Together with the *no-crossing* rule, this equivalence reduces the players strategy space, whenever an offer from the other market side is on the market: buyers can either accept the current market ask, submit a lower bid, or refrain from submitting at all; sellers can either accept the current market bid, submit a higher ask, or refrain from submitting at all.

Since all players on one market side submit their offers simultaneously, - as mentioned before - situations can arise in which two or more players hold the current market offer. If the current market offers match in such cases, obviously, two situations can arise: either the number of players holding the current market offers are equal or they are unequal on the two market sides. In other words, it can happen that the number of players, who have offered to trade at some price, is equal to the number of players, who accept a trade at that price. In this case all players will trade. But, it can also happen that more or less players offer to trade at some price than players accept a trade at that price. In such cases the maximum possible number of contracts are made, where each of the players on the market side with a greater number of equal offers has an equal probability of trading. Thus, the rationing (or *tie-breaking*) in such cases is simply accomplished by using a random draw with uniform probability amongst the players, who have submitted equal offers, on the market side with the greater number of equal offers.

III. The Simplified Alternating Double Auction Market - A Game Theoretic Analysis

III.A. Introduction

Assume the demand and supply curves in an alternating double auction market are *normal*, i.e. a monotonically falling demand and a monotonically rising supply curve. Classical market analysis assumes all trades will occur at the *market clearing price* p*, which lies on the intersection of the demand and supply curves. However, since we have a finite number of players and all values are integers, the intersection of the demand and supply curves must not necessarily be a point. Thus, a range of prices can exist (a *market clearing price range*), in which classical market analysis would expect the *competitive equilibrium*.

The question whether trading at a price in the intersection of the demand and supply curves, i.e. at some market clearing price, also represents a game theoretic equilibrium of the game cannot be answered easily, even though the alternating double auction is a well-defined, finite game. The considerable number of players and the large size of their strategy spaces induce such an enormous game, that a comprehensive comparison of all possible states is a tedious task, especially in the incomplete information setting as described in the previous chapter.

To be able to analyze the situation in such a market game theoretically, it seems sensible and necessary to simplify the game in a number of ways. These simplifications are described in detail below. In the comments of the descriptions below and in the discussions of the last section of this chapter, we argue that a comparison of the result of the theoretic analysis of the simplified alternating double auction market game and the experimental results presented later can be useful, in spite of the simplifications of the model. This seems especially so, because the theoretic analysis leads to general results that are quite intuitive and well in line with the experimental observations.

In the following sections we derive a general game theoretic solution for all simplified alternating double auction market games. We will show that, although multiple equilibria can exist, all equilibria of a specific type, which we call *impatience equilibria*, lead to a unique outcome of the game. The equilibrium outcome is always efficient in the sense that all non-extra-marginal traders, but only these, trade. The equilibrium price that is realized lies either just inside or just outside the market clearing price range, depending solely on the relationship of the player valuations of the marginal and the best extra-marginal traders in the market. Thus, the general game theoretic solution of the simplified alternating double auction market game leads to outcomes that are compatible or at least very *close* to the predictions of competitive market clearing. Since this is true even in markets with very few traders, the analysis seems to give insight into a long standing "scientific mystery" (SMITH 1982), namely the question why experimental double auction markets with very few traders tend to converge to the competitive market clearing price.

All in all, the game theoretic analysis that follows can be seen as a first step towards a better understanding of the strategic situation and interaction in double auction markets. We find that this holds, even though the rules of the alternating double auction market game are very distinct from the rules of other double auction markets in some aspects. After all, the presented analysis is the first complete game theoretic examination of a continuous double auction institution with complete information. (See FRIEDMAN 1993.) Since the dynamics of these institutions present the main difficulty involved in the game theoretic analysis, it seems plausible to assume that - at least - the dynamic aspects of the analysis presented here can be a starting point for the theoretic evaluation of other double auction market institutions.

Doubtlessly, however, the issue is so broad and complex that generalizations from the presented results should be only attempted with great caution, especially since a number of more or less restrictive simplifications were needed to obtain these results. In the following paragraphs, we describe the simplifications used here in more detail. After deriving the results of the game theoretic analysis in the sections III.B., III.C, III.D., and III.E., we summarize the results and return to the discussion of the restrictions in the section III.F.

The first straightforward simplification used in the game theoretic analysis of the simplified alternating double auction market game is to limit the number of items each of the players may trade to one, i.e. $q = 1 \ \forall \ B_i \in B$ and $\forall \ S_j \in S$. Obviously, such a simplification eliminates some inter-temporal strategic possibilities of the players. A player, for example, cannot choose a strategy which establishes any kind of trade-off between the prices he accepts, the number of contracts he makes, and the timing of these contracts. Nevertheless, it seems acceptable to postpone the analysis of markets with multi-unit traders and to start by only examining the limit case with $q = 1$. We resume the discussion of the effects of limiting the number of valuations per trader to one in the last section of this chapter.

In addition to reducing the number of valuations per player to one, we will also assume that all redemption values and unit costs are unequal, i.e. $v_h \neq v_k$ for all v_h, $v_k \in V$, with h and k $= 1 \ldots n_B$ and h $\neq$ k, and $c_h \neq c_k$ for all $c_h, c_k \in C$, with h and k $= 1 \ldots n_S$ and h $\neq$ k. Due to this and to the fact that $q = 1$, we can also simplify the notation, by indexing buyers in falling redemption values and sellers in rising unit costs. This means that B_i's redemption v_i is equal to the i-th highest redemption value in the ordered set of redemption values V. Analogously, S_j's unit cost c_j is equal to the j-th lowest unit cost in the ordered set of unit costs C.

The second simplification is to set the number of cycles without new offers after which the market period terminates to one. As explained above, with $z = 1$, a market period will terminate after the first cycle in which none of the players submits a new offer. This is a merely technical simplification which does not change the basic character of the game. Instead of giving a full proof, let us just present the following two notes: First, increasing z does not change the ultimatum power of the opening market side. If all players on the opening market side refrain from submitting offers and z = 1, at least one of the players on the second moving market side will have to submit an offer in order to keep the market period from terminating. If z is increased, this effect is postponed to a later subgame, but not invalidated. Second, increasing z does not change the informational setting. One could assume that cycles without new offers can be used by the players to signal their commitment to a *tough* bargaining position. But, since there is no immediate cost to such a signal, it cannot be considered as a credible threat. Thus, there seems to be no obvious informational gain with z > 1.

The third simplification is that we assume that a trading bonus β is paid to each contracting party, i.e. the buyer and the seller, for each trade. For merely technical reasons we will assume that β is greater than the smallest money unit ($\beta > 1$) and equal for both parties.

The fourth simplification we will make is by far more critical than the previous ones. We will abandon the incomplete information setting and examine a complete information setting instead. The analyzed information setting is complete, but imperfect, because each player has all information about the values and offers of all other players, but not all information sets are singletons: In each offer cycle the offers of each market side are made independently and simultaneously by the players on that market side. Thus, when deciding on an offer to submit, a player does not have the information which offers will be submitted at the same time by the other players on his market side. After the offers are submitted, however, all players are informed and all other information is common knowledge at all times. Since the simultaneity of the decisions only exists in each half of an offer cycle, i.e. each time the players on one market side decide, and since all information is available after those decisions were made, a game in extensive form such as the simplified alternating double auction market game is sometimes called a *game of almost perfect information* (see for example TIROLE 1990). The important feature a game with almost perfect information is that a proper subgame begins after each stage with simultaneous moves. For this reason it seems adequate to use the concept of subgame perfect equilibria (SELTEN 1965 and SELTEN 1975) for the analysis of the simplified alternating double auction market game.

This serious modification is not primarily a technical simplification, although it, admittedly, does make the analysis somewhat easier. The main reason to examine the complete information setting is to be able to better connect the game theoretic analysis to experimental data. Two features of most double auction market experiments let the complete information setting seem *closer* to the experimental setup than the incomplete information setting.

First, in most such experiments - including our experiment -, subjects were not informed of the distribution from which the redemption values and the unit costs were drawn. In fact, in many cases, as in our case, these values were not drawn randomly at all, but were simply selected to induce a specific form of the demand and supply curves. Second, the market periods were repeated identically in almost all such experiments. In every re-run of the market period, subjects seem to learn to evaluate their own *position in the market* a little better. This is not only underlined by the behavior in the experimental market, but also by the subjects' comments in the post-experimental interviews. Even though we admit that such a *learning* remains an unproven speculation, we must note that the assumption that subjects in such experiments form beliefs (or have priors) on the distribution of values and then choose optimal strategies in the incomplete information game is also quite speculative. Furthermore, there are clues that subjects in experiments in which a basic incomplete information game is played repeatedly seem to have a tendency to move towards the equilibrium of the corresponding complete information game (e.g. COX, SHACHAT, and WALKER 1997). JORDAN (1991) shows that, for an adjustment to what is predicted in the complete information game, Baysian players do not necessarily have to learn the entire informational basis of the game. In our specific case, it generally seems to be sufficient for subjects to learn the range in which the marginal values and the best extra-marginal values lie. After the game theoretic analysis in the following sections, we return to this discussion in the last section of this chapter.

Summarizing, the simplified alternating double auction market game, that we examine in the next sections, is a game of almost perfect information. Each player is strictly associated with the role of a buyer or a seller and can either buy or sell one unit of the hypothetical good, correspondingly. All redemption values and all unit costs are unequal. A trading bonus $\beta > 1$ is paid to each of the contracting parties of a trade. The market period terminates after the first offer cycle in which no new offers were submitted. We will generally assume that the players are risk-neutral. All other features of the game are just as in the general game described in the previous chapter.

III.B. Definitions and Lemmas

The simplified alternating double auction market game, like any double auction market game with static demand and supply, consists of a sequence of similar market subgames in which the number of possible trades decreases as the sequence runs. For example, once the first trade has taken place in the original (full) market, the game moves to a subgame in which the number of potentially possible trades is - at least - one less than in the original market. Many of the definitions that follow are valid both in the original (full) market as well as in the markets which evolve as subgames of the original game, i.e. after a number of trades. In most cases, the instance of each of the defined general terms, in a specific case, depends on the configuration of the subgame market, e.g. which players with which player valuations are present. It should be noted, however, that often we will not explicitly formulate the dependence, when it is self-evident or is clear in the context.

In the game we study not only the number of possible trades decreases, as one subgame market evolves from the other, but also the number of players who can possibly trade, because each player can only trade once. Additionally, since each player in these markets only has a single player valuation and since each of these valuation is unique by assumption, a player can be identified by his role and by the order of his valuation. Thus, buyers will be denoted with B_1, B_2, ..., B_{nB}, where B_1 has the highest redemption value and B_{nB} the lowest (n_B being the total number of buyers). The sellers will be denoted with S_1, S_2, ..., S_{nS}, where S_1 has the lowest unit cost and S_{nS} the highest (n_S being the total number of sellers). To simplify the terminology, we define and use the terms *better* and *best* in a number of ways. This is especially helpful, because it generalizes the relationship between two offers, values, or players, no matter whether bids or asks, redemption values or unit costs, or buyers or sellers are compared.

Definition: *Better Valuation*, *Better Player*, and *Better Offer*

The redemption value v_i or the buyer holding v_i in a subgame of a simplified alternating double auction market game is *better* than the redemption value v_j or the buyer holding v_j, respectively, if $v_i > v_j$. The unit cost c_i or the seller holding c_i is *better* than the unit cost c_j or the seller holding c_j, respectively, if $c_i < c_j$.

The bid b_i is *better* than the bid b_j, if $b_i > b_j$. The ask a_i is *better* than the ask a_j, if $a_i < a_j$. In this case, both b_i and a_i are *better* offers.

Definition: *Best Valuation*, *Best Player*, and *Best Offer*

The greatest redemption value, the buyer holding this value, and the greatest bid in a subgame of a simplified alternating double auction market game are the *best redemption value*, the *best buyer*, and the *best bid*, correspondingly. The smallest unit cost, the seller holding this value, and the smallest ask are the *best unit cost*, the *best seller*, and the *best ask*, correspondingly.

The best redemption value and the best unit cost are the *best player valuations*. The best buyer and the best seller are the *best players* (or *best traders*). The best bid and the best ask are the *best offers*.

The best valuations and best players sometimes play a role in the analysis. But, it will quickly be clear, that the identity of a player, generally, is of little importance to the analysis. The important feature is the relative position of his valuation in the market demand or supply. To identify these relative positions, it will generally suffice to use the concept of *marginal, intra-marginal,* and *extra-marginal valuations*, which originate in the traditional market literature. Figure III.1 displays some of these concepts for different constellations of the market demand and supply.

Definition: *Marginal, Intra-Marginal,* and *Extra-Marginal* Player Valuations and Players

Let B_i with $i = 1 \ldots n_B$ be the buyers and S_j with $j = 1 \ldots n_S$ the sellers in a subgame of a simplified alternating double auction market game, each group ordered and indexed starting from the best player.

The *marginal player valuations* are v_m (*marginal redemption value*) and c_m (*marginal unit cost*) with the property that $v_m - c_m$ is the smallest existing non-negative difference between a redemption value and a unit cost of the same index.

The *intra-marginal player valuations* are all $v_i > v_m$ with $i = 1 \ldots (m - 1)$ and all $c_j < c_m$ with $j = 1 \ldots (m - 1)$. The *extra-marginal player valuations* are all $v_i < v_m$ with $i = (m + 1) \ldots n_B$ and all $c_j > c_m$ with $j = (m + 1) \ldots n_S$.

A player with a marginal, intra-marginal, or extra-marginal player valuation is called a *marginal*, an *intra-marginal*, or an *extra-marginal player*, correspondingly.

Two things should be noted. First, if $v_1 < c_1$, i.e. if not even the best players can trade with one another, neither marginal nor intra-marginal players or valuations exist. This is simply the case of a demand and a supply curve which do not intersect. Second, the definitions of marginal, extra-, and intra-marginal players all depend on the fact that each player only has a single and unique player valuation, which allows for an unambiguous ordering of the players by their valuations.

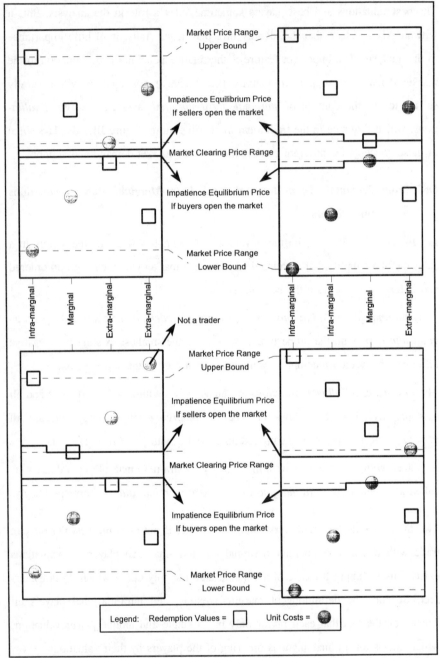

Figure III.1 - Display of demand and supply for all possible constellations of the innermost valuations.

Definition: *Market Clearing Price Range*

Let v_m and v_{m+1} be the marginal and best extra-marginal redemption value and c_m and c_{m+1} be the marginal and best extra-marginal unit cost in a subgame of a simplified alternating double auction market game. If v_m, v_{m+1}, c_m, and c_{m+1} exist, the *market clearing price range* in that subgame is the interval of non-negative integers in the range of and including the boundaries $[\underline{p}^*, \bar{p}^*]$, where the lower bound is $\underline{p}^* = \max(c_m, v_{m+1})$ and the upper bound is $\bar{p}^* = \min(v_m, c_{m+1})$. If $c_m = v_{m+1}$, $\underline{p}^* = v_{m+1}$. If $v_m = c_{m+1}$, $\bar{p}^* = c_{m+1}$. If v_{m+1} does not exist, $\underline{p}^* = c_m$. If c_{m+1} does not exist, $\bar{p}^* = v_m$.

Obviously, the market clearing price range is the interval between (and including) the smallest and greatest possible market clearing prices, $\underline{p}^*$ and $\bar{p}^*$. The smallest possible market clearing price $\underline{p}^*$ is the maximum of the marginal unit cost and the greatest extra-marginal redemption value. The greatest possible market clearing price $\bar{p}^*$ is the minimum of the marginal redemption value and the smallest extra-marginal unit cost. Graphically, when examining the demand and supply schedules, $\underline{p}^*$ and $\bar{p}^*$ represent the *innermost valuations*, i.e. those valuations that enclose the intersection of the two schedules.

Definition: *Innermost Valuations* **and** *Innermost Players*

The *innermost valuations* in a subgame of a simplified alternating double auction market game are those player valuations that specify the upper bound $\underline{p}^*$ and the lower bound $\bar{p}^*$ of the market clearing price range in that subgame. The *lower* innermost valuation is c_m, either if v_{m+1} does not exist or if $c_m > v_{m+1}$; otherwise it is v_{m+1}. The *upper* innermost valuation is v_m, either if c_{m+1} does not exist or if $v_m < c_{m+1}$; otherwise it is c_{m+1}.

The *innermost players* are the players holding the innermost valuations.

Definition: *Market Price Range*

The *market price range* in a subgame of a simplified alternating double auction market game is the range of non-negative integers between and including the lowest unit cost and the highest redemption value of those players who are present in that subgame.

Since due to the *no-loss* rule no buyer can pay a price greater than the best redemption value and no seller can accept a price smaller than the best unit cost, the market price range delineates the range of prices in which trades are possible. Up to this point we allow for markets in which some or all of the players have valuations that lie outside the market price range. Since no matter which strategy such players choose, it will neither lead to a trade nor to a non-zero payoff, we will ignore them in the analysis, assuming that they cannot influence the outcome of a simplified alternating double auction market game (or subgame). The following definitions will be used to limit the analysis to markets in which trades are possible and all players can potentially be involved in a trade.

Definition: *Trader*

A player in a subgame of a simplified alternating double auction market game is a *trader*, if his player valuation (redemption value or unit cost) lies within the market price range of that subgame, i.e. if he potentially can make a contract to buy or to sell in that subgame.

Note that any player who has already traded is no longer considered a trader in the subgames that follow. But, also some of those players who have not yet traded may not be considered traders. This is the case for any buyer who has a redemption value smaller than the lowest unit cost of all present sellers or any seller who has a unit cost greater than the highest redemption value of all present buyers. Note that either of these conditions can already be fulfilled at market opening. Similarly, a trader can *drop out* of the market price range and no longer be considered a trader, because his only (or last) potential contract partner is no longer a trader, i.e. has made a contract with some other trader.

Definition: *m-Trade Market*

A subgame of a simplified alternating double auction market game is an *m-trade market*, if it contains exactly m players on each market side who are intra-marginal or marginal traders.

Note that in any m-trade market, with m > 0, at least one trade is possible, since at least one marginal buyer and one marginal seller exist. Note also that, due to the definition of marginal valuations, a market cannot exist, in which there is a greater number of non-extra-marginal traders on one market side than on the other. Thus, in any possible market the number of non-extra-marginal traders is equal on both market sides. The number of total traders on each market side, however, can be unequal, since an arbitrary number of extra-marginal traders can be present on each market side of an m-trade market.

Definition: *μ-Maximum-Trade Market*

A subgame of a simplified alternating double auction market game is an *μ-maximum-trade market*, if μ is the maximum possible number of contracts that can be made in the subgame.

The concept of the μ-maximum-trade markets will only be used in the induction proofs of the propositions for m-trade markets of an arbitrary size (section III.E.). The reason why the concept is useful is that any μ-maximum-trade market is converted into a (smaller) ν-maximum-trade market, with ν < μ, after a trade occurs. Note that the analogous is not true with m-trade markets: an m-trade market may convert into another (smaller) m-trade market, with less traders in total, but the same number m of non-extra-marginal traders. For example, consider a 1-trade market with an extra-marginal trader on each market side trades. This market is a 2-maximum-trade market, since two trades are possible. When an extra-marginal trader in the original market trades, two traders are left: an originally marginal trader and an originally extra-marginal trader. Since, by definition of traders, the originally extra-marginal trader must have been able to trade with the marginal trader on the other market side, the extra-marginal trader will be a marginal trader in the second market. Thus, the second market is again a 2-trade market. But, since the maximum number of possible trades decreases, after the first considered trade, the second market is a 1-maximum-trade market.

After having established a terminology for the structure of valuations and players in the game, a suitable terminology for the dynamic structure of the decision process in the game will be presented below. Each simplified alternating double auction market game begins with decision nodes in which the traders of the opening market side can decide to submit offers independently and simultaneously. After these decisions are made, the traders of the other market side can decide to submit offers. This alternation between decision nodes of the opening and the second moving market sides is repeated until the game ends with decision nodes of the second moving market side. As described in the previous sections, the decision nodes of first moving and second moving market sides together constitute one complete offer cycle. However, it is often helpful to examine each half of an offer cycle separately, i.e. the decision nodes of only one market side. To simplify the terminology, we introduce the term *decision stage* for the decision nodes of one market side.

Definition: *Decision Stage*

A *decision stage* in a simplified alternating double auction market game consists of half of an offer cycle, i.e. all adjacent decision nodes of the game, in which the traders of one market side independently and simultaneously can decide to submit their offers.

A crucial situation in the game arises when one market side in its decision stage has the option to terminate the market period immediately. For example, in a 1-trade market without extra-marginal traders, in which the buyer is the first mover (i.e. he submits first in every offer cycle), situations can arise in which the seller can choose an action which terminates the market period. This will be the case, every time the buyer has refrained from submitting a new bid. By choosing not to submit a new ask, the seller will force market period termination. This leads to the following definition of a *termination stage*.

Definition: *Termination Stage*

A *termination stage* is a decision stage of a simplified alternating double auction market game in which the deciding traders would inevitably bring on market period termination after the current offer cycle, if they all refrain from submitting new offers.

On first sight it seems, that a decision stage of the opening market side cannot be a termination stage, since all offer cycles end (and thus, the game ends) with a decision stage of the second moving market side. But, this is not true. A decision stage of the opening market side is a termination stage, if the traders on the other market side simply cannot submit any new offers in that cycle. For instance, assume the buyers are first movers and the sellers second movers in every offer cycle. Now take a situation in which each of the sellers has already submitted her lowest feasible ask in an earlier offer cycle, but market ask and bid do not yet match. Due to the *ask-bid spread reduction* rule, this means that sellers will not be able to submit new asks in any following offer cycle, unless the buyers accept the current market ask. In this case, the new offer cycle begins with a buyers' termination stage, because if all buyers decide not to submit a new (higher) bid, the market period will end inevitably. For the analysis of the game, it is important to note, that the described case is the only case in which the opening market side is in a termination stage. This result is contained in the following Lemma 1. Before we can proceed to the lemma, however, we need to define the *best possible offer* of a market side.

Definition: *Best Possible Offer*

The *best possible offer* of a market side in a subgame of the simplified alternating double auction market game is

a] the greatest bid that can be submitted in any decision stage of the buyers in that subgame, i.e. the greatest redemption value of all traders present in that subgame,

b] the smallest ask that can be submitted in any decision stage of the sellers in that subgame, i.e. the lowest unit cost of all traders present in that subgame.

Lemma 1

The opening market side in a simplified alternating double auction market game is in a termination stage if and only if the current market offer of the second moving market side is equal to the second moving market side's best possible offer.

Proof of Lemma 1

Assume we are in a termination stage of the opening market side, but the second moving market side's current offer is not equal to their best possible offer. Now assume that all traders on the opening market side refrain from submitting an offer. According to the definition of the termination stage this should inevitably lead to the termination of the market period. But before the market period can terminate, the traders on the second moving market side must make their decisions, since all offer cycles end with their decisions. Because the best possible offer has not been reached, at least one of the traders on the second moving market side still can choose to submit a new (better) offer. Thus, obviously, we are not in a termination stage of the opening market side.

Now, assume that the current market offer of the second moving market side is equal to the best possible offer and that the traders of the opening market side are in a decision stage which is not a termination stage. Since we are assuming that this is not a termination stage, the market period should not terminate necessarily, even if all traders of the opening market side refrain from submitting new offers. But, since the traders on the other market side will not be able to submit new offers, evidently, the market period will end, if none of the opening traders has submitted a new offer. Hence, we are in a termination stage. ∎

If traders are in a termination stage, they can choose to either accept the current market offer of the other market side (i.e. submit an offer that matches that offer), submit a new offer which reduces the ask-bid spread, or terminate the market period. However, if the market bid is equal to the market ask plus one, they can only choose between accepting the market offer of the other market side and refraining to submit. This subset of the set of termination stages, in which there is a *take-it-or-leave-it* choice, plays an important role, because in such stages at least one trader must make a payoff relevant choice.

Definition: *Take-It-Or-Leave-It Stage*

A *take-it-or-leave-it stage* is a termination stage of a simplified alternating double auction market game, in which at least one of the deciding traders has the alternative to accept the other market side's current offer, but the only other alternative of all deciding traders is to refrain from submitting an offer.

A take-it-or-leave-it stage, thus, is a decision stage with an ultimatum character. Although a trade or the market period termination *can* occur at almost any point in the game, one of the two *must* occur in a take-it-or-leave-it stage. To be able to use the backward induction argument, it is important to show that every path of decisions in a market is finite. The following Lemma 2 gives exactly this result, by establishing that either a trade or a market termination will occur in every market: if not any earlier, then at some point in a take-it-or-leave-it stage.

Lemma 2

Let G' be a subgame of a simplified alternating double auction market game that begins with the decision stage d'. Every play of G' starting in d' that does not end with market period termination leads to a contract. A finite upper bound exists for the length of the plays in G'.

Proof of Lemma 2

As soon as no new offers are submitted in an offer cycle the market period terminates. Thus, as long as the market period does not terminate, at least one new offer must be submitted in every offer cycle. In any given offer cycle that follows the decision stage d' on the path of play, in which the market offers do not match, the current market bid $\bar{b}$ must be smaller than the current market ask $\bar{a}$, due to the *no-crossing* rule. Because of the *ask-bid-spread-reduction* rule, every new offer brings $\bar{b}$ closer to $\bar{a}$. Since all offers are integers, in the course of this ask-bid-spread-reduction, one of three situations can arise: First, $\bar{a} - \bar{b} = 1$. Second, $\bar{b}$ reaches the best possible bid. Third, $\bar{a}$ reaches the best possible ask.

In the first case, if any trader on the opening market side submits a new offer, a trade occurs before a take-it-or-leave-it stage is reached. If all traders of the opening market side refrain from submitting offers, however, a take-it-or-leave-it stage of the second moving market side is reached. This take-it-or-leave-it stage must either lead to market period termination or to a contract.

In the second case, buyers can no longer submit new bids. Thus, to prevent the market period from termination, in each of the following offer cycles at least one of the sellers must submit a new ask. Due to the *ask-bid-spread-reduction* rule, $\bar{a}$ will be decreased until it reaches $\bar{b} + 1$. In the offer cycle after this, a take-it-or-leave-it stage is reached: If buyers have opened the market, they cannot submit new bids, so the activity moves on to the sellers, who are now in a take-it-or-leave-it stage. If sellers opened the market, they are in a take-it-or-leave-it stage immediately.

In the third case, everything is absolutely analogous to the second case with the one difference that sellers cannot submit new asks and the market phase ends with buyers in a take-it-or-leave-it stage.

Quite obviously, a finite upper bound must exist for the length of the plays in G', because there are a finite number of players, each with a finite number of possible actions in each decision stage. (Remember that the smallest money unit is fixed throughout the game.) As the play of G' proceeds, the number of feasible actions of each player is reduced, due to the *ask-bid-spread-reduction* rule. Due to the *no-loss* and the *no-crossing* rule, the set of actions of each player is thus eventually reduced to either only one action, namely having to refrain from submitting, or to two actions, namely accepting or refraining from submitting. At this point either a trade must occur or the market period terminates. ∎

Lemma 2 establishes that any subgame of a simplified alternating double auction market game must either end with market period termination or lead to a trade. Since the total number of possible trades in any market is limited, this result also verifies that the entire game must be finite. Once no more trades are possible, the game must end with market period termination.

The game, however, could also end with market period termination, before all possible trades have occurred. Obviously, if this happens, the play does not lead to market efficiency. The issue is examined in the following Lemma 3. It establishes that as long as at least one trade is feasible (i.e. mutually profitable for both traders), the market period should not terminate in equilibrium. This result is important, because it verifies that a necessary condition for subgame perfect equilibria of the game is that all profitable trades occur. This means that in a subgame perfect equilibrium the entire market rent must be exploited and market inefficiency, due to the shortcoming of trades, cannot prevail.

Before we introduce Lemma 3, it seems necessary to make a few notes on the structure of strategies in the game. Since the simplified alternating double auction market game is a game of almost perfect information in extensive form and since it follows by Lemma 2 that the game is finite, a subgame perfect equilibrium in behavior strategies always exists.

We normally let x_{T_i} denote the behavior strategy of the trader T_i for the game. As usual, the behavior strategy x_{T_i} of each trader T_i specifies a probability distribution over all feasible actions of T_i at each information set of T_i, i.e. x_{T_i} assigns a local strategy to each information set of T_i. Since every information set of each trader T_i is part of some decision stage d of the T_i's market side, we say, that trader T_i's strategy assigns the local strategy $l_{T_i}(d)$ to the decision stage d of the game. In order to keep the terminology as simple as possible, we may sometimes say, that the strategy x_{T_i} of each trader T_i assigns a local strategy $l_{T_i}(d)$ to some decision stage d, even though only the traders on one market side can decide in d. Analogously, we may say, that a strategy combination of the traders T_i assigns a local strategy $l_{T_i}(d)$ to some decision stage d, although not every trader can decide in d. We use these simplifying formulations in those cases, in which it is either clear or of no importance which market side actually decides in the decision stage d.

Lemma 3

Let G be a simplified alternating double auction market game. Let d be a decision stage of G. Let X be a combination of strategies in G with the following properties:

a] The local strategies $l_{T_i}(d)$ assigned by X to the decision stage d always result in immediate market period termination, if d is reached.

b] At the decision stage d, at least one trade is feasible (i.e. at least the two best traders B_1 and S_1 could trade at some price that would not incur a loss on either of them).

X cannot be a subgame perfect equilibrium of the game.

Proof of Lemma 3

By definition the last decision stage d of the game is a termination stage. Thus, two situations are possible in d: either (1) the opening market side is in a termination stage or (2) the second moving market side is in a termination stage. Let us assume the buyers are on the opening market side and the sellers are on the second moving market side. The opposite case can be proven analogously.

Let us first examine (1). From Lemma 1 it follows, that d can only be a termination stage of the opening market side, if the current market offer of the second moving market side is equal to the second moving market side's best possible offer. Thus, if d is a termination stage of the buyers, then the current market ask is equal to the best possible offer of the sellers, i.e. is equal to c_1. Since the market period terminates after d, if d is reached, the strategy of each buyer, that is contained in X, must have specified to refrain from submitting a new bid in d. Specifically, B_1 also refrains from submitting a new bid in d. Thus, the market period terminates and B_1 receives a payoff of 0. But, B_1 also has the alternative in d to accept the current market ask c_1, which would have led to a trade and a payoff of at least $\beta > 1$. Hence, given the strategies of all other buyers, who all refrain from submitting a new bid in d, B_1's strategy in X cannot be a best reply.

Now let us examine (2). In this case the second moving sellers are in a termination stage at d, in which they all refrain from submitting new offers. Instead of also refraining to submit a new offer, the best seller S_1 could have submitted an offer equal to his own valuation c_1. Consider a combination of strategies X' which is identical with X except that S_1's strategy in X' specifies submitting the ask c_1 in d, instead of refraining to submit as in X. Obviously, c_1 is the second moving market side's best possible offer. Thus, the decision stage d' of the buyers that would then follow d on the play realized by X', would be a termination stage of the buyers. With this we are back in case (1) of the proof, this time, however, with the strategy combination X' at the decision stage d'. From the examination of case (1) we know that X' can only be a subgame perfect equilibrium if the market period does not terminate without a contract. Since c_1 is the best possible offer of the sellers, any contract that follows must be a contract with S_1. Thus, by deviating in this way from her strategy in X to the strategy in X', S_1 knows that - if the buyers strategies in X' (and, thus, also in X) are best replies - she will trade at c_1. From this trade S_1 receives a payoff of $\beta > 1$, while sticking to the original strategy in X pays 0, since the market period terminates after d. Hence, X cannot be a subgame perfect equilibrium, because of one of the following reasons: either the buyer strategies in X are not best replies in the subgame that starts d', i.e. after S_1's submission of c_1 at d (i.e. they do not specify trading at the price c_1 at some point after d), or S_1's strategy in X is not a best reply.   ■

To simplify the analysis, we will break the game down into smaller parts, each ending either with a contract or with market period termination. We can do so, because from Lemma 2 we know that any subgame of a simplified alternating double auction market game must either lead to market period termination or to a contract. Since the type of interaction in each of these smaller parts of the game is quite similar, this segmentation of the game will prove to be very useful.

Definition: *Game Segment*

Let G' be a subgame of a simplified alternating double auction market game that begins with the decision stage d'. The *game segment* I' of the subgame G' is the part of G' that begins with the first decision stage d' and contains all decision stages up to and including the decision stages before the next contract or before market period termination, i.e. the game segment begins at d' and ends either with market period termination or with a matching of market ask and market bid.

Definition: *t-th Market Phase* **and** *Opening Market Phase*

The game segment I' of any subgame G' that begins with the decision stage that immediately follows the (t-1)-the matching of market ask and bid in a simplified alternating double auction market game is also called the *t-th market phase* P_t of the game, given the history of play. The game segment that begins with the very first decision stage of the game is called the *opening market phase* or the *first market phase* P_1 of the game.

Note that which t-th market phase is reached depends on the play of the game up to (t-1)-th matching of asks and bids. In general, a number of different t-th market phases will be possible, but only one is reached by any given play. Thus, we will use the term market phase for a specific instance of a generic situation in the game. Note also, that a market phase usually is only the first part of a subgame. It contains the complete subgame, only if all possible plays in the subgame lead to market termination (e.g. only two traders are left in the subgame or no more trades are possible in the subgame). However, it is sometimes very useful to examine the entire subgame that begins with the first decision stage of some market phase. Thus, we introduce the concept of a *t-th market phase subgame*.

Definition: *t-th Market Phase Subgame*

The subgame of a simplified alternating double auction market game G that begins with the first decision stage of some t-th market phase P_t, in which the market clearing price range has the lower bound $\underline{p}^*$ and the upper bound $\bar{p}^*$, is called the *t-th market phase subgame* $G_t(\underline{p}^*,\bar{p}^*)$ of G, given the history of play.

It should be noted, once again, that numerous t-th market phase subgames can be contained in a simplified alternating double auction market game. For example, let us consider a 2-trade market without extra-marginal traders. Let us assume that the best traders, B_1 and S_1, trade in the first market phase. Now some second market phase subgame $G_2(c_2,v_2)$ is reached, in which the two marginal traders, B_2 and S_2, are present. This is true for any second market phase subgame reached after the trade considered, i.e. all second market phases that can be reached are of the same type as $G_2(c_2,v_2)$. But, note that which branch of the game tree is reached, i.e. exactly which second market phase subgame is reached, depends on the history of play up to the trade, that has a very complicated structure in a simplified alternating double auction market game. Specifically, many different combinations of offers can have been submitted in many different sequences, each resulting in a different history of play, but all leading to the considered trade.

Nevertheless, we will show in the following sections, that it suffices for the analysis to know that the second market phase reached is of the same type as $G_2(c_2,v_2)$. In other words, the information concerning the market clearing price range separates the classes of second market phase subgames in the example above sufficiently. In fact, sometimes we will not even need to examine both of the innermost valuations in the $(t+1)$-the market phase subgame. In such cases, we may denote a $(t+1)$-the market phase subgame with $G_{t+1}(.,\bar{p}^*)$ or $G_{t+1}(\underline{p}^*,.)$, if only the upper innermost valuation $\bar{p}^*$ or the lower innermost valuation $\underline{p}^*$, correspondingly, is decisive for the analysis.

Within each specific market phase subgame there are a number of game segments, of which the market phase itself is the largest. In general we will use the concept of game segments for the definitions and proofs. Game segments can start anywhere in the market phase, but always include all decision stages up to market period termination or the next matching of market ask and bid. Often it will suffice to examine only the choices made by players in a game segment, instead of examining the entire strategy combination.

Definition: *Segment Strategy* **and** *Pure Segment Strategy*

A *segment strategy* $x'_{Ti}(I')$ of a trader T_i for a game segment I' assigns a local strategy $l'(d')$ to every decision stage d' of T_i's market side in I'. The segment strategy $x''_{Ti}(I')$ is called a *pure segment strategy*, if $x''_{Ti}(I')$ assigns a pure local strategy $l''(d')$ to every decision stage d' of T_i's market side in I'.

A segment strategy is simply a plan of actions for the game segment. Often in the analysis, we will examine a player's complete strategies (his complete plan of actions) that has the property of inducing a certain segment strategy for a certain game segment. This will allow us to concentrate on the choice of actions in a game segment. One important type of action which can be chosen in a game segment is *price enforcement*, which we formally define in the following. The intuitive meaning of price enforcement is that the traders of one market side attempt to force the traders of the other market side into accepting the current market offer by signaling commitment to this and only this offer. The commitment to the current market offer is signaled by refraining to submit any new offers and, thus, by threatening to let the market period terminate without any further contracts. We will later see that the threat is only credible under certain circumstances. If the threat is credible, we say a market side is *credibly enforcing a price*.

Price enforcement, obviously, is a form of joint (but not explicitly cooperative) action of the traders on one market side. Thus, it is necessary to examine not a single strategy of a trader on the price enforcing market side, but the collection of strategies of the traders on that market side. We introduce a number of needed definitions, before price enforcement is defined.

Definition: *Segment Strategy Combination*

A *segment strategy combination* $X'(I')$ is a collection of segment strategies for the game segment I' that contains a segment strategy $x'_{Ti}(I')$ for each trader who is present in the game segment I' of a simplified alternating double auction market game.

Definition: *Market Side Strategy Combination*

A *market side strategy combination* $X(M)$ of a market side M is a collection of strategies that contains a strategy x_{Ti} for each trader T_i in M.

Definition: *Market Side Segment Strategy Combination*

A *market side segment strategy combination* $X'(I',M)$ of a market side M is a collection of segment strategies for the game segment I' that contains a segment strategy $x'_{Ti}(I')$ for each trader T_i in M.

Definition: *Enforcement Segment Strategy*

A segment strategy $x'_{Ti}(I')$ of a trader T_i for a game segment I' of a simplified alternating double auction market game is called a *enforcement segment strategy*, if $x'_{Ti}(I')$ prescribes refraining to submit an offer in every decision stage of T_i's market side in I'.

Definition: *Price Enforcement*

The market side segment strategy combination $X'(I',M)$ of the market side M in a simplified alternating double auction market game is said to *enforce a price p* in the game segment I' that begins with the decision stage d', if

a] p is the current market offer of the market side M in the decision stage d',

b] p is in the market price range in the decision stage d', and

c] all segment strategies $x'_{Ti}(I')$ contained in the market side segment strategy combination $X'(I',M)$ of the market side M are enforcement segment strategies for I'.

Price enforcement is a property of a market side segment strategy combination, i.e. a property of the joint plan of actions of all traders on one market side. Next, we examine the properties of the reactions of the traders on the other market side to price enforcement.

Definition: *Market Side Best Reply Combination*

Let G be a simplified alternating double auction market game. Let X(M) be the market side strategy combination of the market side M and $X(\bar{M})$ be the market side strategy combination of the other market side $\bar{M}$ in G. X(M) is a *market side best reply combination* of the market side M in G, if each strategy x_{Ti} contained in X(M) is a best reply strategy of the trader T_i in M, given $(x_{T1}, ..., x_{Ti-1}, x_{Ti+1}, ..., x_{Tn})$, in all subgames of G.

The basic notion of a market side best reply combination is that the traders of one market side M choose best reply strategies given some arbitrary, but fixed, market side strategy combination of the other market side M̄. In other words, we let the traders of M optimize their strategy choices, while holding the strategy combination of M̄ fixed. Market side best reply combinations have two - quite obvious - features. On one hand, if both market sides are using market side best reply combinations simultaneously, the entire market is in equilibrium. On the other hand, if one market side is not using a market side best reply combination, the entire market cannot be in a subgame perfect equilibrium.

Lemma 4

Let G be a simplified alternating double auction market game. Let M be the one and M̄ be the other market side in G. Let X be a combination of strategies for G, consisting of the market side segment strategy combinations X(M) and X(M̄) for the market sides M and M̄, correspondingly. Assume X has the following properties:

a] The market side segment strategy combination X'(I',M) of the market side M induced by X on the game segment I' enforces the price p.

b] All segment strategies $x'_{Ti}(I')$ contained in the market side segment strategy combination X'(I',M̄) of the other market side M̄ induced by X on the game segment I' are pure segment strategies.

X(M̄) can be a market side best reply combination, only if any play realized by X on I' ends with a trade at the price p, i.e. only if the segment strategy $x'_{Ti}(I')$ of at least one trader T_i in M̄ specifies accepting the price p in some decision stage of I' that is reached before market period termination.

Proof of Lemma 4

Since p is enforced by the market side M in the game segment I', all segment strategies of the traders on M are segment enforcement strategies, i.e. none of these traders will submit a new offer in I'. Thus, in any play realized by X on I' only two outcomes are possible: either the market period terminates without any further contracts or a contract is made at the price p. Thus, we will only have to show that, if a play realized by X on I' leads to market period termination, then $X(\bar{M})$ cannot be a market side best reply combination of $\bar{M}$.

Let us assume that the play realized by X on I' ends without a trade, i.e. with market period termination. Since, by definition of price enforcement, p is in the market price range in d', at least the best trader T_1 on the other market side $\bar{M}$ can trade at p in d' without a loss. If T_1 trades at the price p in the decision stage d', his payoff is no smaller than the bonus $\beta > 1$. But, since we are assuming that the play realized by X on I' ends without a trade, obviously, T_1's payoff in this play is 0. Thus, at least T_1 has an incentive to deviate from his original segment strategy $x'_{T1}(I')$ to some other segment strategy $x''_{T1}(I')$, which prescribes accepting the price p in some decision stage d'' in the game segment I'. This decision stage d'' can either be equal to or follow after d', but must be reached before market period termination. Hence, if a play realized by X on I', leads to market period termination, there is at least one trader in $\bar{M}$, who is not using a best reply strategy in the subgame starting with d', given the strategies of all other traders in X.   ■

Using the concept of market side best reply combination, two corollaries to the Lemma 4 can be derived, which will both be useful in special cases of price enforcement. Corollary 1 is concerned with cases in which, due to the *no-loss* rule, only a single trader on the *responding* side can accept the enforced price. It follows directly from Lemma 4, that in such cases, this trader's strategy, that is contained in any market side best reply combination of his market side, *must* prescribe accepting this price at some point before market period termination.

Corollary 1

Let G be a simplified alternating double auction market game. Let M be the one and $\bar{M}$ be the other market side in G. Let X be a combination of strategies for G, consisting of the market side segment strategy combinations X(M) and $X(\bar{M})$ for the market sides M and $\bar{M}$, correspondingly. Assume that the market side segment strategy combination $X'(I',M)$ of the market side M induced by X on the game segment I' enforces the price p.

If T_i is the only trader in $\bar{M}$, who makes no loss by trading at the price p in I', $X(\bar{M})$ can be a market side best reply combination, only if any play realized by X on I' ends with a trade involving T_i at the price p, i.e. only if the segment strategy $x'_{T_i}(I')$ of T_i specifies accepting the price p in some decision stage of I' that is reached before market period termination.

Proof of Corollary 1

The proof follows immediately from Lemma 4. Since $X(\bar{M})$ can be a market side best reply combination, only if at least one of the traders on the market side $\bar{M}$ chooses a strategy which prescribes trading at p in I' and since T_i is the only trader who can do so, $X(\bar{M})$ can be a market side best reply combination, only if T_i accepts p in some decision stage of I' before market period termination. ∎

Another special case is also of some importance. Imagine the best buyer has submitted a bid which is greater than the second highest redemption value. Assume further that some seller exists, who can accept this bid, but no lower bids. Obviously, this seller should try to trade in this market phase, since this is her only chance to trade at all. Generally, when a trader's only potential trading counterpart is certain to trade in the current game segment, the trader will try to trade in this game segment, because he will not be able trade later.

Corollary 2

Let G be a simplified alternating double auction market game. Let M be the one and $\bar{M}$ be the other market side in G. Let X be a combination of strategies for G, consisting of the market side segment strategy combinations $X(M)$ and $X(\bar{M})$ for the market sides M and $\bar{M}$, correspondingly. Assume X has the following properties:

a] The market side segment strategy combination $X'(I',M)$ of the market side M induced by X on the game segment I' enforces the price p.

b] All segment strategies $x'_{Ti}(I')$ contained in the market side segment strategy combination $X'(I',\bar{M})$ of the other market side $\bar{M}$ induced by X on the game segment I' are pure segment strategies.

Let T_i be a trader in $\bar{M}$, who makes no loss by trading at the price p in I', but who cannot trade in any subgame following I'. $X(\bar{M})$ can be a market side best reply combination, only if the segment strategy $x'_{Ti}(I')$ of T_i in $\bar{M}$ specifies accepting the price p in some decision stage of I' that is reached before market period termination.

Proof of Corollary 2

It follows by Lemma 4, that $X(\bar{M})$ can be a market side best reply combination, only if at least one of the traders on the market side $\bar{M}$ chooses a strategy which prescribes trading at p in I'. Thus, T_i knows that I' will definitely end with a trade at p. T_i, however, also knows that he cannot trade in any following subgame. Hence, only two outcomes are possible for T_i in this game: he either trades now at p or does not trade at all. T_i's payoff in the latter case is 0. In the former case T_i's expected payoff will be no smaller than $\frac{1}{k} \beta$, with $\beta > 1$, if T_i ties with k - 1 other traders in accepting the offer p. Since the number of traders is finite, $\frac{1}{k} \beta > 0$. Thus, submitting an offer equal to p in I' has a greater expected payoff than not trading at all, no matter with how many other traders T_i ties when accepting p in I'.   ■

In the cases examined in Lemma 4 and the two following corollaries, traders facing price enforcement must accept the enforced price before market period termination. It is important to note, that it makes no difference whether the enforced price is accepted immediately or after a number of decision stages, in which other offers are made by the market side facing the price enforcement. As long as they do not let the market period terminate before the contract is made, the traders facing the price enforcement can choose to submit any feasible offer. Since they are facing price enforcement, however, the offers that do not lead to a trade are irrelevant to the outcome and, thus, irrelevant to the payoffs. In order to be able to specify the strategies properly in some of the propositions of the next section, it sometimes seems sensible to assume that the enforced price is accepted immediately. Thus, we introduce the concept of *acceptance segment strategies*.

Definition: *Acceptance Segment Strategy*

A pure strategy $x'_{Ti}(I')$ of a trader T_i for a game segment I' of a simplified alternating double auction market game is called *acceptance segment strategy*, if $x'_{Ti}(I')$ prescribes:

a] Accept the current market offer p of the other market side in every decision stage of I', in which p exists and a contract at the price p does not lead to a negative payoff.

b] Submit the best possible offer in every decision stage of I', in which the *ask-bid-spread-reduction* rule and the *no-crossing* rule allow the submission.

c] Refrain from submitting an offer in any other decision stage of I'.

Up to this point we have focused on the market side facing price enforcement: note that the best reply property established by Lemma 4 and its corollaries only concerns the market side facing the price enforcement. Neither the Lemma nor the corollaries specify whether the market side strategy combination of the price enforcing market side is a market side best reply combination or not. This is a much more complex problem, with which we will deal in the proofs of the propositions in the next sections. One aspect of the problem, however, will be the examined in the next lemma. With Lemma 5 we will show that a price enforcing market side strategy combination can only be induced by a subgame perfect equilibrium, if the commitment to the enforced price is *credible*. The credibility of the commitment strongly depends on the sequence of actions in the game. The opening market side is in a better position to enforce a price, simply because, when it commits to a price enforcement, the second moving market side is forced into a series of termination stages that - on the longest possible play - end with a take-it-or-leave-it stage of the second moving market side. In contrast, as a result of Lemma 1, the second moving market side can only force the opening traders into such a series of termination stages, if its best possible offer has been submitted.

Definition: *Credible Price Enforcement*

The market side segment strategy combination $X'(I',M)$ of the market side M in a simplified alternating double auction market game that enforces a price p in the game segment I', beginning with the decision stage d', is said to *credibly enforce the price p* , if either

a] M is the opening market side or

b] M is the second moving market side and p is the best possible offer of M in d'.

Lemma 5

Let G be a simplified alternating double auction market game. Let M be the one and $\bar{M}$ be the other market side in G. Let X be a combination of strategies for G, consisting of the market side segment strategy combinations $X(M)$ and $X(\bar{M})$ for the market sides M and $\bar{M}$, correspondingly. Assume X has the following properties:

a] The market side segment strategy combination $X'(I',M)$ of the market side M induced by X on the game segment I' enforces the price p.

b] All segment strategies $x'_{Ti}(I')$ contained in the market side segment strategy combination $X'(I',\bar{M})$ of the other market side $\bar{M}$ induced by X on the game segment I' are pure segment strategies.

X can be a subgame perfect equilibrium of G, only if p is credibly enforced by M.

Proof of Lemma 5

Since p is enforced by the market side M in the game segment I', all segment strategies of the traders in M are segment enforcement strategies, i.e. none of these traders will submit a new offer in I'. Thus, we only have to show is that, if the price enforcement by M is not credible, a decision stage d″ exists in I′, in which a trader in M has an incentive to deviate from his enforcement segment strategy, given the strategies of all other traders.

Any game segment of a simplified alternating double auction market game contains a take-it-or-leave-it stage. This is so by definition, because a game segment contains all decision stages of the game up to and including the decision stages before the next contract or before market period termination. Thus, the game segment with the smallest number of decision stages is a game segment which begins in a take-it-or-leave-it stage and only contains this stage. Any game segment that does not begin with a take-it-or-leave-it stage, must at least contain one take-it-or-leave-it stage, because the traders always have the option to reduce the ask-bid-spread until a take-it-or-leave-it stage is reached.

Let $\bar{a}'$ denote the current market ask and $\bar{b}'$ denote the current market bid in the first decision stage d′ with which the game segment I′ begins. Since all possible paths of ask-bid-spread reduction are contained in I′, for every possible combination of the market ask $\bar{a}''$ and the market bid $\bar{b}''$, that has the properties (i) $\bar{a}'' \leq \bar{a}'$, (ii) $\bar{b}'' \geq \bar{b}'$, (iii) $\bar{a}'' - \bar{b}'' = 1$, and (iv) both $\bar{a}'$ and $\bar{b}'$ in the market price range of d′, at least one take-it-or-leave-it stage is contained in I′. Hence, the examined game segment I′ must contain a take-it-or-leave-it stage d″, in which the current market offer of M is p, the current market offer of $\bar{M}$ is $\bar{p}$, and the equation $|p - \bar{p}| = 1$ is true.

Assume the price enforcement by the market side M is not credible. This can only be the case, if M is the second moving market side and p is not the best possible offer of M. Thus, and since it follows by Lemma 1 that the opening market side can only be in a termination stage, if the second moving market side's current offer is its best possible offer, the take-it-or-leave-it stage d″ must be a termination stage of the second moving market side M. Because the traders in M are all using enforcement segment strategies, i.e. none of them submits a new offer in d″, the market period terminates immediately after d″ and all traders receive a payoff of 0. However, since p is not the best possible offer of M, the best trader T_1 in M could submit a new better offer in d'' without breaking the *no-loss* rule. This new offer of T_1 is equal to $\bar{p}$, because $|p - \bar{p}| = 1$. By submitting $\bar{p}$, T_1 would trade at the price $\bar{p}$ in d'' and receive a payoff no smaller than the bonus $\beta > 1$. Hence, at least T_1 has an incentive to deviate from his enforcement segment strategy in the take-it-or-leave-it stage d''. Note that, no matter whether the play realized by X in I' does or does not reach d'', X cannot be a subgame perfect equilibrium of G, because the segment enforcement strategy of T_1 is not a best reply in all subgames of G. ∎

Neither Lemma 4 nor Lemma 5 address the issue of existence of subgame perfect equilibria with price enforcement. Both only give us necessary conditions for such equilibria. Nevertheless, they are central to the analysis, because they reduce the number of possible candidates for subgame perfect equilibria substantially. Price enforcement is an intuitively attractive behavior in a double auction market, because, as it follows by Lemma 4, at least one of the traders on the other market side must accept a contract at the enforced price. In equilibrium, however, this type of behavior can only persist, if the threat of market period termination is credible.

Clearly, there may be a number of different equilibria, in each of which a different trader facing the credibly enforced price accepts it. For this reason, it is possible that in some situations a coordination problem arises between the traders on the market side facing price enforcement. The coordination problem could arise, because each of these traders prefers not to accept the enforced price, but if none of these trader accepts, each one of them is better off by accepting. Obviously, in such cases, a number of equilibria can exist, in which traders facing price enforcement do not use pure strategies in response and which do not necessarily realize a play that leads to a trade at the credibly enforced price.

We will, however, restrict the following analysis to those cases - as in Lemma 3, 4 and 5 - in which the traders use pure strategies on the equilibrium path. Actually, we will demand more than pureness of the strategies on the equilibrium path. The concept of *quasi-pure subgame perfect equilibria*, which we will use for the analysis, examines those cases in which all traders use a pure local strategy for the first decision stage of each market phase of the game and a pure segment strategy in a distinct game segment of every market phase. In each market phase, this game segment begins with the decision stage that immediately follows the first decision stage of the market phase, given the traders' local strategies for the first decision stage. The traders' segment strategies for the other game segments of the market phase, i.e. for those game segments that are not reached after the first decision stage of the market phase, are not restricted to being pure. Examining the game tree, this means that the segment strategies for only one of the branches that begin in the first decision stage of each market phase are pure segment strategies.

Note that this concept ensures that pure local strategies are used in all decision stages reached on the equilibrium path. Note also, however, that this is true in any market phase subgame. Thus, if a game contains more than one market phase, and if some of the market phases are not reached on the equilibrium path, the traders' strategies induce pure segment strategies also on some game segments in market phases that are not reached on the equilibrium path.

Definition: *Quasi-Pure Subgame Perfect Equilibrium*

Let G be a simplified alternating double auction market game. Let P be the set of all market phases P_t contained in G. Let $d(P_t)$ denote the first decision stage of the market phase P_t.

A *quasi-pure subgame perfect equilibrium* X of the game G is a subgame perfect equilibrium of G with the following properties:

a] For all $P_t \in P$, each strategy x_{Ti}, contained in X, assigns a pure local strategy $l_{Ti}(d(P_t))$ of the trader T_i to the first decision stage $d(P_t)$ of the market phase P_t.

b] Let $d'(P_t)$ denote the decision stage that immediately follows $d(P_t)$, given the $l_{T_i}(d(P_t))$ assigned by X to $d(P_t)$. Let $I'(d'(P_t))$ denote the game segment that begins with $d'(P_t)$. For all $P_t \in P$, each strategy x_{T_i}, contained in X, induces a pure segment strategy $x'_{T_i}(I'(d'(P_t)))$ of trader T_i on the game segment $I'(d'(P_t))$.

In the Propositions of the following sections we will show that multiple quasi-pure subgame perfect equilibrium outcomes of a simplified alternating double auction market game exist. The multiplicity of these equilibrium outcomes results from indifferences of traders concerning the sequence of their actions under certain circumstances. If we restrict the analysis to those quasi-pure subgame perfect equilibria, however, which fulfill a certain *impatience* condition, we can show that all equilibria lead to a unique outcome.

The impatience condition we use is quite simple: if a trader has a number of alternative best reply strategies, he should choose the one that specifies submitting the best offer as early as possible in each market phase of the game. Thus, the idea is to select those quasi-pure subgame perfect equilibria in which no trader delays the submission of his best offer in a market phase, unless delaying the submission leads to a greater payoff than submitting the best offer immediately.

Even though this concept of *impatience equilibria* may seem quite restrictive on first sight, the analysis in the following sections shows, that only a few quasi-pure subgame perfect equilibria are sorted out by the impatience criterium. In the case of 1-trade markets with extra-marginal traders on only the opening market side and in the case of 2-trade markets without extra-marginal traders, we examine quasi-pure subgame perfect equilibria that are not impatience equilibria in order to give an insight into the selection problem.

From a behavioral point of view, we find it quite convincing that a buyer, for example, who plans to bid $\bar{p}$ at some point during a market phase, does so as early as possible, if he is certain that doing so does not lead to a smaller payoff than delaying the submission. In fact, it will become evident in the next sections, that extra-marginal traders, who do not trade in any quasi-pure subgame perfect equilibrium, submit their best offers immediately in equilibria that fulfill the impatience condition. Again, from a behavioral point of view, this seems quite sensible. Because, if such an extra-marginal trader believes that there is a small probability for the other traders to make a mistake, by submitting his best offer immediately, he at least has a small chance of holding the market offer and, thus, a small chance of trading. KAMECKE (1998) finds support for this conjecture in his simple two-bidder english auction experiment. There subjects, who knew that their valuation was lower than the valuation of their opponent, tended to be very competitive in bidding. This led to more competitive equilibrium plays, in which the extra-marginal buyer bid all the way up to his redemption value, than to cooperative plays, in which he refrained from bidding high, in order to leave more gains for his opponent.

It seems similarly reasonable for a trader, who can trade at exactly the same price in an earlier or in a later market phase, to choose to do so early. The early trade not only shortens the duration of the market, but also minimizes the number of decision stages preceding the trade and, thus, the number of decision stages, in which some other trader could mistakenly deviate from his equilibrium strategy and incur a loss or a foregone profit on the initially (theoretically) indifferent trader.

Finally, a learning argument also supports the concept of impatience, if we assume that players in a corresponding game with incomplete information learn the relative positions of their valuations in course of the identical repetitions of such markets. In such cases, missing a trade in an early market leads to bidding higher and earlier - or asking lower and earlier - with the effect that traders tend to *jump* to their best offers in early decision stages in order to avoid being left out of trading.

Before we proceed to the definition of *impatience equilibria*, a note on the paths of play in quasi-pure subgame perfect equilibria seems necessary. As noted before, any play in such equilibria involves only of pure local strategies of the traders. Different plays, however, can be realized in equilibrium, since random draws may occur, if a number of traders tie at some trading price. But, note that in a given market phase equilibrium all plays are identical up to the very end of the market phase, when a random draw may occur. In other words, since local strategies are pure and the only random draw is at the end of the market phase, there is no uncertainty concerning the actions of the traders within a given market phase.

Definition: *Impatience Equilibrium*

Let G be a simplified alternating double auction market game. Let $X = (x_{T1}, ..., x_{Ti-1}, x_{Ti}, x_{Ti+1}, ..., x_{Tn})$ be a combination of strategies for G.

X is an *impatience equilibrium* of G, if

a] X is a quasi-pure subgame perfect equilibrium of G, and

b] no strategy combination X′ with the following properties exists:

b.1] $X' = (x_{T1}, ..., x_{Ti-1}, x'_{Ti}, x_{Ti+1}, ..., x_{Tn})$ and $x'_{Ti} \neq x_{Ti}$.

b.2] x'_{Ti} is a best reply strategy of T_i in all subgames of G, given the strategies $(x_{T1}, ..., x_{Ti-1}, x_{Ti+1}, ..., x_{Tn})$.

b.3]   Some t-th market phase subgame $G_t(.,.)$ exists, in which the trader T_i has not yet traded and in which one of the following conditions is true:

b.3.1]   On the plays realized by X in the first market phase P_t of $G_t(.,.)$, T_i does not submit an offer, but T_i does submit an offer on the plays realized by X′ in P_t.

b.3.2]   The best offer $\bar{p}'_{Ti}(P_t)$ that is submitted by T_i on the plays realized by X′ in the first market phase P_t of G_t is better than the best offer $\bar{p}_{Ti}(P_t)$ submitted by T_i on the plays realized by X in P_t.

b.3.3]   The best offer $\bar{p}'_{Ti}(P_t)$ that is submitted by T_i on the plays realized by X′ in the first market phase P_t of G_t is equal to the best offer $\bar{p}_{Ti}(P_t)$ submitted by T_i on the plays realized by X in P_t, but $\bar{p}_{Ti}'(P_t)$ is submitted after a smaller number of offer cycles than $p_{Ti}(P_t)$.

Definition: *Main Subgame* **with respect to a quasi-pure subgame perfect equilibrium**

Let G be a simplified alternating double auction market game. Let X be a quasi-pure subgame perfect equilibrium of G. Let d be the very first decision stage in G. Let d' be the decision stage that immediately follows d, given the choices of the traders prescribed by X in d. The subgame G' that begins with the decision stage d' is the *main subgame* of G with respect to the quasi-pure subgame perfect equilibrium X of G.

The following analysis will concentrate on specifying the necessary conditions for quasi-pure subgame perfect equilibria and impatience equilibria of the simplified alternating double auction market game. Generally, the necessary condition is that trading occurs at certain prices and - sometimes - in a certain sequence. Thus, the question, that is of interest to our analysis, is which type of market outcomes can be expected in such equilibria. From an institutional point of view, it is especially interesting to see, whether the traditional market clearing price can emerge as the outcome of a quasi-pure subgame perfect equilibrium path. We show that, the equilibrium outcomes are all *close to* or on the boundary of the market clearing price range, even with very little competition in the market. But, we also show that, the most trades occur just outside the market clearing price range in all impatience equilibria of the game.

III.C. 1-Trade Markets

A 1-trade market contains a single non-extra-marginal buyer and a single non-extra-marginal seller. Obviously, both of these must be marginal traders. Any other trader in a 1-trade market must be an extra-marginal trader. We begin by analyzing 1-trade markets without extra-marginal traders. The we examine the cases in which extra-marginal traders are present only on one of the market sides and, finally, the case of 1-trade markets with extra-marginal traders present on both market sides.

III.C.1. 1-Trade Markets Without Extra-Marginal Traders

In Proposition C.1.1 an impatience equilibrium of a 1-trade simplified alternating double auction market game without extra-marginal traders is specified. This establishes the existence of such equilibria of the game. Proposition C.1.2 shows that the equilibrium outcome is unique, i.e. that all quasi-pure subgame perfect equilibria and all impatience equilibria of the game must lead to the same equilibrium outcome.

Proposition C.1.1

Let G be a 1-trade simplified alternating double auction market game without extra-marginal traders. An impatience equilibrium of G exists with the property that in equilibrium a trade at the valuation of the second moving trader is made.

Proof of Proposition C.1.1

Let us assume that the buyer B_1 opens the market and the seller S_1 is on the second moving market side. The opposite case can be proven analogously.

Construction of an impatience equilibrium X of the game G

Consider a strategy combination X, that contains the two strategies x_{B1} and x_{S1} of the traders, with the following properties:

a] In the very first decision stage of the game d, x_{B1} prescribes submitting c_1, i.e. B_1 submits a bid equal to S_1's unit cost. Let d' denote the decision stage that immediately follows this choice. Let I' and G' denote the game segment and the subgame, correspondingly, that begin with d'.

b] x_{B1} induces the enforcement segment strategy $x'_{B1}(I')$ on the game segment I', i.e. B_1 refrains from submitting a new bid in all decision stages in I'.

c] x_{S1} induces the acceptance segment strategy $x'_{S1}(I')$ on the game segment I', i.e. S_1 immediately accepts the current market bid in all decision stages in I'.

d] In each other subgame of the game, except G', the combination of subgame strategies induced by X represents some arbitrary subgame perfect equilibrium of the subgame.

We start by splitting the game up into a number of subgames at the very first decision stage of B_1. Each of these subgames begins with a different first bid of B_1 (one of them begins with B_1 refraining to submit a first offer) and has at least one subgame perfect equilibrium in behavior strategies. In each of these subgames, except G', we assume with d] that X induces some arbitrary subgame perfect equilibrium. To prove that X is a subgame perfect equilibrium of the entire game, we have to show that the combination of the segment strategies $X'(I')$, that contains $x'_{B1}(I')$ and $x'_{S1}(I')$, is a subgame perfect equilibrium of G' and that B_1 has no reason to deviate from submitting c_1 in the very first decision stage d of G.

Subgame perfectness in the main subgame G'

First note that any subgame of G' is a game segment of G', because only two traders are in the market and only one trade is possible. This is also true for the largest game segment in G', namely, I', which is identical to G'. For this reason, the segment strategies contained in $X'(I')$ are complete strategies for G' and $X'(I')$ can qualify as an equilibrium of G'. Thus, to show that $X'(I')$ is a subgame perfect equilibrium of G', it suffices to show that: No game segment I'' exists in G', in which either B_1 or S_1 has an incentive to deviate from the game segment strategy combination induced by $X'(I')$ on I''.

Let I'' be some arbitrary game segment of G' that begins with the decision stage d''. Let $\tilde{b}''$ be the current market bid in d''. Due to the *ask-bid-spread-reduction* rule $\tilde{b}''$ cannot be smaller than the market bid c_1 in the first decision stage d' of G'. Thus, $\tilde{b}''$ must be in the market price range of d''. Since $\tilde{b}''$ is in the market price range of d'' and B_1 is on the opening market side and B_1 refrains from submitting a new bid in all his decision stages in I', it follows that S_1 faces the credibly enforced price $\tilde{b}''$ in I''. S_1 is the only trader on the second market side. Thus, it follows by

Corollary 1, that S_1's strategy in any market side best reply combination of the sellers (of which only one is present) prescribes accepting a trade at the enforced price $\tilde{b}''$ in I''. Note that for S_1's payoff in I'', given the strategy of B_1, it is irrelevant, whether S_1 accepts immediately or not. Thus, the acceptance segment strategy $x'_{S1}(I')$ of S_1 maximizes S_1's payoff in any game segment I'' of G', given the enforcement segment strategy $x'_{B1}(I')$ of B_1.

Since S_1 accepts any price that is credibly enforced by B_1 in I'', B_1 maximizes his payoff in I'' by enforcing the lowest possible bid in I''. The lowest possible bid, that B_1 can enforce in I'' is $\tilde{b}''$, because - due to the *ask-bid-spread-reduction* rule - no bid in I'' can be lower than the current market bid $\tilde{b}''$ in d''. Thus, $x'_{B1}(I')$ maximizes B_1's payoff in any game segment I'' of G', given the acceptance segment strategy $x'_{S1}(I')$ of S_1. Hence, for any possible I'' in G', neither B_1 nor S_1 have an incentive to deviate from their strategies contained in $X'(I')$. Thus, it follows that $X'(I')$ is a subgame perfect equilibrium of G'.

Optimality of choices in the first decision stage d

Now we have to prove that B_1 has no reason to deviate from submitting c_1 in the very first decision stage d of the game. As shown above, if B_1 submits c_1 in d, he trades at the price c_1 in G'. Due to the *no-loss* rule, c_1 is the lowest price at which a trade is possible. Hence, B_1's equilibrium payoff in any subgame other than G', that is reached by deviating in d, cannot be greater than the equilibrium payoff he achieves in G'.

Quasi-pureness and impatience

X is a quasi-pure subgame perfect equilibrium of the game, because the local strategy of the buyer in d is pure and the segment strategies of both traders in I' are pure segment strategies. But, X is also an impatience equilibrium of the game. This is so, because B_1's best offer on the play realized by any other strategy combination $\bar{X} = (\bar{x}_{B1}, x_{S1})$ which leads to a payoff for B_1 equal to his equilibrium payoff in X, must be c_1; otherwise B_1 would trade at a higher price. But, since x_{B1} prescribes submitting c_1 in the very first decision stage, no $\bar{x}_{B1}$ can specify submitting c_1 any earlier. Similarly, every alternative strategy of S_1, with the same payoff as x_{S1}, must prescribe submitting c_1, but cannot prescribe doing so any earlier than x_{S1} does. ∎

Proposition C.1.2

Let G be a 1-trade simplified alternating double auction market game without extra-marginal traders. Any combination of strategies that is a quasi-pure subgame perfect equilibrium of G - on the equilibrium path - leads to a trade at a price equal to the player valuation of the second moving trader.

Proof of Proposition C.1.2

We must only show that strategy combinations leading to other outcomes on cannot be quasi-pure subgame perfect equilibria. It follows by Lemma 3 that market period termination without a trade cannot be an equilibrium outcome. The only other possible outcome of the game is that a trade at some other price occurs.

Consider some price p unequal to the second mover's valuation at which a trade occurs. Due to the *no-loss* rule, it is true for all feasible prices p, that trading at p gives the opening trader a smaller and the second moving trader a greater payoff than trading at the second mover's valuation. Since a trade occurs at the price p, there must have been some decision stage d of the opening trader in which he submitted p. By Lemma 1 and since p is unequal to the second mover's valuation it follows that d was not a termination stage and, thus, not a take-it-or-leave-it stage of the opening trader. Hence, in the subgame starting at that stage, the opening trader also had the alternative to credibly enforce another price p*, which would have led to a greater payoff (e.g. his own current market offer at that time). By Corollary 1 it follows that any best reply strategy of the second moving trader must prescribe accepting the enforced price p*. Obviously, as long as p is unequal to the second mover's valuation, there is always some p* which - by enforcement - leads to a more profitable trade for the opening trader. ∎

III.C.2. 1-Trade Markets With Extra-Marginal Traders on Only One Market Side

As mentioned above, a 1-trade market contains a marginal buyer and a marginal seller. Additionally, we now allow for extra-marginal traders on one of the market sides, i.e. either extra-marginal buyers or extra-marginal sellers are present in the market. In Proposition C.2.1 an impatience equilibrium of the game is specified for the case that the extra-marginal traders are on the second moving market side. This establishes the existence of such equilibria of the game in that case. Proposition C.2.2 shows that the equilibrium outcome is unique, i.e. that all quasi-pure subgame perfect equilibria and all impatience equilibria of the game - in that case - must lead to the same equilibrium outcome.

In Proposition C.2.3 two different quasi-pure subgame perfect equilibria of the game are specified for the case that the extra-marginal traders are on the opening market side. The first is also an impatience equilibrium, while the second is not. This establishes the existence of such equilibria. Proposition C.2.4 shows that all quasi-pure subgame perfect equilibria must lead to one of the outcomes specified in Proposition C.2.3, where all impatience equilibria must lead to the outcome of the first equilibrium specified in Proposition C.2.3.

Proposition C.2.1

Let G be a 1-trade simplified alternating double auction market game with extra-marginal traders only on the second moving market side. An impatience equilibrium of G exists with the property that in equilibrium a trade occurs at the valuation of the marginal trader on the second moving market side.

Proof of Proposition C.2.1

Assume that the buyer is on the opening market side, i.e. that the extra-marginal traders are sellers. The opposite case can be proven analogously. Thus, the marginal buyer is B_1. The marginal seller is S_1 and the extra-marginal sellers are S_j, with

$j = 2 \dots n_S$. Obviously, the best extra-marginal seller is S_2 with the unit cost c_2.

Construction of an impatience equilibrium X of the game G

Consider a strategy combination X, that contains the strategies x_{B1} and x_{Sj}, with $j = 1 \dots n_S$, with the following properties:

a] In the very first decision stage of the game d, x_{B1} prescribes submitting c_1, i.e. B_1 submits a bid equal to the marginal seller S_1's unit cost. Let d' denote the decision stage that immediately follows this choice. Let I' and G' denote the game segment and the subgame, correspondingly, that begin with d'.

b] x_{B1} induces the enforcement segment strategy $x'_{B1}(I')$ on the game segment I', i.e. B_1 refrains from submitting a new bid in all decision stages in I'.

c] All x_{Sj} induce the acceptance segment strategies $x'_{Sj}(I')$ on the game segment I', i.e. the marginal seller S_1 immediately accepts the current market bid in every decision stage of I' and each of the extra-marginal sellers immediately accepts the current market bid in those decision stages in I', in which the *no-loss* rule allows accepting, submits her best ask in those decision stages in I', in which the *ask-bid-spread-reduction* rule and the *no-crossing* rule allow the submission, and otherwise, refrains from submitting a new ask.

d] In each other subgame of the game, except G', the combination of subgame strategies induced by X represents some arbitrary subgame perfect equilibrium of the subgame.

We start by splitting the game up into a number of subgames at the very first decision stage of B_1. Each of these subgames begins with a different first bid of B_1 (one of them begins with B_1 refraining to submit a first offer). Each of these subgames has at least one subgame perfect equilibrium in behavior strategies. In each of these subgames, except G', we assume with d] that X induces some arbitrary subgame perfect equilibrium. To prove that X is a subgame perfect equilibrium of the entire game, we have to show that the combination of the segment strategies

$X'(I')$, that contains $x'_{B1}(I')$ and $x'_{Sj}(I')$, with $j = 1 \ldots n_S$, is a subgame perfect equilibrium of G' and that B_1 has no reason to deviate from submitting c_1 in the very first decision stage d of the game.

Subgame perfectness in the main subgame G'

First note that any subgame of G' is a game segment of G', because only one trade is possible. This is also true for the largest game segment in G', namely, I', which is identical to G'. For this reason, the segment strategies contained in $X'(I')$ are complete strategies for G' and $X'(I')$ can qualify as an equilibrium of G'. Thus, to show that $X'(I')$ is a subgame perfect equilibrium of G', it suffices to show that: No game segment I'' exists in G', in which one of the traders has an incentive to deviate from the game segment strategy combination induced by $X'(I')$ on I''.

Let I'' be some arbitrary game segment of G' that begins with the decision stage d''. Let $\bar{b}''$ be the current market bid in d''. Due to the *ask-bid-spread-reduction* rule $\bar{b}''$ cannot be smaller than the market bid c_1 in the first decision stage d' of G'. Thus, $\bar{b}''$ must be in the market price range of d''. Since $\bar{b}''$ is in the market price range of d'' and B_1 is on the opening market side and B_1 refrains from submitting a new bid in all his decision stages in I'', it follows that the sellers face the credibly enforced price $\bar{b}''$ in I''. No seller can trade after I'', because only one trade is possible in the game. Thus, it follows by Corollary 2, that, in any market side best reply combination of the sellers, each seller, who makes no loss by trading at $\bar{b}''$, must accept a trade at the enforced price $\bar{b}''$. If any number k of sellers, with $k > 1$ can accept a trade at $\bar{b}''$, each of them should accept immediately to have a $\frac{1}{k}$ chance of trading at the price $\bar{b}''$, because each other one also accepts immediately. If one of them would delay the acceptance, she would be certain not to trade at all. If only S_1 can accept $\bar{b}''$, it is irrelevant for her payoff in I'', whether she accepts a trade at $\bar{b}''$ immediately or not, given B_1's strategy. Thus, the acceptance segment strategy $x'_{S1}(I')$ maximizes S_1's payoff in any game segment I'' of G', given the enforcement segment strategy of B_1 and given the acceptance segment strategies of the other sellers. If an extra-marginal seller S_j, with $j = 2 \ldots n_S$ cannot accept $\bar{b}''$, because $\bar{b}'' < c_j$, all her segment strategies in I'' lead to a payoff of 0, given the strategies of the other traders. Thus, by submitting her best ask, S_j achieves a no smaller payoff in I'' than with any other segment strategy. Hence, the acceptance segment strategy $x'_{Sj}(I')$ maximizes each S_j's payoff in any game segment I'' of G', given the enforcement segment strategy of B_1 and given the acceptance segment strategies of the other sellers.

Since at least one seller accepts any price that is credibly enforced by B_1 in I'', B_1 maximizes his payoff in I'' by enforcing the lowest possible bid in I''. The lowest possible bid, that B_1 can enforce in I'' is $\bar{b}''$, because - due to the *ask-bid-spread-reduction* rule - no bid in I'' can be smaller than the current market bid $\bar{b}''$ in d''. Thus, $x'_{B1}(I')$ maximizes B_1's payoff in any game segment I'' of G', given the acceptance segment strategies of the sellers. Hence, for any possible I'' in G', neither of the buyer nor the sellers have an incentive to deviate from their strategies contained in $X'(I')$. Thus, it follows that $X'(I')$ is a subgame perfect equilibrium of G'.

Optimality of choices in the first decision stage d

Now we have to prove that B_1 has no reason to deviate from submitting c_1 in the very first decision stage d of the game. As shown above, if B_1 submits c_1 in d, he trades at the price c_1 in G'. Due to the *no-loss* rule, c_1 is the highest price at which a trade is possible. Hence, B_1's equilibrium payoff in any subgame other than G', that is reached by deviating in d, cannot be greater than the equilibrium payoff he achieves in G'.

Quasi-pureness and impatience

X is a quasi-pure subgame perfect equilibrium of the game, because the local strategy of the buyer in d is pure and the segment strategies of all traders in I' are pure segment strategies. But, X is also an impatience equilibrium of the game. This is so, because B_1's best offer on the play realized by any other strategy combination $\bar{X} = (\bar{x}_{B1}, x_{S1}, ..., x_{SnS})$ which leads to a payoff for B_1 equal to his equilibrium payoff in X, must be c_1; otherwise B_1 would trade at a higher price. But, since x_{B1}

prescribes submitting c_1 in the very first decision stage, no $\bar{x}_{B1}$ can specify submitting c_1 any earlier. Similarly, every alternative strategy of the marginal seller S_1, with the same payoff as x_{S1}, must prescribe submitting c_1, but cannot prescribe doing so any earlier than x_{S1} does. Lastly, every alternative strategy of any extra-marginal seller S_j, with $j = 2 \dots n_S$, also leads to a payoff of 0, given the other strategies. Thus, each S_j must submit her best offer c_j as soon as possible in an impatience equilibrium. This is true in X.   ∎

Proposition C.2.2

Let G be a 1-trade simplified alternating double auction market game with extra-marginal traders only on the second moving market side. Any combination of strategies that is a quasi-pure subgame perfect equilibrium of G - on the equilibrium path - leads to a trade at a price equal to the player valuation of the marginal trader of the second moving market side.

Proof of Proposition C.2.2

We must only show that strategy combinations leading to other outcomes cannot be quasi-pure subgame perfect equilibria. Since only one trade is possible (i.e. only one pair of traders can trade), the only other possible outcomes of the game are: either market period termination without a trade or a trade at some other price. It follows by Lemma 3 that market period termination without a trade cannot be an equilibrium outcome, so we only have to check for the latter case.

We give no formal proof for the proposition, because the proof completely analogous to the Proof of Proposition C.1.2: Since the trader on the opening market side has no competition, it immediately follows by Lemma 5 that this trader can choose to credibly enforce any price within the market price range without fearing to be overbid or underasked. Thus, he enforces the best possible offer of the other market side. The marginal trader on the second moving market side must then accept this enforced price, because he is the only one who can. (This can also easily be seen by checking the proof to the Proposition C.2.1.)   ∎

Proposition C.2.3

Let G be a 1-trade simplified alternating double auction market game with extra-marginal traders only on the opening moving market side.

(1)   An impatience equilibrium of G exists with the property that in equilibrium the two marginal traders trade at a price, which lies just outside the range of feasible prices for the best extra-marginal trader (i.e. exactly one money unit greater than the best extra-marginal buyer's redemption value or exactly one money unit smaller than the best extra-marginal seller's unit cost).

(2)   Another quasi-pure subgame perfect equilibrium of G exists, that is not an impatience equilibrium, with the property that in equilibrium the two marginal traders trade at a price equal to the best extra-marginal trader's player valuation.

Proof of Proposition C.2.3 - Case (1)

Assume that the buyers are on the opening market side, i.e. that the extra-marginal traders are buyers. The opposite case can be proven analogously. Thus, the marginal buyer is B_1 and the extra-marginal buyers are B_i, with $i = 2 \ldots n_B$. Obviously, the best extra-marginal buyer is B_2 with the redemption value v_2. The seller is S_1. We must show that an impatience equilibrium exists in which B_1 and S_1 trade at the price $v_2 + 1$.

Construction of an impatience equilibrium X of the game G

Consider a strategy combination X, that contains the strategies x_{Bi}, with $i = 1 \ldots n_B$, and x_{S1} of the traders, with the following properties:

a]   In the very first decision stage d of the game, x_{B1} prescribes submitting $v_2 + 1$ and each x_{Bi}, with $i = 2 \ldots n_B$, prescribes submitting v_i, i.e. the marginal buyer B_1 submits a bid equal to the best extra-marginal buyer B_2's redemption value plus one and each extra-marginal buyer submits a bid equal to his own redemption value. Let d' denote the decision stage that immediately follows these choices. Let I' and G' denote the game segment and the subgame, correspondingly, that begin with d'.

b] All x_{Bi}, with $i = 1 \dots n_B$, induce the enforcement segment strategies $x'_{Bi}(I')$ on the game segment I', i.e. all buyers refrain from submitting in all decision stages in I'.

c] x_{S1} induces the acceptance segment strategy $x'_{S1}(I')$ on the game segment I', i.e. S_1 immediately accepts the current market bid in all decision stages in I'.

d] In each other subgame of the game, except G', the combination of subgame strategies induced by X represents some arbitrary subgame perfect equilibrium of the subgame.

We start by splitting the game up into a number of subgames at the very first decision stage of the buyers. This first decision stage is not a single decision node, but the combination of the decision nodes of all B_i. Thus, each of the subgames that follow the first decision stage begins with a different combination of first bids of the buyers (one of them begins with all buyers refraining to submit a first offer). Each of these subgames has at least one subgame perfect equilibrium in behavior strategies. In each of these subgames, except G', we assume with d] that X induces some arbitrary subgame perfect equilibrium. To prove that X is a quasi-pure subgame perfect equilibrium of the entire game, we have to show that the combination of the segment strategies $X'(I')$, that contains $x'_{Bi}(I')$, with $i = 1 \dots n_B$, and $x'_{S1}(I')$, is a subgame perfect equilibrium of G' and that none of the buyers B_i has an incentive to deviate from submitting the bids specified in a] in the very first decision stage d of the game.

Subgame perfectness in the main subgame G'

First note that any subgame of G' is a game segment of G', because only one trade is possible. This is also true for the largest game segment in G', namely, I', which is identical to G'. For this reason, the segment strategies contained in $X'(I')$ are complete strategies for G' and $X'(I')$ can qualify as an equilibrium of G'. Thus, to show that $X'(I')$ is a subgame perfect equilibrium of G', it suffices to show that: No game segment I'' exists in G', in which one of the traders has an incentive to deviate from the game segment strategy combination induced by $X'(I')$ on I''.

Let I'' be some arbitrary game segment of G' that begins with the decision stage d''. Let $\bar{b}''$ be the current market bid in d''. Due to the *ask-bid-spread-reduction* rule $\bar{b}''$ cannot be smaller than the market bid $v_2 + 1$ in the first decision stage d' of G'. Thus, $\bar{b}''$ must be in the market price range of d''. Since $\bar{b}''$ is in the market price range of d'' and the buyers are on the opening market side and all buyers refrain from submitting new bids in all their decision stages in I'', it follows that S_1 faces the credibly enforced price $\bar{b}''$ in I''. S_1 is the only trader on the second market side. Thus, it follows by Corollary 1, that S_1's strategy in any market side best reply combination of the sellers prescribes accepting a trade at the enforced price $\bar{b}''$ in I''. Note that for S_1's payoff in I'', given the strategies of the buyers, it is irrelevant, whether S_1 accepts immediately or not. Thus, the acceptance segment strategy $x'_{S1}(I')$ maximizes S_1's payoff in any game segment I'' of G', given the enforcement segment strategies of the buyers.

Since S_1 accepts any price that is credibly enforced by the buyers in I'' and since the extra-marginal buyers B_i, with $i = 2 \ldots n_B$, have no choice but to refrain from submitting new bids, B_1 maximizes his payoff in I'' by enforcing the lowest possible bid in I''. The lowest possible bid, that B_1 can enforce in I'' is $\bar{b}''$, because - due to the *ask-bid-spread-reduction* rule - no bid in I'' can be lower than the current market bid $\bar{b}''$ in d''. Thus, $x'_{B1}(I')$ maximizes B_1's payoff in any game segment I'' of G', given the acceptance segment strategy $x'_{S1}(I')$ of S_1 and the enforcement segment strategies of the extra-marginal buyers.

Finally, since all segment strategy of the extra-marginal buyers B_i, with $i = 2 \ldots n_B$, in any I'' lead to payoffs of 0, each $x'_{Bi}(I')$ maximizes B_i's payoff in any game segment of G', given the strategies of the others. Hence, for any possible I'' in G', neither the buyers nor S_1 have an incentive to deviate from their strategies contained in $X'(I')$. Thus, it follows that $X'(I')$ is a subgame perfect equilibrium of G'.

Optimality of choices in the first decision stage d

Now we have to prove that no buyers has a reason to deviate from the choice specified in a] in the very first decision stage d of the game. Let us begin with B_1. Given the strategies of the extra-marginal buyers B_i, with $i = 2 \ldots n_B$, and of S_1, if B_1 submits $v_2 + 1$ in d, he trades at the price $v_2 + 1$ in G'. With this, he achieves a payoff $v_1 - (v_2 + 1) + \beta$. If he deviates by submitting a higher bid, he earns less, since - even in the very best case - he trades at price greater than $v_2 + 1$. If he deviates by submitting a bid smaller than v_2 (or deviates by refraining to submit a bid), one of two outcomes are possible: either he does not trade at all (e.g. S_1 accepts B_2's bid at v_2) or he later overbids B_2's opening bid v_2. In the former case, B_1 has a payoff of 0, which is smaller than the payoff of immediately bidding $v_2 + 1$. In the latter case, since overbidding means bidding at least $v_2 + 1$, B_1 can at maximum have a payoff equal to that of immediately bidding $v_2 + 1$. Finally, if B_1 deviates by also bidding v_2, one of two outcomes are possible: either B_1 later overbids v_2 or B_1 and B_2 remain in a tie at v_2 and each of them has a ½ chance of trading with S_1. In the former case, since overbidding means bidding at least $v_2 + 1$, B_1 at maximum has a payoff equal to that of immediately bidding $v_2 + 1$. In the latter case, B_1 has a maximum expected payoff of $\frac{1}{2}(v_1 - v_2 + \beta)$, which is smaller than the payoff of immediately bidding $v_2 + 1$. This is so, because $v_1 > v_2$ together with $\beta > 1$ implies that $v_1 - v_2 \geq 1 > 2 - \beta$, which means $v_1 - v_2 > 2 - \beta \Leftrightarrow v_1 - (v_2 + 1) + \beta > \frac{1}{2}(v_1 - v_2 + \beta)$. Thus, deviating from the opening bid $v_2 + 1$ cannot lead to a higher payoff for B_1, given the other strategies.

The question left to answer is whether any extra-marginal buyer B_i, with $i = 2 \ldots n_B$ has an incentive to deviate from his opening bid v_i, given the strategies of the others. Clearly, this is not the case, since, given the strategies of the others, no extra-marginal buyer B_i is able to trade in this game. Thus, all of B_i's strategies lead to a payoff of 0 and any one is a best reply strategy. Since neither of the buyers has an incentive to deviate from his choice in d and since X'(I') is a subgame perfect equilibrium of G', X is a subgame perfect equilibrium of the entire game.

Quasi-pureness and impatience

X is a quasi-pure subgame perfect equilibrium of the game, because the local strategies of the buyers in d are pure and the segment strategies of all traders in I' are pure segment strategies. But, X is also an impatience equilibrium of the game. This is so, because B_1's best offer on the play realized by any other strategy combination $\bar{X} = (\bar{x}_{B1}, x_{B2}, ..., x_{BnB}, x_{S1})$ which leads to a payoff for B_1 equal to his equilibrium payoff in X, must be $v_2 + 1$; otherwise B_1 would trade at a different price or not at all. But, since x_{B1} prescribes submitting $v_2 + 1$ in the very first decision stage, no $\bar{x}_{B1}$ can specify submitting $v_2 + 1$ any earlier. Similarly, every alternative strategy of the marginal seller S_1, with the same payoff as x_{S1}, must prescribe submitting $v_2 + 1$, but cannot prescribe doing so any earlier than x_{S1} does. Lastly, every alternative strategy of any extra-marginal buyer B_i, with i = 2 ... n_B, also leads to a payoff of 0, given the other strategies. Thus, each B_i must submit his best offer v_i as soon as possible in an impatience equilibrium. This is true in X. ∎

Proof of Proposition C.2.3 - Case (2)

Again, we assume that the buyers are on the opening market side, i.e. that the extra-marginal traders are buyers. The opposite case can be proven analogously. Thus, the marginal buyer is B_1 and the extra-marginal buyers are B_i, with i = 2 ... n_B. Obviously, the best extra-marginal buyer is B_2 with the redemption value v_2. The seller is S_1.

We must show that a quasi-pure subgame perfect equilibrium of G exists, that is not an impatience equilibrium, in which B_1 and S_1 trade at the price v_2. Let us, again, start by splitting the game up into a number of subgames at the game's very first decision stage of the buyers d. This first decision stage is not a single decision node, but the combination of the decision nodes of all B_i. Thus, each of the subgames that follows the first decision stage begins with a different combination of first bids of the buyers (one of them begins with all buyers refraining to submit a first offer). Each of these subgames has at least one subgame perfect equilibrium in behavior strategies. For two of these subgames G_1 and G_2, however, we specify pure subgame strategies of the traders.

G_1 is the subgame that begins after the following combination of decisions in d: the marginal buyer B_1 submits v_2; the best extra-marginal buyer B_2 submits $v_2 - 1$; each other extra-marginal buyer B_i, with $i = 3 \ldots n_B$ submits v_i. Let d_1 denote the first decision stage in G_1, i.e. the decision stage of the seller S_1 that immediately follows d in G_1. Let I_1 denote the game segment that begins with d_1. We will see that G_1 is the main subgame of G.

G_2 is the subgame of G that begins after the following combination of decisions in d: the marginal buyer B_1 submits v_2; the best extra-marginal buyer B_2 submits v_2; each other extra-marginal buyer B_i, with $i = 3 \ldots n_B$ submits v_i. Let d_2 denote the first decision stage in G_2, i.e. the decision stage of the seller S_1 that immediately follows d in G_2. Let I_2 denote the game segment that begins with d_2. We will call G_2 the *threat subgame* of G.

Construction of a quasi-pure subgame perfect equilibrium X of the game G

Now, consider a strategy combination X, that contains the strategies x_{Bi}, with $i = 1 \ldots n_B$, and x_{S1} of the traders, with the following properties:

a]   In the very first decision stage d of the game, x_{B1} prescribes submitting v_2, i.e. the marginal buyer B_1 submits a bid equal to the best extra-marginal buyer B_2's redemption value. x_{B2} prescribes submitting $v_2 - 1$ in d, i.e. the best extra-marginal buyer B_2 submits a bid equal to his own redemption value minus one in d. Each x_{Bi}, with $i = 3 \ldots n_B$, prescribes submitting v_i in d, i.e. each other extra-marginal buyer submits a bid equal to his own redemption value in d.

b.1]   All x_{Bi}, with $i = 1 \ldots n_B$, induce the enforcement segment strategies $x_{1-Bi}(I_1)$ on the game segment I_1, i.e. all buyers refrain from submitting in all decision stages in I_1.

b.2] x_{B1} induces the segment strategy $x_{2-B1}(I_2)$ on the game segment I_2 that pre-scribes the following: (a) Submit $v_2 + 1$ in any decision stage of I_2, in which the current market bid is equal to v_2. (b) Refrain from submitting a bid in any decision stage of I_2, in which the current market bid is greater than v_2. All x_{Bi}, with $i = 2 \ldots n_B$, induce the enforcement segment strategies $x_{2-Bi}(I_2)$ on the game segment I_2, i.e. all extra-marginal buyers refrain from submitting in all decision stages in I_2.

c.1] x_{S1} induces the acceptance segment strategy $x_{1-S1}(I_1)$ on the game segment I_1, i.e. S_1 immediately accepts the current market bid in all decision stages in I_1.

c.2] x_{S1} induces the segment strategy $x_{2-S1}(I_2)$ on the game segment I_2 that pre-scribes the following: (a) Submit an ask equal to $v_2 + 1$ in any decision stage of I_2, in which the current market bid is equal to v_2 and either the current market ask is greater than $v_2 + 1$ or no current market ask exists. (b) Submit an ask equal to v_2 in any decision stage of I_2, in which the current market bid is equal to v_2 and the current market ask is equal to $v_2 + 1$. (c) Accept a trade at the current market bid in any decision stage of I_2, in which the current market bid is greater than v_2.

d] In each other subgame of the game, except G_1 and G_2, the combination of subgame strategies induced by X represents some arbitrary subgame perfect equilibrium of the subgame.

The following table III.1 displays the plays that are realized by X, depending on whether B_2 bids $v_2 - 1$ in d (in which case the subgame G_1 is reached) or bids v_2 in d (in which case the subgame G_2 is reached). Given the prescriptions of X, B_2 bids $v_2 - 1$ and the main subgame G_1 is reached. The threat subgame G_2, that is not reached in equilibrium, allows B_1 to threaten to overbid a tie at v_2 in d.

Table III.1 - Plays realized by X depending on the first bid of B_2

subgame	offer cycle - decision stage	B_1's bid	B_2's bid	B_i's bid ($i = 3 \ldots n_B$)	S_1's ask
	if B_2 submits $v_2 - 1$ in the very first decision stage d				
G_1	1 - buyers	v_2	$v_2 - 1$	v_i	
	1 - sellers				v_2
	B_1 trades with S_1 at the price v_2				
	if B_2 submits v_2 in the very first decision stage d				
G_2	1 - buyers	v_2	v_2	v_i	
	1 - sellers				$v_2 + 1$
	2 - buyers	$v_2 + 1$	-	-	
	B_1 trades with S_1 at the price $v_2 + 1$				

In each of the subgames, that follow the first decision stage d of the buyers, except G_1 and G_2, we assume with d] that X induces some arbitrary subgame perfect equilibrium. To prove that X is a subgame perfect equilibrium of the entire game, we first have to show that the combination of the segment strategies $X_1(I_1)$, that contains $x_{1-Bi}(I_1)$, with $i = 1 \ldots n_B$, and $x_{1-S1}(I_1)$, is a subgame perfect equilibrium of G_1. Next we show that the combination of the segment strategies $X_2(I_2)$, that contains $x_{2-Bi}(I_2)$, with $i = 1 \ldots n_B$, and $x_{2-S1}(I_2)$, is a subgame perfect equilibrium of G_2. Last, we show that none of the buyers B_i has an incentive to deviate from submitting the bids specified in a] in the very first decision stage d.

Before we begin, note that any subgame of G_1 or G_2 is a game segment of G_1 or G_2, correspondingly, because only one trade is possible. This is also true for the largest game segments in G_1 and in G_2, namely, I_1 and I_2, which are identical to G_1 and G_2, correspondingly. For this reason, the segment strategies contained in $X_1(I_1)$ and $X_2(I_2)$ are complete strategies for G_1 and G_2, correspondingly, and $X_1(I_1)$ and $X_2(I_2)$ can qualify as equilibria of G_1 or G_2, correspondingly.

Subgame perfectness in the main subgame G_1

To show that $X_1(I_1)$ is a subgame perfect equilibrium of G_1, it suffices to show that: No game segment I_1'' exists in G_1, in which one of the traders has an incentive to deviate from the game segment strategy combination induced by $X_1(I_1)$ on I_1.

Let I_1'' be some arbitrary game segment of G_1 that begins with the decision stage d_1''. Let $\bar{b}_1''$ be the current market bid in d_1''. Due to the *ask-bid-spread-reduction* rule $\bar{b}_1''$ cannot be smaller than the market bid v_2 in the first decision stage d_1 of G_1. Thus, $\bar{b}_1''$ must be in the market price range of d_1''. Since $\bar{b}_1''$ is in the market price range of d_1'' and the buyers are on the opening market side and all buyers refrain from submitting new bids in all their decision stages in I_1'', it follows that S_1 faces the credibly enforced price $\bar{b}_1''$ in I_1''. S_1 is the only trader on the second market side. Thus, it follows by Corollary 1, that S_1's strategy in any market side best reply combination of the sellers prescribes accepting a trade at the enforced price $\bar{b}_1''$ in I_1''. Note that for S_1's payoff in I_1'', given the strategies of the buyers, it is irrelevant, whether S_1 accepts immediately or not. Thus, the acceptance segment strategy $x_{1-S1}(I_1)$ maximizes S_1's payoff in any game segment I_1'' of G_1, given the enforcement segment strategies of the buyers.

The marginal buyer B_1 is the only buyer who can hold the current market bid $\bar{b}_1''$ in the game segment I_1'', because of the following reason: Since B_1 is the only buyer who holds the current market bid v_2 in the first decision stage d_1 of the subgame G_1, for another buyer B_i, with $i = 2 \ldots n_B$, to hold $\bar{b}_1''$, it is necessary that B_i has overbid v_2 at some decision stage before d_1''. Due to the *no-loss* rule, however, no buyer - except B_1 himself - can overbid v_2. Thus, any $\bar{b}_1''$ is held by B_1 alone. Since S_1 accepts any price that is credibly enforced by the buyers in I_1'' and since B_1 alone holds $\bar{b}_1''$, B_1 maximizes his payoff in I_1'' by enforcing the lowest possible bid in I_1''. The lowest possible bid, that B_1 can enforce in I_1'' is $\bar{b}_1''$, because - due to the *ask-bid-spread-reduction* rule - no bid in I_1'' can be lower than the current market bid $\bar{b}_1''$ in d_1''. Thus, $x_{1-B1}(I_1)$ maximizes B_1's payoff in any game segment I_1'' of G_1, given the acceptance segment strategy $x_{1-S1}(I_1)$ of S_1

and the enforcement segment strategy of B_2.

Finally, since all segment strategies of the extra-marginal buyers B_i, with $i = 2 \ldots$ n_B, in any I_1'' lead to payoffs of 0, each $x_{1-Bi}(I_1)$ maximizes B_i's payoff in any game segment of G_1, given the strategies of the others. Hence, for any possible I_1'' in G_1, neither the buyers nor S_1 have an incentive to deviate from their strategies contained in $X_1(I_1)$. It follows that $X_1(I_1)$ is a subgame perfect equilibrium of G_1.

Subgame perfectness in the threat subgame G_2

Just as before, to show that $X_2(I_2)$ is a subgame perfect equilibrium of G_2, it suffices to show that: No game segment I_2'' exists in G_2, in which one of the traders has an incentive to deviate from the game segment strategy combination induced by $X_2(I_2)$ on I_2''.

Let I_2'' be some arbitrary game segment of G_2 that begins with the decision stage d_2''. Let $\bar{b}_2''$ be the current market bid in d_2''. Due to the *ask-bid-spread-reduction* rule $\bar{b}_2''$ cannot be smaller than the market bid v_2 in the first decision stage d_2 of G_2. Thus and since no B_i, with $i = 2 \ldots n_B$, can overbid v_2, $\bar{b}_2''$ must either be greater than v_2 and held only by B_1 or equal to v_2 and held by both B_1 and B_2. (Remember that B_1 and B_2 jointly hold the market bid v_2 in the first decision stage d_2 of the subgame G_2.) Hence, any game segment I_2'' of G_2 is either of type A, in which $\bar{b}_2'' = v_2$ is held by B_1 and B_2, or of type B, in which $\bar{b}_2'' > v_2$ is held by B_1 alone. Let I_{2-A}'' denote an arbitrary game segment of type A in G_2 and I_{2-B}'' denote an arbitrary game segment of type B in G_2.

Subgame perfectness in the threat subgame G_2: Game segments I_{2-B}'' of G_2 in which B_1 alone holds the current market bid $\bar{b}_{2-B}'' > v_2$

Let us first examine some game segment I_{2-B}'' that begins with the decision stage d_{2-B}'', in which the current market bid is $\bar{b}_{2-B}'' > v_2$. Since $\bar{b}_{2-B}''$ is in the market price range of d_{2-B}'' and the buyers open the market and all refrain from submitting new bids in I_{2-B}'', $\bar{b}_{2-B}''$ is a credibly enforced price. Thus, it follows by Corollary 1, that S_1's strategy in any market side best reply combination of the sellers prescribes accepting a trade at the enforced price $\tilde{b}_{2\text{-}B}''$ in $I_{2\text{-}B}''$. Note that for S_1's payoff in $I_{2\text{-}B}''$, given the strategies of the buyers, it is irrelevant, whether S_1 accepts $\tilde{b}_{2\text{-}B}'' > v_2$ immediately or not. Thus, accepting $\tilde{b}_{2\text{-}B}''$ maximizes S_1's payoff in any game segment $I_{2\text{-}B}''$ of G_2 and S_1 has no reason to deviate from her segment strategy $x_{2\text{-}S1}(I_2)$ in any $I_{2\text{-}B}''$.

B_1, too, has no reason to deviate from his segment strategy $x_{2\text{-}B1}(I_2)$ in any $I_{2\text{-}B}''$. Since S_1 accepts any price that is credibly enforced by the buyers in $I_{2\text{-}B}''$ and since B_1 alone holds $\tilde{b}_{2\text{-}B}'' > v_2$, B_1 maximizes his payoff in $I_{2\text{-}B}''$ by enforcing the lowest possible bid in $I_{2\text{-}B}''$. The lowest possible bid, that B_1 can enforce in $I_{2\text{-}B}''$ is $\tilde{b}_{2\text{-}B}''$, because - due to the *ask-bid-spread-reduction* rule - no bid in $I_{2\text{-}B}''$ can be lower than the current market bid $\tilde{b}_{2\text{-}B}''$ in any $d_{2\text{-}B}''$. Thus, $x_{2\text{-}B1}(I_2)$ maximizes B_1's payoff in any game segment $I_{2\text{-}B}''$ of G_2, given the segment strategies of the other traders.

Finally, since all segment strategies of the extra-marginal buyers B_i, with $i = 2 \dots$ n_B, in any $I_{2\text{-}B}''$ lead to payoffs of 0, each $x_{2\text{-}Bi}(I_2)$ maximizes B_i's payoff in any game segment $I_{2\text{-}B}''$ of G_2, given the strategies of the others. Hence, for any possible $I_{2\text{-}B}''$ in G_2, neither the buyers nor S_1 have an incentive to deviate from their strategies contained in $X_2(I_2)$.

Subgame perfectness in the threat subgame G_2: Game segments $I_{2\text{-}A}''$ of G_2 in which B_1 and B_2 together hold the current market bid $\tilde{b}_{2\text{-}A}'' = v_2$

Now, let us examine some game segment $I_{2\text{-}A}''$ that begins with the decision stage $d_{2\text{-}A}''$, in which the current market bid is $\tilde{b}_{2\text{-}A}'' = v_2$. We only have to examine whether the traders have an incentive to deviate from their segment strategies in those decision stages of $I_{2\text{-}A}''$, in which the current market bid is equal to v_2. This is so, because any other decision stage of $I_{2\text{-}A}''$, with a current market bid greater than v_2, is part of some game segment $I_{2\text{-}B}''$ of type B, in which - as we have shown above - the traders have no incentive to deviate from their strategies. Since any decision stage of $I_{2\text{-}A}''$, in which the current market bid is equal to v_2, is the first decision stage $d_{2\text{-}A}'$ of some other game segment $I_{2\text{-}A}'$ of type A in G_2, it suffices to show that neither of the buyers nor the seller have an incentive to deviate from the choice induced by their segment strategies in the first decision stage d_{2-A}' of any game segment I_{2-A}' of type A in G_2.

First, assume that the examined arbitrary decision stage d_{2-A}' is a decision stage of the buyers. Obviously, none of the extra-marginal buyers B_i, with $i = 2 \ldots n_B$ has a choice but to refrain from submitting a bid in any d_{2-A}', since no B_i can overbid the current market bid $\tilde{b}_{2-A}' = v_2$ in d_{2-A}'. Thus, the enforcement segment strategy $x_{2-Bi}(I_2)$ maximizes each extra-marginal buyer's payoff in any game segment I_{2-A}'.

B_1 has no incentive to deviate from his segment strategy's prescription for d_{2-A}' either, i.e. B_1 cannot achieve a greater payoff than he achieves by overbidding v_2 with $v_2 + 1$ in d_{2-A}'. If B_1 overbids v_2 with $v_2 + 1$ in d_{2-A}', a game segment of type B is reached, in which S_1 accepts a trade at $v_2 + 1$. If B_1 overbids v_2 with a bid $p > v_2 + 1$ in d_{2-A}', another game segment of type B is reached, in which S_1 accepts a trade at p. But trading at p results in a smaller payoff for B_1 than trading at $v_2 + 1$, since $p > v_2 + 1$. If B_1 refrains from bidding in d_{2-A}', two outcomes are possible, given S_1's segment strategy: (i) If the current market ask $\tilde{a}_{2-A}'$ in d_{2-A}' is equal to $v_2 + 1$, S_1 accepts a trade at v_2 in her next decision stage, which is a take-it-or-leave-it stage. This leads to an expected payoff of $\frac{1}{2}(v_1 - v_2 + \beta)$ for B_1, since either he or B_2 trade with S_1 at the price v_2. (ii) If the current market ask $\tilde{a}_{2-A}'$ in d_{2-A}' is greater than $v_2 + 1$ (or not specified yet), S_1 submits the ask $v_2 + 1$ in her next decision stage, after which the next game segment of type A is reached. Thus, this outcome takes us to the beginning of this paragraph, i.e. to some other decision stage of the buyers in which the current market bid is v_2. By recursion (and since the game is finite), it becomes evident that, if B_1 refrains to submit a bid in d_{2-A}', the game segment I_{2-A}' must end either with B_1 trading at some price $p \geq v_2 + 1$ with probability 1 or with B_1 trading at v_2 with probability $\frac{1}{2}$. Thus, if B_1 refrains to submit a bid in d_{2-A}', he can achieve a no greater payoff than he achieves by bidding $v_2 + 1$ in d_{2-A}'. Since a bid greater than $v_2 + 1$ leads to a smaller payoff than the bid $v_2 + 1$ and since refraining to bid can at maximum lead to a payoff equal to the payoff of bidding $v_2 + 1$, B_1 has no incentive to deviate from his segment strategy in any first decision stage d_{2-A}' of any game segment I_{2-A}' of G_2.

Now, assume that the examined arbitrary decision stage d_{2-A}' is a decision stage of the seller. We have to distinguish two cases: (i) The current market ask $\tilde{a}_{2-A}'$ in d_{2-A}' is greater than $v_2 + 1$ (or not specified yet). (ii) The market ask $\tilde{a}_{2-A}'$ in d_{2-A}' is equal to $v_2 + 1$.

Let us begin with the second case (ii). Since the current market ask $\tilde{a}_{2-A}' = v_2 + 1$ and the current market bid $\tilde{b}_{2-A}' = v_2$ in d_{2-A}', d_{2-A}' is a take-it-or-leave-it stage of S_1 and no decision stage of the buyers follows d_{2-A}'. Thus, S_1 can only choose to either accept a trade at v_2 or to let the market period terminate without a trade, by refraining to submit v_2. Obviously, x_{2-S1}, which prescribes submitting an ask equal to v_2, i.e. accepting a trade at the price v_2, maximizes S_1's payoff.

In the case (i), the current market ask a_{2-A}' in d_{2-A}' is either greater than $v_2 + 1$ or not specified yet. If S_1 submits v_2, she trades at v_2, since $\tilde{b}_{2-A}' = v_2$. If S_1 submits any ask $p = v_2 + 1$, a decision stage of the buyers is reached, in which B_1 submits $v_2 + 1$ and a trade at $p = v_2 + 1$ occurs. If $p > v_2 + 1$, again a decision stage of the buyers is reached, in which B_1 submits $v_2 + 1$. After this submission, a game segment of type B is reached, in which B_1 enforces $v_2 + 1$ and S_1 accepts a trade at $v_2 + 1$. Thus, given the segment strategies of the buyers, any game segment I_{2-A}', that begins with a decision stage d_{2-A}', with $\tilde{a}_{2-A}'$ not yet specified or $\tilde{a}_{2-A}' > v_2 + 1$, results in a trade at $v_2 + 1$, if S_1 submits an ask $p > v_2$. Since, this is true for any $p > v_2$, S_1 has no reason to deviate from her segment strategy that prescribes submitting $v_2 + 1$. Finally, if S_1 refrains from submitting an offer in d_{2-A}', either the period terminates or (if d_{2-A}' is not a termination stage, which it is, unless $d_{2-A}' = d_2$) a decision stage of the buyers is reached, in which B_1 submits $v_2 + 1$. Again, after B_1 submits $v_2 + 1$, a game segment of type B is reached, in which B_1 enforces $v_2 + 1$ and S_1 accepts a trade at $v_2 + 1$. Thus, submitting an ask smaller than $v_2 + 1$ leads to a smaller payoff, while submitting an ask greater than $v_2 + 1$ and refraining to submit an ask in d_{2-A}' lead to no greater payoffs than the payoff of submitting $v_2 + 1$ for S_1 in I_{2-A}', given the segment strategies of the buyers.

Summarizing, we have shown that neither a game segment of type A nor of type B exists in G_2 in which one of the traders has an incentive to deviate from his segment strategy. Since all game segments in G_2 are of either type A or B, no game segment in G_2 exists, in which one of the traders has an incentive to deviate from his segment strategy contained in $X_2(I_2)$. Hence, it follows that $X_2(I_2)$ is a subgame perfect equilibrium of G_2.

Optimality of choices in the first decision stage d

Now we have to prove that none of the buyers has a reason to deviate from his choice specified in a] in the very first decision stage d of the game. Let us begin with B_1. Given the strategies of the extra-marginal buyers B_i, with $i = 2 \ldots n_B$, and of S_1, if B_1 submits v_2 in d, he trades at the price v_2 in G_1. Thus, his payoff is $v_1 - v_2 + \beta$. If he deviates by submitting a higher bid, he earns less, since - even in the very best case - he trades at price greater than v_2. If he deviates by submitting a bid smaller than $v_2 - 1$ (or deviates by refraining to submit a bid), one of two outcomes are possible: either he does not trade at all (e.g. S_1 accepts B_2's bid at $v_2 - 1$) or he later overbids B_2's opening bid $v_2 - 1$. In the former case, B_1 has a payoff of 0, which is smaller than the payoff of immediately bidding v_2. In the latter case, since overbidding means bidding at least v_2, B_1 can at maximum have a payoff equal to that of immediately bidding v_2. Finally, if B_1 deviates by also bidding $v_2 - 1$, one of two outcomes are possible: either B_1 later overbids the opening bid $v_2 - 1$ or B_1 and B_2 remain in a tie at $v_2 - 1$ and each of them has a ½ chance of trading with S_1. In the former case, since overbidding means bidding at least v_2, B_1 at maximum has a payoff equal to that of immediately bidding v_2. In the latter case, B_1 has an expected payoff of $\frac{1}{2}(v_1 - (v_2 - 1) + \beta)$, which is smaller than the payoff of immediately bidding v_2, because $v_1 > v_2$ together with $\beta > 1$ implies $v_1 - v_2 \geq 1 > 1 - \beta$ and $v_1 - v_2 > 1 - \beta \Leftrightarrow v_1 - v_2 + \beta > \frac{1}{2}(v_1 - (v_2 - 1) + \beta)$.

Hence, no deviation from the opening bid v_2 can lead to a higher payoff for B_1, given the other strategies.

Now let us examine the choice of B_2. Given the strategies of the other traders, B_2 cannot trade in this game and all his strategies lead to a payoff of 0. If B_2 submits $v_2 - 1$ in d, the subgame G_1 is reached in which he does not trade. If he deviates by submitting v_2, the only higher bid that he can submit, the subgame G_2 is reached, in which he is overbid by B_1 and, thus, does not trade. Finally, if he deviates by submitting a bid smaller than $v_2 - 1$ (or deviates by refraining to submit a bid), he does not trade either. Thus, no deviation from the opening bid $v_2 - 1$ can lead to a trade and, thus, to a payoff greater than 0 for B_2, given the other strategies.

The question left to answer is whether any other extra-marginal buyer B_i, with $i = 3 \ldots n_B$ has an incentive to deviate from his opening bid v_i, given the strategies of the others. Obviously, this is not the case, since, given the strategies of the others, no extra-marginal buyer B_i is able to trade in this game. Thus, all of B_i's strategies lead to a payoff of 0 and any one is a best reply strategy. Since neither of the buyers has an incentive to deviate from his choice in d and since $X_1(I_2)$ and $X_2(G_2)$ are subgame perfect equilibria of G_1 and G_2, correspondingly, X is a subgame perfect equilibrium of the entire game.

Quasi-pureness, but no impatience

X is a quasi-pure subgame perfect equilibrium of the game, because the local strategies of the buyers in d are pure and the segment strategies of all traders in I_1 are pure segment strategies. But, X is not an impatience equilibrium of the game. This is so, because B_2's strategy x_{B2} in X could be replaced by a different strategy $\bar{x}_{B2}$, which prescribes submitting v_2 in d, instead of submitting $v_2 - 1$. Since the subgame G_2 is reached on the play realized by the strategy combination $\bar{X} = (x_{B1}, \bar{x}_{B2}, x_{B3}, \ldots, x_{BnB}, x_{S1})$, $\bar{X}$ leads to a payoff for B_2 that is equal to his equilibrium payoff in X, namely 0. But, B_2 submits a better offer on the play realized by $\bar{X}$ than on the play realized by X. Thus, X cannot be an impatience equilibrium of the game. ∎

Proposition C.2.4

Let G be a 1-trade simplified alternating double auction market game with extra-marginal traders only on the opening market side. Any combination of strategies that is a quasi-pure subgame perfect equilibrium of the game - on equilibrium path -

(1)   either leads to a trade involving the two marginal traders at a price which lies just outside the range of feasible prices for the best extra-marginal trader (i.e. exactly one money unit greater than the best extra-marginal buyer's redemption value or exactly one money unit smaller than the best extra-marginal seller's unit cost),

(2)   or leads to a trade involving the two marginal traders at a price equal to the player valuation of the best extra-marginal trader.

All impatience equilibria of G are of the type described in (1).

Proof of Proposition C.2.4

We must first show that strategy combinations leading to other outcomes cannot be quasi-pure subgame perfect equilibria. Once that is established, we must show that type (2) equilibria cannot be impatience equilibria. Since only one trade is possible (i.e. only one pair of traders can trade), the only other possible outcomes of the game are: either market period termination without a trade or a trade at some other price. It follows by Lemma 3 that market period termination without a trade cannot be an equilibrium outcome, so we only have to check for the latter case.

Assume that the buyers are on the opening market side, i.e. that the extra-marginal traders are buyers. (The opposite case can be proven analogously.) It follows that the marginal buyer is B_1 and the extra-marginal buyers are B_i, with $i = 2 \dots n_B$. Obviously, the best extra-marginal buyer is B_2 with the redemption value v_2. The seller is S_1. According to the proposition, in every quasi-pure subgame perfect equilibrium of the game, a trade occurs between B_1 and S_1 either at the price $v_2 + 1$ or at the price v_2.

A trade occurs at $p > v_2 + 1$

Consider some price $p > v_2 + 1$. Due to the *no-loss* rule, only B_1 can trade at p. Trading at p, however, gives B_1 a smaller payoff than trading at $v_2 + 1$. If a trade occurs at the price p, in some decision stage d' of the buyers, B_1 must have submitted p. By Lemma 1 and since $p \neq c_1$, it follows that d' was not a termination stage and thus, not a take-it-or-leave-it stage for B_1. Since the buyers open the market and since no extra-marginal buyer B_i, with $i = 2 \dots n_B$ can overbid $p > v_2$, it follows that in the game segment I' beginning with d', B_1 also had the alternative to credibly enforce some other price $p^* < p$ and $p^* \geq v_2 + 1$. Note that B_1 can choose to do this alone, because the extra-marginal buyers B_i have no choice, but to refrain from bidding. Since S_1 is alone on the second moving market side, it follows by Corollary 1 that S_1's strategy must prescribe accepting a trade at the credibly enforced price p^* in any market side best reply combination of the sellers. Thus, as long as $p > v_2 + 1$, there is always some $p^* < p$, which B_1 can credibly enforce and which leads to a more profitable trade for B_1.

A trade occurs at $p < v_2 + 1$

Now, consider some price $p < v_2 + 1$. If a trade occurs at the price p, the current market bid $\bar{b}'$ in some decision stage d' must have been equal to p. In d' some number k of buyers, with $1 \leq k \leq n_B$, held the current market bid $\bar{b}'$. Additionally, S_1 must have accepted $\bar{b}'$ in the game segment I' beginning with d'. Obviously, some decision stage d'' must have existed, after $\bar{b}'$ was first submitted, but before S_1 accepted $\bar{b}'$, in which S_1 had the alternative to submit an ask $\bar{a}'' = \bar{b}' + 1$. Let I'' denote the game segment and G'' denote the subgame that begins with d'', i.e. with S_1's submission of $\bar{b}' + 1$. If any buyer accepts a trade at $\bar{b}' + 1$ in G'', S_1 receives a higher payoff than on the original play that led to a trade at $\bar{b}'$, because she trades at a higher price. Thus, if any buyer is willing to accept a trade at the price $\bar{b}' + 1$ in G'', S_1 has an incentive to deviate from her original strategy. Hence, to prove that a quasi-pure subgame perfect equilibrium cannot lead to any price $p < v_2 + 1$, it is sufficient to show that at least one of the buyers prefers to trade at $\bar{b}' + 1$ in G'' for any $\bar{b}' < v_2 + 1$, i.e. overbids any current market bid $\bar{b}' < v_2 + 1$. Showing that S_1 is able to trade at $\bar{b}' + 1$ in G'' is sufficient, because, for any strategy combination that realizes a play that leads to a trade at a price $\bar{b}' < v_2 + 1$, an alternative strategy combination exists, in which S_1's strategy choice is altered in such a way, that the realized play reaches the subgame G'', in which S_1 achieves a greater payoff.

Since $\bar{b}' \leq v_2$, there are at least 2 buyers, who can hold the market bid $\bar{b}'$ in d'. Therefore, if $k \geq 1$ buyers hold $\bar{b}'$ in d', three situations are possible: (i) $k > 1$ (i.e. more than one buyer holds $\bar{b}'$) and B_1 is in the group of buyers holding $\bar{b}'$. (ii) $k \geq 1$ (i.e. one or more buyers holds $\bar{b}'$), but B_1 is not in the group of buyers holding $\bar{b}'$. (iii) B_1 is the only buyer, who holds $\bar{b}'$.

A trade occurs at p $<$ v_2 + 1
Case (i) - B_1 is in the group of buyers holding $\bar{b}' = p < v_2 + 1$

Let us first examine case (i), i.e. B_1 and a number of extra-marginal buyers together hold the current market bid $\bar{b}'$ in d'. If none of the k buyers, who jointly hold $\bar{b}'$, choose to overbid $\bar{b}'$ in I'', B_1 has an expected payoff of $\frac{1}{k}(v_1 - \bar{b}' + \beta)$. But, keeping all other buyers' strategies constant, B_1 can deviate from his strategy by overbidding $\bar{b}'$ with $\bar{b}' + 1$, i.e. by accepting S_1's offer and trading at $\bar{b}' + 1$. In this case, B_1's payoff is $v_1 - (\bar{b}' + 1) + \beta$. Thus, B_1 has an incentive to overbid as long as: $\frac{1}{k}(v_1 - \bar{b}' + \beta) < v_1 - (\bar{b}' + 1) + \beta \Leftrightarrow v_1 - \bar{b}' > \frac{k}{k-1} - \beta$. Because $\frac{k}{k-1} \leq 2$, for $k > 1$, and $\beta > 1$, it follows that $\frac{k}{k-1} - \beta < 1$. This implies $v_1 - \bar{b}' > 1 \Leftrightarrow v_1 \geq \bar{b}'$. Hence, as long as B_1 is in a tie at a current market offer $\bar{b}' < v_2 + 1 \leq v_1$, he has an incentive to overbid $\bar{b}'$ and to trade at $\bar{b}' + 1$.

A trade occurs at p $<$ v_2 + 1
Case (ii) - B_1 is not in the group of buyers holding $\bar{b}' = p < v_2 + 1$

Let us now examine (ii), i.e. $k \geq 1$ buyers - excluding B_1 - hold the current market bid $\bar{b}'$ in d''. In this case, B_1 has incentive to overbid $\bar{b}'$ and to accept S_1's offer and trade at $\bar{b}' + 1$. If B_1 refrains from submitting a bid in all decision stages of I'', his payoff is 0, since he does not trade in I″ and cannot trade later. But, if B_1 over-bids $\tilde{b}'$ and trades at $\tilde{b}' + 1$ with S_1 in I″, he achieves a payoff $v_1 - (\tilde{b}' + 1) + \beta$, that is greater than 0 for any $b' < v_2 + 1 \leq v_1$. Hence, as long as B_1 is not in the group of buyers, who have tied at a current market offer $\tilde{b}' < v_2 + 1 \leq v_1$, he has an incentive to overbid $\tilde{b}'$ and to trade with S_1 at $\tilde{b}' + 1$.

A trade occurs at $p < v_2 + 1$
Case (iii) - B_1 is the only buyer holding $\tilde{b}' = p < v_2 + 1$

Finally, in case (iii) B_1 alone holds $\tilde{b}'$. Now B_2 has an incentive to overbid any $\tilde{b}' < v_2$, because B_2 has a payoff of 0, if he does not overbid $\tilde{b}'$. If B_2, however, overbids $\tilde{b}'$, he can achieve a payoff $v_2 - (\tilde{b}' + 1) + \beta > 0$, for any $\tilde{b}' < v_2$. Thus, as long as B_1 alone holds $\tilde{b}'$ and $\tilde{b}' < v_2$, B_2 has an incentive to overbid $\tilde{b}'$ and to trade with S_1 at $\tilde{b}' + 1$.

Trading at $p = v_2$ cannot be the outcome of an impatience equilibrium

Summarizing, we have shown that neither a trade at a price $p > v_2 + 1$ nor at a price $p < v_2$ can be the equilibrium outcome of a quasi-pure subgame perfect equilibrium of the game. We have also shown that a trade at the price v_2 can be an equilibrium outcome, only if the marginal buyer B_1 alone held the current market bid at v_2. This is so, because, if the best extra-marginal buyer B_2 alone holds v_2 or if B_1 and B_2 tie at v_2, B_1 has an incentive to overbid v_2 with $v_2 + 1$. Thus, the only candidates left for equilibrium outcomes are a trade at the price $v_2 + 1$, involving the two marginal traders, or a trade at the price v_2, also involving the two marginal traders. The only thing left to show is that the latter type of outcome cannot be the outcome of an impatience equilibrium.

Assume X is a quasi-pure subgame perfect equilibrium of the game, which leads to a trade involving B_1 and S_1 at v_2 and in which B_1 alone bids v_2 in some decision stage d and holds the current market bid at v_2 alone in the decision stages following d. X is not an impatience equilibrium, because B_2's strategy x_{B2} in X could be replaced by a different strategy $\bar{x}_{B2}$, which prescribes submitting v_2 in d, instead of submitting $v_2 - 1$. Since X is a quasi-pure subgame perfect equilibrium of the game, it follows that B_2 cannot achieve a higher payoff by deviating from his strategy in this way. Thus, the strategy combination $\bar{X} = (x_{B1}, \bar{x}_{B2}, x_{B3}, ..., x_{BnB}, x_{S1})$ leads to a payoff for B_2 that is equal to his equilibrium payoff in X, namely 0. But, B_2 submits a better offer on the play realized by $\bar{X}$ than on the play realized by X. Thus, X cannot be an impatience equilibrium of the game. Hence, all impatience equilibria of the game must lead to a trade at the price $v_2 + 1$.   ∎

III.C.3. 1-Trade Markets With Extra-Marginal Traders on Both Market Sides

As in all 1-trade markets, a 1-trade market with extra-marginal traders on both market sides contains two marginal traders, one on each market side. Additionally, at least one extra-marginal trader is present on each market side of these markets. The major difference to the smaller markets is that two market phases are possible, i.e. two contracts can possibly be made. This is true, because each of the extra-marginal traders - by definition of traders - can trade with the marginal trader on the other market side. As before, we index the buyers in an order from the highest to the lowest redemption value and the sellers from the lowest to the highest unit cost. Thus, traders are denoted by B_1, ..., B_{nB}, and S_1, ..., S_{nS}, where B_1 and S_1 are the marginal traders and B_2 and S_2 are the best extra-marginal traders. As always, the player valuations are indexed correspondingly.

In Proposition C.3.1 an impatience equilibrium of the game is specified. This establishes the existence of such equilibria of the game. The equilibrium in Proposition C.3.1 is not unique, but Proposition C.3.2 shows that the equilibrium outcome is unique, i.e. that all impatience equilibria of the game must lead to the same equilibrium outcome.

To be able to specify a unique outcome in the necessary condition, we need the following technical assumption:

Assumption A1

Let B_x and S_x be the best extra-marginal traders in a simplified alternating double auction market game. The unit cost c_x of the best extra-marginal seller S_x is at least two money units greater than the redemption value v_x of the best extra-marginal buyer, i.e. $c_x - v_x > 2$.

This assumption is crucial, because it disallows a number of indifferences, which all arise from the fact, that we have an environment with a discrete money unit. The assumption, however, is merely technical, because the assumed property can be achieved in any given game by a simple redefinition of the smallest money unit. Given that the smallest money unit can be chosen arbitrarily, it suffices to have the property $c_x > v_x$. This is always the case, since B_x and S_x are extra-marginal traders. Nevertheless, we later discuss the effects of the cases $c_x - v_x \leq 2$ in a 1-trade market with extra-marginal traders on both market sides.

Proposition C.3.1

Let G be a 1-trade simplified alternating double auction market game with at least one extra-marginal trader on each market side. Let the Assumption A1 be true in G. An impatience equilibrium of G exists with the property that in equilibrium a trade occurs at a price which lies just outside the range of feasible prices for the best extra-marginal trader of the opening market side (i.e. exactly one money unit greater than the best extra-marginal buyer's redemption value or exactly one money unit smaller than the best extra-marginal seller's unit cost).

Proof of Proposition C.3.1

Let us assume the buyers open the market and the sellers are on the second moving market side. The opposite case can be proven analogously.

Since two trades are possible, certain plays of the game G lead to a second market phase. Obviously, which second market phase is reached, depends on the history of play. Any second market phase that can be reached, however, is either a 1-trade market without extra-marginal traders or a 1-trade market with extra-marginal traders on only one market side. This is true, because at least one of the marginal traders, B_1 or S_1, must be involved in any trade. Thus, in any second market phase that can evolve one of the marginal traders is not present. This means that none of the extra-marginal traders on the other market side can trade in those second market phases, because the only trader, with whom they could have possibly traded, is not in the market anymore.

The truncated game $\bar{G}$

Since all possible second market phase subgames are either 1-trade markets without extra-marginal traders or 1-trade markets with extra-marginal traders on only one market side, we can use the results of the propositions C.1.2, C.2.2, and C.2.4 to examine these subgames. We assume that the strategies of the traders for G induce such subgame strategies in each of the second market phase subgames without extra-marginal traders that have the properties specified in the Proof of Proposition C.1.1. We further assume that the strategies of the traders for G induce such subgame strategies in each of the second market phase subgames with extra-marginal traders only on the second moving market side or only on the opening market side that have the properties specified in the Proof of Proposition C.2.1 or C.2.3 (case a]), correspondingly. Thus, for each of these second phases, we can conclude the following: First, the strategy combination for G induces an impatience equilibrium in each second market phase subgame of G. Second, if a second market phase is reached on a play in G, the outcome corresponds to the outcome specified either in Proposition C.1.2, in Proposition C.2.2, or in Proposition C.2.4.

Given the assumption above, we can analyze the truncated game $\bar{G}$ instead of the original game G, where $\bar{G}$ is constructed in the following manner: $\bar{G}$ is identical to G, except that every second market phase subgame of G is replaced by an end node in $\bar{G}$, which has the impatience equilibrium payoffs of the corresponding second market phase subgame of G. Four types of second market phases subgames are possible in G: (1) If S_1 trades in the first market phase and only one extra-marginal seller exists, some $G_2(c_2, v_1)$ evolves, in which the traders B_1 and S_2 are present and trade at the price c_2. (2) If S_1 trades in the first market phase and more than one extra-marginal seller exist, some $G_2(c_3, v_1)$ evolves, in which the traders B_1 and S_j, with $j = 2 \ldots n_S$, are present and B_1 and S_2 trade at the price c_2. (Remember that we are assuming that the buyers are on the opening market side.) (3) If B_1 trades in the first market phase and only one extra-marginal buyer exists, some $G_2(c_1, v_2)$ evolves, in which the traders B_2 and S_1 are present and trade at the price c_1. (4) If B_1 trades in the first market phase and more than one extra-marginal buyer exist, some $G_2(c_1, v_3)$ evolves, in which the traders B_i, with $i = 2 \ldots n_B$, and S_1 are present and B_2 and S_2 trade at the price $v_3 + 1$. These results, that all follow by the propositions C.1.2, C.2.2, and C.2.4, supply all equilibrium payoffs we need to construct the truncated game $\bar{G}$, as explained above. Obviously, an impatience equilibrium of $\bar{G}$ is expanded to an impatience equilibrium of G, if the strategies are expanded to cover the second market phases as specified above. Thus, in the rest of the proof, we analyze the truncated game $\bar{G}$.

Construction of an impatience equilibrium X of the truncated game $\bar{G}$

Consider a strategy combination X for the truncated game $\bar{G}$, that contains the strategies x_{Bi}, with $i = 1 \ldots n_B$, and x_{Sj}, with $j = 1 \ldots n_S$, of the traders, with the following properties:

a] In the very first decision stage d of $\bar{G}$, x_{B1} prescribes submitting $v_2 + 1$ and each x_{Bi}, with $i = 2 \ldots n_B$, prescribes submitting v_i, i.e. the marginal buyer B_1 submits a bid equal to the best extra-marginal buyer B_2's redemption value plus one and each extra-marginal buyer submits a bid equal to his own redemption value. Let d' denote the decision stage that immediately follows these choices. Let I' and $\bar{G}'$ denote the game segment and the subgame, correspondingly, that begin with d'.

b] All x_{Bi}, with $i = 1 \ldots n_B$, induce the enforcement segment strategies $x'_{Bi}(I')$ on the game segment I', i.e. all buyers refrain from submitting in all decision stages in I'.

c] All x_{Sj} induce the acceptance segment strategies $x'_{Sj}(I')$ on the game segment I', i.e. the marginal seller S_1 immediately accepts the current market bid in every decision stage of I' and each of the extra-marginal sellers immediately accepts the current market bid in those decision stages in I', in which the *no-loss* rule allows accepting, submits her best ask in those decision stages in I', in which the *ask-bid-spread-reduction* rule and the *no-crossing* rule allow the submission, and otherwise, refrains from submitting a new ask.

d] In each other subgame of the game, except $\bar{G}'$, the combination of subgame strategies induced by X represents some arbitrary subgame perfect equilibrium of the subgame.

We start by splitting up the truncated game $\bar{G}$ into a number of subgames after the very first decision stage. This first decision stage is not a single decision node, but the combination of the decision nodes of the buyers. Thus, each of the subgames that follow the first decision stage begins with a different combination of first bids of the buyers (one of them begins with all buyers refraining to submit a first offer). Each of these subgames has at least one subgame perfect equilibrium in behavior strategies. In each of these subgames, except $\bar{G}'$, we assume with d] that X induces some arbitrary subgame perfect equilibrium. To prove that X is a quasi-pure subgame perfect equilibrium of $\bar{G}$, we only have to show that the combination of the segment strategies $X'(I')$, that contains $x'_{Bi}(I')$, with $i = 1 \ldots n_B$, and $x'_{Sj}(I')$, with $j = 1 \ldots n_S$, is a subgame perfect equilibrium of $\bar{G}'$ and that none of the buyers has an incentive to deviate from submitting the bids specified in a] in the very first decision stage d of the game.

Subgame perfectness in the main subgame $\bar{G}'$

First note that any subgame of $\bar{G}'$ is a game segment of $\bar{G}'$, because the truncated game $\bar{G}$ has no second market phases. This is also true for the largest game segment in $\bar{G}'$, namely, I', which is identical to $\bar{G}'$. For this reason, the segment strategies contained in $X'(I')$ are complete strategies for $\bar{G}'$ and $X'(I')$ can qualify as an equilibrium of $\bar{G}'$. To show that $X'(I')$ is a subgame perfect equilibrium of $\bar{G}'$, it suffices to show that: (*) No game segment I'' exists in $\bar{G}'$, in which one of the traders has an incentive to deviate from the game segment strategy combination induced by $X'(I')$ on I''.

Let I'' be some arbitrary game segment of $\bar{G}'$ beginning with the decision stage d''. Let $\tilde{b}''$ be the current market bid in d''. Due to the *ask-bid-spread-reduction* rule $\tilde{b}''$ cannot be smaller than the market bid $v_2 + 1$ in the first decision stage d' of $\bar{G}'$. Thus, $\tilde{b}''$ must be in the market price range of d''. Since $\tilde{b}''$ is in the market price range of d'' and the buyers are on the opening market side and all buyers refrain from submitting new bids in all their decision stages in I', it follows that the sellers face the credibly enforced price $\tilde{b}''$ in I''.

Since $\tilde{b}'' \geq v_2 + 1$ is true anywhere in $\bar{G}'$, no extra-marginal buyer B_i, with $i = 2 \ldots n_B$, can trade in I''. This, however, means that only two outcomes are possible in $\bar{G}'$: either B_1 trades in I'' or the market period terminates without any trades at all. Thus, $\bar{G}'$ does not contain any end nodes that correspond either to second market phase subgames $G_2(c_2, v_1)$ or to $G_2(c_3, v_1)$, in which the best extra-marginal seller S_2 can trade. Similarly, if $n_S > 2$, none of the other extra-marginal sellers S_j, with $j = 3 \ldots n_S$, is able to trade in a second market phase subgame. Hence, if at all, an extra-marginal seller S_j, with $j = 2 \ldots n_S$, can only trade in the first market phase, i.e. in I''.

Since all extra-marginal sellers know that they can only trade in the first market phase, it follows by Corollary 2, that, in any market side best reply combination of the sellers, each extra-marginal seller, who makes no loss by trading at $\tilde{b}''$, must accept a trade at the enforced price $\tilde{b}''$. If any number k of sellers, with $k > 1$ can accept a trade at $\bar{b}''$, each of them should accept immediately to have a $\frac{1}{k}$ chance of trading at the price $\bar{b}''$, because each other one also accepts immediately. If one of them would delay the acceptance, she would be certain not to trade at all. If an extra-marginal seller S_j, with $j = 2 \dots n_S$ cannot accept $\bar{b}''$, because $\bar{b}'' < c_j$, all her segment strategies in I'' lead to a payoff of 0, given the strategies of the other traders. Thus, by submitting her best ask, S_j achieves a no smaller payoff in I'' than with any other segment strategy. Hence, the acceptance segment strategy $x'_{Sj}(I')$ maximizes each S_j's payoff in any game segment I'' of G', given the enforcement segment strategy of B_1 and given the acceptance segment strategies of the other sellers.

In contrast, $\bar{G}'$ does contain end nodes that correspond to second market phase subgames $G_2(c_1,v_2)$ and $G_2(c_1,v_3)$, in which S_1 trades with B_2 either at the price c_1 or at the price $v_3 + 1$. However, since $v_2 + 1 > c_1$ and $v_2 + 1 > v_3 + 1$, S_1 trades at a higher price by accepting any $\bar{b}'' \geq v_2 + 1$ in I'' and, thus, always prefers trading in I'' to trading in a second market phase subgame. Because S_1 prefers trading in I'' to trading in a second market phase subgame and because at least one extra-marginal seller S_j, with $j = 2 \dots n_S$, accepts any current market bid $\bar{b}'' \geq c_2$ immediately, S_1 maximizes her payoff by also accepting the enforced current market bid $\bar{b}'' \geq c_2$ immediately. If only S_1 can accept $\bar{b}''$, because $\bar{b}'' < c_2$, it is irrelevant for her payoff in I'', whether she accepts a trade at $\bar{b}''$ immediately or not, given the buyers' strategies. Thus, the acceptance segment strategy $x'_{S1}(I')$ maximizes S_1's payoff in any game segment I'', given the other segment strategies.

Since at least one seller accepts any price that is credibly enforced by the buyers in I'' and since the extra-marginal buyers B_i, with $i = 2 \dots n_B$, have no choice but to refrain from submitting new bids in I'', B_1 maximizes his payoff in I'' by enforcing the lowest possible bid in I''. The lowest possible bid, that B_1 can enforce in I'' is $\bar{b}''$, because - due to the *ask-bid-spread-reduction* rule - no bid in I'' can be lower than the current market bid $\bar{b}''$ in d''. Thus, $x'_{B1}(I')$ maximizes B_1's payoff in any game segment I'' of $\bar{G}'$, given the acceptance segment strategies of the sellers and given the enforcement segment strategies of the extra-marginal buyers.

Finally, since all segment strategies of the extra-marginal buyers B_i, with $i = 2 \dots n_B$, in any I'' lead to a payoff of 0, given the strategies of the others, each $x'_{Bi}(I')$ maximizes each B_i's payoff in any game segment of $\bar{G}'$. Hence, for any possible I'' in $\bar{G}'$, neither a buyer nor a seller has an incentive to deviate from his strategy contained in $X'(I')$. Thus, it follows that $X'(I')$ is a subgame perfect equilibrium of $\bar{G}'$.

Optimality of choices in the first decision stage d

Now we have to prove that none of the buyers has a reason to deviate from his choice specified in a] in the very first decision stage d of the truncated game $\bar{G}$. Let us begin with B_1. As shown above and given the strategies of the others, if B_1 submits $v_2 + 1$ in d, he trades at the price $v_2 + 1$ in $\bar{G}'$. This gives him a payoff of $v_1 - (v_2 + 1) + \beta$. If he deviates by submitting a bid greater than $v_2 + 1$, he earns less, since - even in the very best case - he trades at price greater than $v_2 + 1$. If B_1 deviates by submitting a bid smaller than $v_2 + 1$ (or deviates by refraining to submit a bid), one of four outcomes are possible, since each extra-marginal buyer B_i, with $i = 2 \dots n_B$, bids v_i in d: (i) B_1 does not trade at all and receives a payoff of 0. (ii) B_1 does not trade in the first market phase, but trades in some second market phase with S_2 at the price c_2 and receives a payoff of $v_1 - c_2 + \beta$. (iii) B_1 and B_2 tie at the market bid v_2 in the first market phase. If B_1 is the *winner* in the tie-breaking, he trades with S_1 at v_2 in the first market phase. If B_1 is not the *winner* in the tie-breaking, he trades with S_2 at c_2 in some second market phase. Thus, B_1's expected payoff in this case is $\frac{1}{2}(v_1 - v_2 + \beta) + \frac{1}{2}(v_1 - c_2 + \beta)$. (iv) B_1 overbids B_2's opening bid v_2 in some later decision stage of the first market phase and trades with S_1. B_1 must bid at least $v_2 + 1$ to overbid v_2, thus his payoff at maximum is $v_1 - (v_2 + 1) + \beta$.

Clearly, B_1's payoff in case (i) is smaller than his payoff from bidding $v_2 + 1$ in d. In case (ii) B_1's payoff is also smaller than his payoff from bidding $v_2 + 1$ in d, since it follows by Assumption A1 that $c_2 > v_2 + 2$, i.e. B_1 pays a higher price c_2 if he trades with S_2 in any second market phase. The same is true for the case (iii), since $c_2 > v_2 + 2 \Rightarrow \frac{1}{2}(v_2 + c_2) > v_2 + 1 \Rightarrow \frac{1}{2}(v_1 - v_2 + \beta) + \frac{1}{2}(v_1 - c_2 + \beta)$ $< v_1 - (v_2 + 1) + \beta$. Finally, the maximum payoff B_1 can achieve in case (iv) is not greater than the payoff of bidding $v_2 + 1$ in d. Thus, no deviation from the opening bid $v_2 + 1$ can lead to a higher payoff for B_1, given the other strategies.

The question left to answer is whether any extra-marginal buyer B_i, with $i = 2 \ldots$ n_B, has an incentive to deviate from his opening bid v_i, given the other strategies. Obviously, this is not the case, since, given the strategies of the others, no extra-marginal buyer B_i is able to trade in this game. Thus, all of B_i's strategies lead to a payoff of 0 and any one is a best reply strategy. Since neither of the buyers has an incentive to deviate from his choice in d and since $X'(I')$ is a subgame perfect equilibrium of $\bar{G}'$, X is a subgame perfect equilibrium of the truncated game $\bar{G}$.

Quasi-pureness and impatience

X is a quasi-pure subgame perfect equilibrium of the game, because the local strategies of the buyers in d are pure and the segment strategies of all traders in I' are pure segment strategies. But, X is also an impatience equilibrium of the game. This is so, because B_1's best offer on the play realized by any strategy combination $\bar{X} = (\bar{x}_{B1}, x_{B2}, \ldots, x_{BnB}, x_{S1}, \ldots, x_{SnS})$, that leads to a payoff for B_1 equal to his equilibrium payoff in X, must be $v_2 + 1$; otherwise B_1 would trade at a different price or not at all. But, since x_{B1} prescribes submitting $v_2 + 1$ in the very first decision stage, no $\bar{x}_{B1}$ can specify submitting $v_2 + 1$ any earlier. Similarly, every alternative strategy of the marginal seller S_1, with the same payoff as x_{S1}, must prescribe submitting $v_2 + 1$, but cannot prescribe doing so any earlier than x_{S1} does. Lastly, every alternative strategy of any extra-marginal buyer B_i, with $i = 2$ $\ldots n_B$, or of any extra-marginal seller S_j, with $j = 2 \ldots n_S$, also leads to a payoff of 0, given the other strategies. Thus, each B_i must submit his best offer v_i and each S_j must submit her best offer c_j as soon as possible in an impatience equilibrium. This is true in X. ∎

Proposition C.3.2

Let G be a 1-trade simplified alternating double auction market game with at least one extra-marginal trader on each market side. Let the Assumption A1 be true in G. Any combination of strategies that is an impatience equilibrium of G - on the equilibrium path - leads to a trade at a price which lies just outside the range of feasible prices for the best extra-marginal trader on the opening market side (i.e. exactly one money unit greater than the best extra-marginal buyer's redemption value or exactly one money unit smaller than the best extra-marginal seller's unit cost).

Proof of Proposition C.3.2

For the entire proof we assume the buyers open the market. The other case can be proven analogously. We must only show that strategy combinations leading to other outcomes cannot be impatience equilibria.

The truncated game $\bar{G}$

Since two trades are possible, certain plays of the game G lead to a second market phase. Which second market phase is reached, depends on the history of play. But, as shown in the Proof of Proposition C.3.1, any second market phase subgame that can be reached is a 1-trade market either without extra-marginal traders or with extra-marginal traders on only one market side. It follows by Proposition C.1.2, that any impatience equilibrium of the second market phase 1-trade markets without extra-marginal traders, $G_2(c_2,v_1)$ and $G_2(c_1,v_2)$, must lead to trade at c_2 or c_1, correspondingly. It follows by Proposition C.2.2 and by Proposition C.2.4, that any impatience equilibrium of the second market phase 1-trade markets with extra-marginal traders on only one market side, $G_2(c_3,v_1)$ and $G_2(c_1,v_3)$, must lead to trade at c_2 or $v_3 + 1$, correspondingly. Thus, any strategy combination for G that leads to another outcome in a second market phase cannot be an impatience equilibrium of G. Hence - as in the Proof of Proposition C.3.1 -, instead of analyzing the original game G, we can analyze the truncated game $\bar{G}$, in which the second market phase subgames of G are replaced with end nodes with the corresponding payoffs of the 1-trade markets $G_2(c_2,v_1)$, $G_2(c_1,v_2)$, $G_2(c_3,v_1)$, and $G_2(c_1,v_3)$.

By Lemma 3 we know that market period termination without a trade cannot be an equilibrium outcome. Thus, the only other possible outcome of $\bar{G}$, that must be checked, is that a trade at some other price than that specified in Proposition C.3.2 occurs in the first market phase. Since the buyers open the market, the only outcome compatible with the proposition is that a trade at the price $v_2 + 1$ occurs in the first market phase and that no second market phase is reached.

A trade occurs at $p > v_2 + 1$

Let us first check the cases in which a trade occurs at a price $p > v_2 + 1$. Due to the *no-loss* rule, only B_1 can trade at p. Trading at p, however, gives B_1 a smaller payoff than trading at $v_2 + 1$. Since a trade occurs at the price p, there must have been some decision stage d' of the buyers in which B_1 submitted p. By Lemma 1 and since $p \neq c_1$, it follows that d' was not a termination stage and thus, not a take-it-or-leave-it stage for B_1. Since the buyers open the market and since no extra-marginal buyer B_i, with $i = 2 \ldots n_B$ can overbid any $p > v_2$, it follows that, in the game segment I' beginning with d', B_1 also had the alternative to credibly enforce some other price $p^* < p$ and $p^* \geq v_2 + 1$. Note that B_1 can choose to do this alone, because the extra-marginal buyers have no choice, but to refrain from bidding. It follows by Lemma 4 that one of the sellers accepts the credibly enforced price p^* in equilibrium. Thus, as long as $p > v_2 + 1$, there is always some $p^* < p$, which B_1 can credibly enforce and which leads to a more profitable trade for B_1.

A trade occurs at $p < v_2 + 1$

Now, consider some price $p < v_2 + 1$. If a trade occurs at the price p, the current market bid $\bar{b}'$ in some decision stage d' must have been equal to p. In d' some number k of buyers, with $1 \leq k \leq n_B$, held the current market bid $\bar{b}'$. Additionally, S_1 must have accepted $\bar{b}'$ in the game segment I' beginning with d'. (Remember that the extra-marginal sellers S_j, with $j = 2 \ldots n_S$, cannot trade at $\bar{b}' < v_2 + 1 <$

c_2, due to the Assumption A1 and to the *no-loss* rule.) Obviously, some decision stage d'' must have existed, after $\bar{b}'$ was first submitted, but before S_1 accepted $\bar{b}'$, in which S_1 had the alternative to submit an ask $\bar{a}'' = \bar{b}' + 1$. Let I'' denote the game segment and G'' denote the subgame that begins with d'', i.e. with S_1's submission of $\bar{b}' + 1$. If any buyer accepts a trade at $\bar{b}' + 1$ in G'', S_1 receives a higher payoff than on the original play that led to a trade at $\bar{b}'$, because she trades at a higher price. Thus, if any buyer is willing to accept a trade at $\bar{b}' + 1$ in G'', S_1 has an incentive to deviate from her original strategy. Hence, to prove that a quasi-pure subgame perfect equilibrium cannot lead to a price $p < v_2 + 1$, it is sufficient to show that at least one of the buyers prefers to trade at $\bar{b}' + 1$ in G'' for any $\bar{b}' < v_2 + 1$, i.e. overbids any current market bid $\bar{b}' < v_2 + 1$. Showing that S_1 is able to trade at $\bar{b}' + 1$ in G'' is sufficient, because, for any strategy combination that realizes a play that leads to a trade at a price $\bar{b}' < v_2 + 1$, an alternative strategy combination exists, in which S_1's strategy choice is altered in such a way, that the realized play reaches the subgame G'', in which S_1 achieves a greater payoff.

Since $\bar{b}' \leq v_2 < v_2 + 1$, there are at least 2 buyers, who can hold the market bid $\bar{b}'$ in d'. Thus, if $k \geq 1$ buyers hold $\bar{b}'$ in d', three situations are possible in d'': (i) $k > 1$ (i.e. more than one buyer holds $\bar{b}'$) and B_1 is in the group of buyers holding $\bar{b}'$. (ii) $k \geq 1$ (i.e. one or more buyers holds $\bar{b}'$), but B_1 is not in the group of buyers holding $\bar{b}'$. (iii) B_1 is the only buyer, who holds $\bar{b}'$.

A trade occurs at $p < v_2 + 1$
Case (i) - B_1 is in the group of buyers holding $\bar{b}' = p < v_2 + 1$

Let us first examine case (i), i.e. B_1 and a number of extra-marginal buyers together hold the current market bid $\bar{b}'$ in d'. If none of the k buyers, who jointly hold $\bar{b}'$, choose to overbid $\bar{b}'$ in I'', B_1 has an expected payoff of $\frac{1}{k}(v_1 - \bar{b}' + \beta) + \frac{k-1}{k}(v_1 - c_2 + \beta)$. The first term represents B_1's payoff of trading with S_1 at $\bar{b}'$ in the first market phase multiplied by the probability of being the *winner* of the tie-breaking rule. The second term represents B_1's payoff of trading in some second market phase subgame with S_2 multiplied by the residual probability. But, keeping all other buyers' strategies constant, B_1 can deviate from his strategy by overbidding $\bar{b}'$ with $\bar{b}' + 1$, i.e. by accepting S_1's offer and trading at $\bar{b}' + 1$. In this case, B_1's payoff is $v_1 - (\bar{b}' + 1) + \beta$. Thus, B_1 has an incentive to overbid as long as:

$$\frac{1}{k}(v_1 - \bar{b}' + \beta) + \frac{k-1}{k}(v_1 - c_2 + \beta) < v_1 - (\bar{b}' + 1) + \beta \Leftrightarrow \bar{b}' < c_2 - \frac{k}{k-1}.$$

Because it follows by Assumption A1 that $v_2 < c_2 - 2$ and because $\frac{k}{k-1} \leq 2$, for $k > 1$, this implies $\bar{b}' < v_2 + 1 \leq c_2 - 2 \leq c_2 - \frac{k}{k-1}$. Hence, as long as B_1 is in a tie at a current market offer $\bar{b}' < v_2 + 1$, he has an incentive to overbid $\bar{b}'$ and to trade at $\bar{b}' + 1$.

A trade occurs at $p < v_2 + 1$
Case (ii) - B_1 is not in the group of buyers holding $\bar{b}' = p < v_2 + 1$

Let us now examine (ii), i.e. $k \geq 1$ buyers - excluding B_1 - hold the current market bid $\bar{b}'$ in d''. In this case, B_1 has an incentive to overbid $\bar{b}'$ and to accept S_1's offer and trade at $\bar{b}' + 1$. If B_1 refrains from submitting a bid in all decision stages of I'', he achieves a payoff $v_1 - c_2 + \beta$ by trading with S_2 in a second market phase subgame. But, if B_1 overbids $\bar{b}'$ and trades at $\bar{b}' + 1$ with S_1 in I'', he achieves a payoff of $v_1 - (\bar{b}' + 1) + \beta$. The former payoff is smaller, because $\bar{b}' < v_2 + 1 \leq c_2 - 2$ is true due to the Assumption A1 and implies $\bar{b}' < c_2 - 1$, which is equivalent to $v_1 - c_2 + \beta < v_1 - (\bar{b}' + 1) + \beta$. Hence, as long as B_1 is not in the group of buyers, who have tied at a current market offer $\bar{b}' < v_2 + 1$, he has an incentive to overbid $\bar{b}'$ and to trade with S_1 at $\bar{b}' + 1$.

A trade occurs at $p < v_2 + 1$
Case (iii) - B_1 is the only buyer holding $\bar{b}' = p < v_2 + 1$

Finally, in case (iii) B_1 alone holds $\bar{b}'$. Now B_2 has an incentive to overbid any $\bar{b}' < v_2$. If B_2 does not overbid $\bar{b}'$, his payoff is 0, since he can neither trade now nor later. (Remember that S_1, the only seller B_2 can trade with, is not present in a second market phase, if a trade at $\bar{b}' < v_2 + 1$ occurs in the first market phase, because no other seller can accept a trade at $\bar{b}'$.) But, if B_2 overbids $\bar{b}'$ with $\bar{b}' + 1$, he trades at $\bar{b}' + 1$ with S_1 in I'' and achieves a payoff $v_2 - (\bar{b}' + 1) + \beta > 0$, for any $\bar{b}' < v_2$. Thus, as long as B_1 alone holds $\bar{b}'$ and $\bar{b}' < v_2$, B_2 has an incentive to overbid $\bar{b}'$ and to trade with S_1 at $\bar{b}' + 1$.

Trading at $p = v_2$ cannot be the outcome of an impatience equilibrium

Summarizing, we have shown that neither a trade at a price $p > v_2 + 1$ nor at a price $p < v_2$ can be the equilibrium outcome of a quasi-pure subgame perfect equilibrium of the game. We have also shown that a trade at the price v_2 can be an equilibrium outcome, only if the marginal buyer B_1 alone held the current market bid at v_2. This is so, because if the best extra-marginal buyer B_2 alone holds v_2 or if B_1 and B_2 tie at v_2, B_1 has an incentive to overbid v_2 with $v_2 + 1$. Thus, the only candidates left for equilibrium outcomes are a trade at the price $v_2 + 1$, involving the two marginal traders, or a trade at the price v_2, also involving the two marginal traders. The only thing left to show is that the latter type of outcome cannot be the outcome of an impatience equilibrium.

Assume X is a quasi-pure subgame perfect equilibrium of the game, which leads to a trade involving B_1 and S_1 at v_2 and in which B_1 alone bids v_2 in some decision stage d and holds the current market bid at v_2 alone in the decision stages following d. X is not an impatience equilibrium, because B_2's strategy x_{B2} in X could be replaced by a different strategy $\bar{x}_{B2}$, which prescribes submitting v_2 in d, instead of submitting a smaller bid or of submitting no bid at all. Since X is a quasi-pure subgame perfect equilibrium of the game, it follows that B_2 cannot achieve a higher payoff by deviating from his strategy in this way. Thus, the strategy combination $\bar{X} = (x_{B1}, \bar{x}_{B2}, x_{B3}, ..., x_{BnB}, x_{S1}, ..., x_{SnS})$ leads to a payoff for B_2 that is equal to his equilibrium payoff in X, namely 0. But, B_2 submits a better offer on the play realized by $\bar{X}$ than on the play realized by X. Thus, X cannot be an impatience equilibrium of the game. Hence, all impatience equilibria must lead to a trade at the price $v_2 + 1$. ∎

Before moving on to the 2-trade markets without extra-marginal traders, a note on the role of the Assumption A1 should be made. The assumption was used in the proofs of Proposition C.3.1 and C.3.2 to exclude the opening marginal trader's indifference between overbidding and not overbidding a market bid at the valuation of the competing best extra-marginal trader. The key to this indifference is that the marginal trader in these markets has a trade-off between contracts with the marginal trader or the best extra-marginal trader on the other market side.

For simplicity, let us assume that only two extra-marginal traders, B_2 and S_2, are in the market and that the buyers open the market. In this case B_1 can choose to either trade with S_1 or with S_2. Assumption A1 simply assures that the price B_1 has to pay to be certain to trade with S_1 (i.e. the price which bars B_2 from the trading with S_1) is sufficiently smaller than the price B_1 would realize in any second market phase in a trade with S_2. From Proposition C.1.1 we know that B_1 pays c_2, if he remains in a second market phase market with S_2. The lowest price that enables B_1 to bar B_2 from trading in the first market phase is $v_2 + 1$. If $c_2 - v_2 < 2$, it follows that $v_2 + 1 \geq c_2$. In such cases it is more costly (in terms of expected payoff) for B_1 to force B_2 out of trading with S_1, than it is for him to take the risk of a tie at v_2. This is so, because in the tie he might be lucky (which he is with probability $\frac{1}{2}$) and trade in the first market phase at a price v_2. If he is not lucky, he trades at $c_2 \leq v_2 + 1$ in a second market phase. In case $c_2 - v_2 = 2$, B_1 is indifferent between tieing with B_2 and overbidding, since the expected payoff of both actions is equal, i.e. $v_1 - (v_2 + 1) + \beta = \frac{1}{2}(v_1 - v_2 + \beta) + \frac{1}{2}(v_1 - (v_2 + 2) + \beta)$.

Obviously, without Assumption A1 a number of other quasi-pure subgame perfect equilibria outcomes are possible. In (some of) these two trades may occur: first a trade between S_1 and B_2 at the price v_2 and then a trade between S_2 and B_1 at a price c_2. Whether or not two trades occur in such cases, however, strongly depends on the random draw in case of a tie. Whenever the marginal trader (in our example B_1) is chosen by the random draw, the market functions *normally*, i.e. only the marginal traders trade. Whenever the extra-marginal trader wins the tie, the market exhibits more trades than in the traditional market clearing and a price discrimination takes place. (But, the two prices are very close to each other.) It should, however, once again be noted that this phenomenon is crucially driven by the fact that we are examining an environment with discrete valuations and prices. By simply choosing an *adequately* small money unit, this indifference phenomenon disappears.

III.D. 2-Trade Markets Without Extra-Marginal Traders

In this section we examine the case of 2-trade markets without extra-marginal traders. Even though these markets are a special case of the general m-trade markets of the next section, we have chosen to examine them separately, in order to demonstrate that a number of quasi-pure subgame perfect equilibria can evolve in such markets that are not impatience equilibria. In contrast, the case of 2-trade markets with extra-marginal traders is not examined in detail (but only as a special case of m-trade markets), because in section C.2. we already addressed the problem of non-impatience equilibria in the similar case of 1-trade markets with extra-marginal traders on only one market side.

In 2-trade markets without extra-marginal traders one can expect that two trades should occur in equilibrium, since this is required by full efficiency. Indeed in the following Proposition D.2 we show that this is one of the necessary conditions for all quasi-pure subgame equilibria of such markets. The surprising result of the analysis, however, is that, generally, both trades do not occur at the same price on the equilibrium path. Proposition D.1 establishes that three different types of quasi-pure subgame perfect equilibria exist in such markets. In two of these equilibria we find price discrimination. In both of these equilibria the intra-marginal trader on the second moving market side is forced to trade at a *worse* price (i.e. higher price, if he is a buyer, or lower price, if she is a seller) than the marginal trader on that side. The two prices, however, are very close to each other, being not more than two money units apart. The two prices are also very close to the market clearing price range, with one price just outside the range and one on the boundary. In the first equilibrium the intra-marginal trader of the second moving market side trades at a price equal to the player valuation of the marginal trader on the second moving market side minus one money unit. Then the marginal trader on the second moving market side trades at his own player valuation. In the second equilibrium the first trade is at a price equal to the marginal trader's valuation minus two, while the second trade is as before. Finally, in the third type of quasi-pure subgame perfect equilibrium, just like in the two other equilibria, the intra-marginal trader on the second moving market side trades first with either one of the traders on the opening market side. In this equilibrium, however, both prices are equal to the player valuation of the marginal trader on the second moving market side.

The second and third type of equilibria evolve because of the indifference of traders between trading in the first market phase and trading in some second market phase. This means that these equilibria can only evolve, if the traders are *patient*, in the sense that they do not trade as soon as possible. Thus, all impatience equilibria of 2-trade markets without extra-marginal traders have an outcome of the first type.

Proposition D.1

Let G be a 2-trade simplified alternating double auction market game without extra-marginal traders.

(1)   An impatience equilibrium of G exists with the following properties:

   (1.1)   Two trades occur in equilibrium.

   (1.2)   The first trade occurs at a price which lies just outside the range of feasible prices for the marginal trader of the second moving market side (i.e. exactly one money unit greater than the marginal buyer's redemption value or exactly one money unit smaller than the marginal seller's unit cost).

   (1.3)   The second trade occurs at a price equal to the player valuation of the marginal trader on the second moving market side.

(2)   If the player valuations of the traders on the second moving market side are
two or more money units apart, a second quasi-pure subgame perfect equilibrium of G exists, that is not an impatience equilibrium, with the same properties (1.1) and (1.3), but with the property (1.2) replaced by (2.2):

(2.2)   The first trade occurs at a price which lies two money units outside the
range of feasible prices for the marginal trader of the second moving
market side (i.e. exactly two money units greater than the marginal
buyer's redemption value or exactly two money units smaller than the
marginal seller's unit cost).

(3)   A third quasi-pure subgame perfect equilibrium of G exists, that is not an
impatience equilibrium, with the same property (1.1), but with the properties
(1.2) and (1.3) replaced by the properties (3.2) and (3.3), correspondingly:

(3.2)   The first trade occurs at a price equal to the player valuation of the
marginal trader on the second moving market side and involves one of
the traders on the opening market side and the intra-marginal trader on
the second moving market side.

(3.3)   The second trade occurs at a price equal to the player valuation of the
marginal trader on the second moving market side and involves one of
the traders on the opening market side and the marginal trader on the
second moving market side.

Proof of Proposition D.1 - Case (1)

For the entire proof, let us assume the buyers B_1 and B_2 open the market and the
sellers S_1 and S_2 are on the second moving market side. The opposite case can be
proven analogously.

The truncated game $\bar{G}$

Since two trades are possible, certain plays of the game G lead to a second market
phase. Obviously, which second market phase is reached, depends on the history of

play. Any second market phase that can be reached, however, is a 1-trade market without extra-marginal traders. Thus, we can use result of the Propositions C.1.2 to examine these subgames. We assume that the strategies of the traders for G induce subgame strategies in each of the second market phase subgames that have the properties specified in the Proof of Proposition C.1.1. Thus, for each of these second phases, we can conclude the following: First, the strategy combination for G induces an impatience equilibrium in each second market phase subgame of G. Second, if a second market phase subgame is reached on a play in G, the outcome corresponds to the outcome specified in Proposition C.1.2.

Given the assumption above, we can analyze the truncated game $\bar{G}$ instead of the original game G, where $\bar{G}$ is constructed in the following manner: $\bar{G}$ is identical to G, except that every second market phase subgame of G is replaced by an end node in $\bar{G}$, which has the impatience equilibrium payoffs of the corresponding second market phase subgame of G. Four types of second market phase subgames are possible in G: (1) If B_1 and S_1 trade in the first market phase, some $G_2(c_2,v_2)$ evolves, in which the traders B_2 and S_2 are present and trade at the price c_2. (Remember that we are assuming that the buyers are on the opening market side.) (2) If B_1 and S_2 trade in the first market phase, some $G_2(c_1,v_2)$ evolves, in which the traders B_2 and S_1 are present and trade at the price c_1. (3) If B_2 and S_1 trade in the first market phase, some $G_2(c_2,v_1)$ evolves, in which the traders B_1 and S_2 are present and trade at the price c_2. (4) If B_2 and S_2 trade in the first market phase, some $G_2(c_1,v_1)$ evolves, in which the traders B_1 and S_1 are present and trade at the price c_1. These results, that all follow by Proposition C.1.1, supply all equilibrium payoffs we need to construct the truncated game $\bar{G}$, as explained above. Obviously, an impatience equilibrium of $\bar{G}$ is expanded to an impatience equilibrium of G, if the strategies are expanded to cover the second market phases as specified above. Thus, in the rest of the proof, we analyze the truncated game $\bar{G}$.

Construction of an impatience equilibrium X of the truncated game $\bar{G}$

Consider a strategy combination X, that contains the four strategies x_{B1}, x_{B2}, x_{S1}, and x_{S2} of the traders, with the following properties:

a] In the very first decision stage d of $\bar{G}$, both x_{B1} and x_{B2} prescribe submitting $c_2 - 1$, i.e. B_1 and B_2 submit bids equal to S_2's unit cost minus one. Let d' denote the decision stage that immediately follows these choices. Let I' and $\bar{G}'$ denote the game segment and the subgame, correspondingly, that begin with d'.

b] x_{B1} and x_{B2} induce the enforcement segment strategies $x'_{B1}(I')$ and $x'_{B2}(I')$ on the game segment I', i.e. B_1 and B_2 refrain from submitting in all decision stages in I'.

c] x_{S1} and x_{S2} induce the acceptance segment strategies $x'_{S1}(I')$ and $x'_{S2}(I')$ on the game segment I', i.e. the marginal seller S_1 immediately accepts the current market bid in every decision stage of I' and S_2 immediately accepts the current market bid in those decision stages in I', in which the *no-loss* rule allows accepting, submits her best ask in those decision stages in I', in which the *ask-bid-spread-reduction* rule and the *no-crossing* rule allow the submission, and otherwise, refrains from submitting a new ask.

d] In each other subgame of the game, except $\bar{G}'$, the combination of subgame strategies induced by X represents some arbitrary subgame perfect equilibrium of the subgame.

The following table III.2 displays the plays that are realized by X in $\bar{G}$. Two plays are possible, because of the random draw at the end of the first market phase of G (which corresponds to the end of the truncated game $\bar{G}$). The plays displayed for the second market phase subgames, depending on whether B_1 or B_2 trades with S_1 in $\bar{G}$, are realized by strategy combinations for these second market phase subgames that correspond to the impatience equilibrium constructed in the Proof of Proposition C.1.1.

Table III.2 - Plays realized by X in $\bar{G}$ and in the second market phase subgames.

subgame	offer cycle - decision stage	B_1's bid	B_2's bid	S_1's ask	S_2's ask
$\bar{G}'$	1 - buyers	$c_2 - 1$	$c_2 - 1$		
	1 - sellers			$c_2 - 1$	c_2
	either B_1 or B_2 trades with S_1 at the price $c_2 - 1$				
	if B_1 trades with S_1 in $\bar{G}'$				
$G_2(c_2,v_2)$	2 - buyers	-	c_2		
	2 - sellers			-	c_2
	B_2 trades with S_2 at the price c_2				
	if B_2 trades with S_1 in $\bar{G}'$				
$G_2(c_2,v_1)$	2 - buyers	c_2	-		
	2 - sellers			-	c_2
	B_1 trades with S_2 at the price c_2				

We start by splitting up the truncated game $\bar{G}$ into a number of subgames at the very first decision stage. This first decision stage is not a single decision node, but the combination of the decision nodes of B_1 and B_2. Thus, each of the subgames that follow the first decision stage begin with a different combination of first bids of B_1 and B_2 (one of them begins with both B_1 and B_2 refraining to submit a first offer). Each of these subgames has at least one subgame perfect equilibrium in behavior strategies. In each of these subgames, except $\bar{G}'$, we assume with d] that X induces some arbitrary subgame perfect equilibrium. To prove that X is a quasi-pure subgame perfect equilibrium of $\bar{G}$, we only have to show that the combination of the segment strategies $X'(I')$, that contains $x'_{B1}(I')$, $x'_{B2}(I')$, $x'_{S1}(I')$, and $x'_{S2}(I')$, is a subgame perfect equilibrium of $\bar{G}'$ and that B_1 and B_2 have no reason to deviate from submitting the bids specified in a] in the very first decision stage d of $\bar{G}$.

Subgame perfectness in the main subgame $\bar{G}'$

First note that any subgame of $\bar{G}'$ is a game segment of $\bar{G}'$, because the truncated game $\bar{G}$ has no second market phases. This is also true for the largest game segment in $\bar{G}'$, namely, I', which is identical to $\bar{G}'$. For this reason, the segment strategies contained in $X'(I')$ are complete strategies for $\bar{G}'$ and $X'(I')$ can qualify as an equilibrium of $\bar{G}'$. To show that $X'(I')$ is a subgame perfect equilibrium of $\bar{G}'$, it suffices to show that: No game segment I'' exists in $\bar{G}'$, in which one of the traders has an incentive to deviate from the game segment strategy combination induced by $X'(I')$ on I''.

Let I'' be some arbitrary game segment of $\bar{G}'$ that begins with the decision stage d''. Let $\bar{b}''$ be the current market bid in d''. Due to the *ask-bid-spread-reduction* rule, $\bar{b}''$ cannot be smaller than the market bid $c_2 - 1$ in the first decision stage d' of $\bar{G}'$. Thus, $\bar{b}''$ must be in the market price range of d''. Since $\bar{b}''$ is in the market price range of d'' and the buyers are on the opening market side and both B_1 and B_2 refrain from submitting new bids in all their decision stages in I', it follows that S_1 and S_2 face the credibly enforced price $\bar{b}''$ in I''.

Since $\bar{b}'' \geq c_2 - 1$ is true anywhere in $\bar{G}'$ and since S_1 knows that she must trade at the price $c_1 \leq c_2 - 1$ in any $G_2(c_1, v_1)$ or $G_2(c_1, v_2)$, S_1 has no incentive to wait for a second market phase. (In fact, if $c_1 < c_2 - 1$, S_1 strictly prefers trading in I'' to trading in $G_2(c_1, v_1)$ or $G_2(c_1, v_2)$.) Thus, since $\bar{b}''$ is credibly enforced in I'' and since S_2 is using an acceptance segment strategy in I', S_1 maximizes her profit by also accepting $\bar{b}''$ immediately. Hence, the acceptance segment strategy $x'_{S1}(I')$ maximizes S_1's payoff in any game segment I'', given the other segment strategies.

Analogously, since $\bar{b}'' \geq c_2 - 1$ is true anywhere in $\bar{G}'$ and since S_2 knows that she must trade at the price c_2 in any $G_2(c_2, v_1)$ or $G_2(c_2, v_2)$, S_2 has no incentive to wait for a second market phase. If $\bar{b}'' > c_2$, then S_2 strictly prefers trading in I'' to trading in a $G_2(c_2, v_1)$ or $G_2(c_2, v_2)$. But, even if $\bar{b}'' = c_2$, S_2 has no reason to deviate from her acceptance segment strategy, since she is indifferent between trading in I'' or in any $G_2(c_2, v_1)$ or $G_2(c_2, v_2)$. If $\bar{b}'' < c_2$, then S_2 has no choice, but to trade in a $G_2(c_2,v_1)$ or $G_2(c_2,v_2)$, since she cannot trade in I″. But, since she cannot trade in I″, if $\bar{b}'' < c_2$, S_2 achieves a no smaller payoff in I″, by submitting her best ask, than by any other segment strategy. Thus, since $\bar{b}''$ is credibly enforced in I″ and since S_1 is using an acceptance segment strategy in I′, S_2 maximizes her profit by also accepting $\bar{b}''$ immediately, whenever possible. Hence, the acceptance segment strategy $x'_{S2}(I')$ maximizes S_2's payoff in any game segment I″, given the other segment strategies.

Since S_1 and S_2 accept any price that is credibly enforced by the buyers in I″ and since B_2 refrains from submitting new bids in I″, B_1 maximizes his payoff in I″ by enforcing the current market bid $\bar{b}''$ in I″. To see why, let us compare B_1's payoff from enforcement to his payoff from overbidding. If B_1 enforces $\bar{b}''$, two cases are possible: (i) Both S_1 and S_2 can accept $\bar{b}'' \geq c_2$, then B_1's payoff is $v_1 - \bar{b}'' + \beta$. (ii) Only S_1 can accept $\bar{b}'' = c_2 - 1$, then B_1's expected payoff is $\frac{1}{2}(v_1 - \bar{b}'' + \beta) + \frac{1}{2}(v_1 - c_2 + \beta)$. (Remember that, if $\bar{b}'' = c_2 - 1$, both buyers must be holding it, because they started off with a tie at $c_2 - 1$ in d. Additionally, the buyer, who does not trade with S_1 in the first market phase, reaches a second market phase subgame $G_2(c_2,.)$, in which he trades with S_2 at c_2.) If B_1 overbids $\bar{b}''$, his payoff is $v_1 - (\bar{b}'' + 1) + \beta$. (Naturally, B_1 could overbid $\bar{b}''$ with more than one money unit, but that would only lead to an even lower payoff.) Obviously, overbidding leads to a lower payoff in case (i). But, overbidding also leads to a lower payoff than the payoff in case (ii), because that case only evolves if $\bar{b}'' = c_2 - 1$. To overbid $\bar{b}'' = c_2 - 1$, B_1 must at least bid c_2, which is greater than the expected price in case (ii) $c_2 - \frac{1}{2}$. Hence, $x'_{B1}(I')$ maximizes B_1's payoff in any game segment I″ of $\bar{G}'$, given the acceptance segment strategies of S_1 and S_2 and given the enforcement segment strategy of B_2.

Finally, the same argument as above can be used to show that $x'_{B2}(I')$ maximizes B_2's payoff in any game segment I″ of $\bar{G}'$, given the acceptance segment strategies of S_1 and S_2 and given the enforcement segment strategy of B_1. Thus, for any possible I″ in $\bar{G}'$, no trader has an incentive to deviate from his strategy contained in $X'(I')$, and it follows that $X'(I')$ is a subgame perfect equilibrium of $\bar{G}'$.

Optimality of choices in the first decision stage d

Now we have to prove that B_1 and B_2 have no reason to deviate from their choices specified in a] in the very first decision stage d of the truncated game $\bar{G}$. As shown above - given the strategies of the others -, if B_i, with $i = 1$ or 2, submits $c_2 - 1$ in d, he either trades at the price $c_2 - 1$ in the first market phase or trades at the price c_2 in some second market phase. Thus, B_i's expected payoff is $v_i - (c_2 - \frac{1}{2}) + \beta$. If B_i deviates by submitting a bid greater than $c_2 - 1$, he earns less, since - even in the very best case - he trades at price c_2 and earns $v_i - c_2 + \beta$. If B_i deviates by submitting a bid smaller than $c_2 - 1$ (or deviates by refraining to submit a bid), one of three outcomes are possible, since the other buyer bids $c_2 - 1$ in d: (i) B_i does not trade at all and receives 0. (ii) B_i does not trade in the first market phase, but trades in some second market phase with S_2 at the price c_2 and receives $v_i - c_2 + \beta$. (iii) B_i overbids the other buyer's opening bid $c_2 - 1$ in some later decision stage of the first market phase and trades with either S_1 or S_2. Since B_i must bid at least c_2 to overbid $c_2 - 1$, his payoff - at maximum - is $v_i - c_2 + \beta$.

Clearly, B_i's payoff in the case (i) is smaller than his payoff from bidding $c_2 - 1$ in d. In the cases (ii) and (iii) B_i's payoff is also smaller than his payoff from bidding $c_2 - 1$ in d, as already explained above. Thus, no deviation from the opening bid $c_2 - 1$ can lead to a higher payoff for either of the buyers B_i, given the other strategies. Since neither B_1 nor B_2 have an incentive to deviate from their choices in d and since $X'(I')$ is a subgame perfect equilibrium of $\bar{G}'$, X is a subgame perfect equilibrium of the truncated game $\bar{G}$.

Quasi-pureness and impatience

X is a quasi-pure subgame perfect equilibrium of the $\bar{G}$, because the local strategies of the buyers in d are pure and the segment strategies of all traders in I' are pure segment strategies. But, X is also an impatience equilibrium of the game. This is so, because each buyer B_i's best offer in the first market phase on the play realized by any strategy combination $\bar{X}$, with $\bar{X} = (\bar{x}_{B1}, x_{B2}, x_{S1}, x_{S2})$, if $i = 1$, and $\bar{X} = (x_{B1}, \bar{x}_{B2}, x_{S1}, x_{S2})$, if $i = 2$, which leads to a payoff for B_i equal to his equilibrium payoff in X, must be $c_2 - 1$; otherwise B_i, would trade at a different expected price or not at all. But, since x_{Bi} prescribes submitting $c_2 - 1$ in the very first decision stage, no $\bar{x}_{Bi}$ can specify submitting $c_2 - 1$ any earlier. Similarly, every alternative strategy of the intra-marginal seller S_1, with the same payoff as x_{S1}, must prescribe submitting $c_2 - 1$, but cannot prescribe doing so any earlier than x_{S1} does. Lastly, no alternative strategy of the marginal seller S_2 in the first market phase can lead to a trade in the first market phase, given the other strategies. Thus, S_2 must submit her best offer c_2 as soon as possible in an impatience equilibrium. This is true in X.   ■

Proof of Proposition D.1 - Case (2)

The proof for case (2) of the Proposition D.1, with a first trade at the price $c_2 - 2$, is completely analogous the proof for case (1), so we refer the reader to that proof. The following comments summarize the steps necessary for the proof.

B_1 and B_2 open the market by both submitting bids at $c_2 - 2$ in the very first decision stage d of the game and then enforce this price. S_1 accepts $c_2 - 2$. The first market phase ends. In the following 1-trade market S_2 is forced to accept a trade at c_2. The main point of the proof, again, is that neither of the buyers has an incentive to overbid the original bid $c_2 - 2$. The expected payoff is $\frac{1}{2}(v_i - (c_2 - 2) + \beta) + \frac{1}{2}(v_i - c_2 + \beta)$, for each buyer B_i, with $i = 1$ or 2, if $c_2 - 2$ is enforced. We have a case of indifference, since this is equal to $(v_i - (c_2 - 1) + \beta)$, the payoff of overbidding $c_2 - 2$ with the bid $c_2 - 1$.

Note, however, that an equilibrium of this type cannot be an impatience equilibrium of the game. This is so, because either buyer's strategy could be replaced by a different strategy, which prescribes submitting $c_2 - 1$ in d, instead of submitting $c_2 - 2$. This alternative strategy combination, in which one of the buyers B_i, with $i = 1$ or 2, bids $c_2 - 1$ in d, while the other buyer B_j, with $j = 1$ or 2 and $j \neq i$, bids c_2 - 2, leads to a payoff for B_i that is equal to his expected payoff in the original equilibrium, namely $(v_i - (c_2 - 1) + \beta)$. But, B_i submits the better offer c_2 - 1 on the play realized by the alternative strategy combination than on the play realized by the equilibrium strategy combination. Thus, an equilibrium strategy combination leading to case (2) cannot be an impatience equilibrium of the game. ■

The proof for case (3) of Proposition D.1, with both trades at the price c_2, is somewhat different than the proofs of the other cases. This equilibrium evolves, because of the indifference of S_2 between trading at the price c_2 in the first and in a second market phase. If S_2 tries to trade as soon as possible, this equilibrium cannot evolve. But, if S_2 waits to trade in some second market phase subgame, then the buyers have no incentive to deviate from a tie at c_2 in the first market phase.

Proof of Proposition D.1 - Case (3)

For the entire proof, let us assume the buyers B_1 and B_2 open the market and the sellers S_1 and S_2 are on the second moving market side. The opposite case can be proven analogously.

The truncated game $\bar{G}$

Since two trades are possible, certain plays of the game G lead to a second market phase. Obviously, which second market phase is reached, depends on the history of play. Any second market phase that can be reached, however, is a 1-trade market without extra-marginal traders. Thus, we can use result of the Propositions C.1.2 to examine these subgames. We assume that the strategies of the traders for G induce subgame strategies in each of the second market phase subgames that have the properties specified in the Proof of Proposition C.1.1. Thus, for each of these second phases, we can conclude the following: First, the strategy combination for G induces an impatience equilibrium in each second market phase subgame of G. Second, if a second market phase subgame is reached on a play in G, the outcome corresponds to the outcome specified in Proposition C.1.2.

Given the assumption above, we can analyze the truncated game $\bar{G}$ instead of the original game G, where $\bar{G}$ is constructed in the following manner: $\bar{G}$ is identical to G, except that every second market phase subgame of G is replaced by an end node in $\bar{G}$, which has the impatience equilibrium payoffs of the corresponding second market phase subgame of G. Four types of second market phase subgames are possible in G: (1) If B_1 and S_1 trade in the first market phase, some $G_2(c_2,v_2)$ evolves, in which the traders B_2 and S_2 are present and trade at the price c_2. (Remember that we are assuming that the buyers are on the opening market side.) (2) If B_1 and S_2 trade in the first market phase, some $G_2(c_1,v_2)$ evolves, in which the traders B_2 and S_1 are present and trade at the price c_1. (3) If B_2 and S_1 trade in the first market phase, some $G_2(c_2,v_1)$ evolves, in which the traders B_1 and S_2 are present and trade at the price c_2. (4) If B_2 and S_2 trade in the first market phase, some $G_2(c_1,v_1)$ evolves, in which the traders B_1 and S_1 are present and trade at the price c_1. These results, that all follow by Proposition C.1.1, supply all equilibrium payoffs we need to construct the truncated game $\bar{G}$, as explained above. Obviously, an impatience equilibrium of $\bar{G}$ is expanded to an impatience equilibrium of G, if the strategies are expanded to cover the second market phases as specified above. Thus, in the rest of the proof, we analyze the truncated game $\bar{G}$.

Construction of a quasi-pure subgame perfect equilibrium X of the truncated game $\bar{G}$

Consider a strategy combination X, that contains the four strategies x_{B1}, x_{B2}, x_{S1}, and x_{S2} of the traders, with the following properties:

a] In the very first decision stage d of $\bar{G}$, both x_{B1} and x_{B2} prescribe submitting c_2, i.e. B_1 and B_2 submit bids equal to S_2's unit cost. Let d' denote the decision stage that immediately follows these choices. Let I' and $\bar{G}'$ denote the game segment and the subgame, correspondingly, that begin with d'.

b] x_{B1} and x_{B2} induce the enforcement segment strategies $x'_{B1}(I')$ and $x'_{B2}(I')$ on the game segment I', i.e. B_1 and B_2 refrain from submitting in all decision stages in I'.

c] x_{S1} induces the acceptance segment strategy $x'_{S1}(I')$ on the game segment I', i.e. S_1 immediately accepts the current market bid in any decision stage of I'. x_{S2} induces a segment strategy $x'_{S2}(I')$ on the game segment I' that prescribes accepting a trade at the current market bid in any decision stage of I', in which the current market bid is greater than c_2, and otherwise, to refrain from submitting a new ask.

d] In each other subgame of the game, except $\bar{G}'$, the combination of subgame strategies induced by X represents some arbitrary subgame perfect equilibrium of the subgame.

The following table III.3 displays the plays that are realized by X in $\bar{G}$. Two plays are possible, due to the random draw at the end of the first market phase of G (that corresponds to the end of the truncated game $\bar{G}$). The plays shown for the second market phase subgames, depending on whether B_1 or B_2 trades with S_1 in $\bar{G}$, are realized by strategy combinations for these second market phase subgames, that correspond to the impatience equilibrium constructed in the Proof of Proposition C.1.1.

Compared to case (1) of the Proposition D.1 - the major modification is due to S_2's segment strategy. Instead of accepting a trade at c_2 in the first market phase, S_2 refrains from submitting an ask and waits for a trade in some second market phase.

For the proof, we once again start by splitting up the truncated game $\bar{G}$ into a number of subgames after the very first decision stage. And again, to prove that X is a quasi-pure subgame perfect equilibrium of $\bar{G}$, we only have to show that the combination of the segment strategies $X'(I')$, that contains $x'_{B1}(I')$, $x'_{B2}(I')$, $x'_{S1}(I')$, and $x'_{S2}(I')$, is a subgame perfect equilibrium of $\bar{G}'$ and that B_1 and B_2 have no reason to deviate from submitting the bids specified in a] in the very first decision stage d of the game.

Table III.3 - Plays realized by X in $\bar{G}$ and in the second market phase subgames.

subgame	offer cycle - decision stage	B_1's bid	B_2's bid	S_1's ask	S_2's ask
$\bar{G}'$	1 - buyers	c_2	c_2		
	1 - sellers			c_2	no offer
	either B_1 or B_2 trades with S_1 at the price c_2				
	if B_1 trades with S_1 in $\bar{G}'$				
$G_2(c_2,v_2)$	2 - buyers	-	c_2		
	2 - sellers			-	c_2
	B_2 trades with S_2 at the price c_2				
	if B_2 trades with S_1 in $\bar{G}'$				
$G_2(c_2,v_1)$	2 - buyers	c_2	-		
	2 - sellers			-	c_2
	B_1 trades with S_2 at the price c_2				

Subgame perfectness in the main subgame $\bar{G}'$

For the same reasons given in the Proof of Proposition D.1 - Case (1), to show that $X'(I')$ is a subgame perfect equilibrium of $\bar{G}'$, it suffices to show that: No game segment I'' exists in $\bar{G}'$, in which one of the traders has an incentive to deviate from the segment strategy combination induced by $X'(I')$ on I''.

Let I'' be some arbitrary game segment of $\bar{G}'$ that begins with the decision stage d''. Let $\bar{b}''$ be the current market bid in d''. Due to the *ask-bid-spread-reduction* rule $\bar{b}''$ cannot be smaller than the market bid c_2 in the first decision stage d' of $\bar{G}'$. Thus, $\bar{b}''$ must be in the market price range of d''. Since $\bar{b}''$ is in the market price range of d'' and the buyers are on the opening market side and both B_1 and B_2 refrain from submitting new bids in all their decision stages in I', it follows that S_1 and S_2 face the credibly enforced price $\bar{b}''$ in I''.

Since $\bar{b}'' \geq c_2$ is true anywhere in $\bar{G}'$ and since S_1 knows that she must trade at the price $c_1 < c_2$ in any $G_2(c_1,v_1)$ or $G_2(c_1,v_2)$, S_1 strictly prefers trading in I'' to trading in any $G_2(c_1,v_1)$ or $G_2(c_1,v_2)$. Thus, since $\bar{b}''$ is credibly enforced in I'' two situations can arise: If $\bar{b}'' = c_2$, then S_2 refrains from accepting and it follows by Lemma 4 that S_1's strategy in any market side best reply combination must prescribe accepting $\bar{b}''$. If $\bar{b}'' > c_2$, S_2 uses an acceptance segment strategy in I' and S_1 maximizes her payoff by also accepting $\bar{b}''$ immediately. Thus, the acceptance segment strategy $x'_{S1}(I')$ maximizes S_1's payoff in any game segment I'', given the other segment strategies.

Analogously, since $\bar{b}'' \geq c_2$ is true anywhere in $\bar{G}'$ and since S_2 knows that she must trade at the price c_2 in any $G_2(c_2,v_1)$ or $G_2(c_2,v_2)$, two situations can arise for S_2: If $\bar{b}'' > c_2$, then S_2 strictly prefers trading in I'' to trading in any $G_2(c_2,v_1)$ or $G_2(c_2,v_2)$. But, if $\bar{b}'' = c_2$, S_2 has no reason to trade in the first market segment, since she is indifferent between trading at the price c_2 in I'' or in any $G_2(c_2,v_1)$ or $G_2(c_2,v_2)$. Thus, since $\bar{b}''$ is credibly enforced in I'' and since S_1 is using an acceptance segment strategy in I', S_2 maximizes her payoff by also accepting $\bar{b}''$ immediately, whenever $\bar{b}'' > c_2$, and refraining from submitting an offer otherwise. Thus, the segment strategy $x'_{S2}(I')$ maximizes S_2's payoff in any game segment I'', given the other segment strategies.

Since either S_1 alone or both S_1 and S_2 accept any price that is credibly enforced by the buyers in I'' and since B_2 refrains from submitting new bids in I'', B_1 maximizes his payoff in I'' by enforcing the current market bid $\bar{b}''$ in I''. To see why, let us compare B_1's payoff from the enforcement of $\bar{b}''$ to his payoff from overbidding. If B_1 enforces $\bar{b}''$ in I'', two cases are possible: (i) Both S_1 and S_2 accept $\bar{b}'' > c_2$, then B_1's payoff is $v_1 - \bar{b}'' + \beta$. (ii) Only S_1 accepts $\bar{b}'' = c_2$, then B_1's expected payoff is $\frac{1}{2}(v_1 - \bar{b}'' + \beta) + \frac{1}{2}(v_1 - c_2 + \beta)$. (Remember that, if $\bar{b}'' = c_2$, both buyers must be holding it, because they started off with a tie at c_2 in d. Additionally, the buyer, who does not trade with S_1 in the first market phase, reaches a second market phase subgame $G_2(c_2,.)$, in which he trades with S_2 at c_2.) If B_1 overbids $\bar{b}''$, his payoff is $v_1 - (\bar{b}'' + 1) + \beta$. (Naturally, B_1 could overbid $\bar{b}''$ with more than one money unit, but that would only lead to an even lower payoff.) Obviously, overbidding leads to a lower payoff in case (i). But, overbidding also leads to a lower payoff than the payoff in case (ii), because that is only the case, if $\bar{b}'' = c_2$. The expected price in that case is c_2 and, thus, smaller than the price in case of overbidding, since $\bar{b}'' + 1 = c_2 + 1$. Hence, $x'_{B1}(I')$ maximizes B_1's payoff in any game segment I'' of $\bar{G}'$, given the acceptance segment strategies of S_1 and S_2 and given the enforcement segment strategy of B_2.

Finally, the same argument as above can be used to show that $x'_{B2}(I')$ maximizes B_2's payoff in any game segment I'' of $\bar{G}'$, given the acceptance segment strategies of S_1 and S_2 and given the enforcement segment strategy of B_1. Thus, for any possible I'' in $\bar{G}'$, no trader has an incentive to deviate from his strategy contained in $X'(I')$, and it follows that $X'(I')$ is a subgame perfect equilibrium of $\bar{G}'$.

Optimality of choices in the first decision stage d

Now we have to prove that B_1 and B_2 have no reason to deviate from their choices specified in a] in the very first decision stage d of the truncated game $\bar{G}$. As shown above - given the strategies of the others -, if B_i, with i = 1 or 2, submits c_2 in d, he either trades at the price c_2 in the first market phase or trades at the price c_2 in some second market phase. This leads to an expected payoff $v_i - c_2 + \beta$. If B_i deviates by submitting a bid greater than c_2, he earns less, since - even in the very best case - he trades at price $c_2 + 1$ and achieves a payoff of $v_i - (c_2 + 1) + \beta$. If he deviates by submitting a bid smaller than c_2 (or deviates by refraining to submit a bid), one of three outcomes are possible, since the other buyer bids c_2 in d: (i) B_i does not trade at all and receives a payoff of 0. (ii) B_i does not trade in the first market phase, but trades in some second market phase with S_2 at the price c_2 and receives a payoff of $v_i - c_2 + \beta$. (iii) B_i overbids the other buyer's opening bid c_2 in some later decision stage of the first market phase and trades with either S_1 or S_2. Since B_i must bid at least $c_2 + 1$ to overbid c_2, his payoff - at maximum - can be $v_i - (c_2 + 1) + \beta$.

Clearly, B_i's payoff in the case (i) is smaller than his payoff from bidding c_2 in d. In the case (ii) B_i's payoff is equal to the expected payoff of bidding c_2 in d. In the case (iii) B_i's payoff is smaller than his expected payoff from bidding c_2 in d. Thus, no deviation from the opening bid c_2 can lead to a higher payoff for B_i given the other strategies. Since neither B_1 nor B_2 have an incentive to deviate from their choices in d and since $X'(I')$ is a subgame perfect equilibrium of $\bar{G}'$, X is a subgame perfect equilibrium of the truncated game $\bar{G}$.

Quasi-pureness, but no impatience

X is a quasi-pure subgame perfect equilibrium of the $\bar{G}$, because the local strategies of the buyers in d are pure and the segment strategies of all traders in I' are pure segment strategies. But, X is not an impatience equilibrium of the game. This is so, because S_2's strategy x_{S2} in X could be replaced by a different strategy $\bar{x}_{S2}$, which prescribes accepting c_2 immediately, i.e. submitting c_2 in d', instead of refraining to submit an ask. Since S_2 trades at c_2 in d' on the play realized by the strategy combination $\bar{X} = (x_{B1}, x_{B2}, x_{S1}, \bar{x}_{S2})$, $\bar{X}$ leads to a payoff for S_2 that is equal to her equilibrium payoff in X, namely $c_2 - c_2 + \beta = \beta$. But, S_2 submits c_2 in an earlier decision stage on the play realized by $\bar{X}$ than on the play realized by X. Thus, X cannot be an impatience equilibrium of the game. ■

Proposition D.2

Let G be a 2-trade simplified alternating double auction market game without extra-marginal traders. Any combination of strategies that is an impatience equilibrium of G - on the equilibrium path - leads to two trades with the first at a price which lies just outside the range of feasible prices for the marginal trader on the second moving market side (i.e. exactly one money unit greater than the marginal buyer's redemption value or exactly one money unit smaller than the marginal seller's unit cost) and the second at a price equal to the player valuation of the marginal trader on the second moving market side.

Proof of Proposition D.2

For the entire proof, we assume the buyers open the market. The other case can be proven analogously. We must only show that strategy combinations leading to other outcomes cannot be impatience equilibria.

The truncated game $\bar{G}$

Since two trades are possible, certain plays of the game G lead to a second market phase. Which second market phase is reached, depends on the history of play. Any second market phase that can be reached, however, is a 1-trade market without extra-marginal traders. It follows by Proposition C.1.2, that any quasi-pure subgame perfect equilibrium of these 1-trade markets without extra-marginal traders must lead to a trade at the player valuation of the second moving trader on the equilibrium path. Thus, any strategy combination for G that leads to another outcome in a second market phase cannot be a quasi-pure subgame perfect equilibrium of G. Hence - as in the Proof of Proposition D.1 -, instead of analyzing the original game G, we can analyze the truncated game $\bar{G}$, in which the second market phase subgames of G are replaced with end nodes with the corresponding payoffs of the 1-trade markets: $G_2(c_2,v_1)$, $G_2(c_1,v_1)$, $G_2(c_2,v_2)$, or $G_2(c_1,v_2)$.

By Lemma 3 we know that market period termination without a trade cannot be an equilibrium outcome. Thus, the only other possible outcome of $\bar{G}$ that must be checked is that a trade at some other price than specified in Proposition D.2 occurs in the first market phase. Since the buyers open the market, the only trading price in the first market phase, that is compatible with the proposition, is the price $c_2 - 1$.

A trade occurs at $p > c_2$

Let us first check the cases in which a trade occurs at a price $p > c_2$. If a trade occurs at the price p, the current market bid $\bar{b}'$ in some decision stage d' must have been equal to p. In d' either B_1, or B_2, or both held the current market bid $\bar{b}'$. Obviously, in equilibrium, the sellers accept a trade at $\bar{b}'$, only if neither of the buyers is planning to overbid $\bar{b}'$ in a later decision stage of the game segment I'.

Additionally, if neither buyer was planning to overbid $\tilde{b}'$ in I', both sellers should have accepted $\tilde{b}'$ in I', if this is an equilibrium. S_1 should have, because trading at $\tilde{b}' > c_2$ has a greater payoff than trading at the price c_1 in any second market phase. S_2 should have, because trading at $\tilde{b}' > c_2$ has a greater payoff than trading at the price c_2 in any second market phase. Thus, in any quasi-pure subgame perfect equilibrium in which a trade at $\tilde{b}' > c_2$ occurred, $\tilde{b}'$ was the credibly enforced current market bid and was accepted by both sellers in I'.

However, if both buyers together held the market bid $\tilde{b}' > c_2$, there must have been some decision stage d'' before d', in which both submitted $\tilde{b}'$, but each of them had an incentive to deviate from his strategy. This is so, because each could have achieved a higher expected payoff by simply refraining to bid, instead of bidding $\tilde{b}'$ in d''. Without loss of generality, let us assume B_1 refrains from bidding $\tilde{b}'$ in d''. In this case, since both sellers accept $\tilde{b}' > c_2$, held by B_2, only one of them trades at $\tilde{b}'$ with B_2. Then B_1 trades with the other seller in some second market phase. Since B_1 reaches either $G_2(c_1,v_1)$ or $G_2(c_2,v_1)$, each with probability ½, the expected price B_1 pays in a second market phase is $\frac{1}{2}c_1 + \frac{1}{2}c_2$. This price is smaller than the price he pays by tieing with B_2 at $\tilde{b}' > c_2$ in d'', since $\frac{1}{2}c_1 + \frac{1}{2}c_2 < c_2 < \tilde{b}'$. Obviously, the same argument can be made for B_2. Thus, in d'' each of the buyers has an incentive to deviate from his strategy and not tie with the other buyer at $\tilde{b}' > c_2$.

If only one of the buyers held the market bid $\tilde{b}' > c_2$ in d'', this buyer also had an incentive to deviate from his strategy. This is so, because he could have chosen to enforce some other price $p^* < \tilde{b}'$ and $p^* > c_2$ in I'. Because of the argument above, the other buyer would not have overbid p^*. Evidently, any $p^* < \tilde{b}'$ leads to a higher payoff than $\tilde{b}'$. Thus, if a strategy combination leads to a current market bid $\tilde{b}' > c_2$, it cannot be a quasi-pure subgame perfect equilibrium of the game, since - no matter whether a buyer alone or both together hold $\tilde{b}'$ - at least one of the buyers has an incentive to deviate from his strategy.

A trade occurs at $p < c_2 - 1$

Now, consider some price $p < c_2 - 1$. If a trade occurs at the price p, the current market bid $\bar{b}'$ in some decision stage d' must have been equal to p. In d' either B_1, or B_2, or both held the current market bid $\bar{b}'$. Additionally, S_1 must have accepted $\bar{b}'$ in the game segment I' beginning with d'. (Remember that S_2 cannot trade at $\bar{b}' < c_2 - 1$, due to the *no-loss* rule.) Obviously, some decision stage d'' must have existed, after $\bar{b}'$ was first submitted, but before S_1 accepted $\bar{b}'$, in which S_1 had the alternative to submit an ask $\bar{a}'' = \bar{b}' + 1$. Let I'' denote the game segment and $\bar{G}''$ denote the subgame that begins with d'', i.e. with S_1's submission of $\bar{b}' + 1$. If any buyer accepts a trade at $\bar{b}' + 1$ in $\bar{G}''$, S_1 receives a greater payoff than on the original play that led to a trade at $\bar{b}'$, because she trades at a higher price. Thus, if any buyer is willing to accept a trade at $\bar{b}' + 1$ in $\bar{G}''$, S_1 has an incentive to deviate from her original strategy. Hence, to prove that a quasi-pure subgame perfect equilibrium cannot lead to a price $p < c_2 - 1$, it is sufficient to show that at least one of the buyers prefers to trade at $\bar{b}' + 1$ with S_1 in $\bar{G}''$ for any $\bar{b}' < c_2 - 1$, i.e. overbids any current market bid $\bar{b}' < c_2 - 1$. Showing that S_1 is able to trade in $\bar{G}''$ is sufficient, because, for any strategy combination that realizes a play that leads to a trade at a price $\bar{b}' < c_2 - 1$, an alternative strategy combination exists, in which S_1's strategy choice is altered in such a way, that the realized play reaches the subgame $\bar{G}''$, in which S_1 achieves a greater payoff.

Two situations are possible in d'': either (i) only one of the buyers holds the current market bid $\bar{b}'$ or (ii) both buyers hold $\bar{b}'$. Let us first examine (i), i.e. only one buyer holds the current market bid $\bar{b}'$ in d''. In this case, the buyer not holding $\bar{b}'$ has an incentive to overbid $\bar{b}'$, i.e. to accept S_1's offer. If B_2 holds $\bar{b}'$ and B_1 does not overbid, B_1 achieves the payoff $v_1 - c_2 + \beta$, by trading with S_2 in some second market phase. Analogously, if B_1 holds $\bar{b}'$ and B_2 does not overbid, then B_2 earns $v_2 - c_2 + \beta$, by trading with S_2 in some second market phase. If the buyer not holding $\bar{b}'$ overbids, by submitting $\bar{b}' + 1$, he trades at $\bar{b}' + 1$ with S_1 in I''. Thus, overbidding leads to a payoff $v_1 - (\bar{b}' + 1) + \beta$ for B_1 and $v_2 - (\bar{b}' + 1) + \beta$ for

B_2. B_1 has an incentive to overbid as long as $\bar{b}' < c_2 - 1$, since $\bar{b}' < c_2 - 1 \Leftrightarrow$ $v_1 - (\bar{b}' + 1) + \beta > v_1 - c_2 + \beta$. Obviously, B_2 also has an incentive to overbid as long as $b' < c_2 - 1$. Thus, any strategy combination that leads to a trade at a current market bid $\bar{b}' < c_2 - 1$, held by only one buyer in I'', cannot be an equilibrium of the game, because the buyer not holding the current market bid $\bar{b}$ has an incentive to overbid $\bar{b}'$ and accept a trade with S_1 at $\bar{b}' + 1$.

Now, let us examine case (ii), i.e. both buyers together hold the current market bid $\bar{b}'$ in d''. If neither buyer chooses to overbid $\bar{b}'$ in I'', each buyer B_i, with $i = 1$ or 2, has an expected payoff $\frac{1}{2}(v_i - \bar{b}' + \beta) + \frac{1}{2}(v_i - c_2 + \beta)$. Keeping B_j's strategy constant, with $j = 1$ or 2 and $j \neq i$, B_i could deviate from his strategy by overbidding $\bar{b}'$ with $\bar{b}' + 1$. B_i then trades at the price $\bar{b}' + 1$ and receives the payoff $v_i - (\bar{b}' + 1) + \beta$. Thus, B_i has an incentive to overbid $\bar{b}'$ with $\bar{b}' + 1$ as long as: $\frac{1}{2}(v_i - \bar{b}' + \beta) + \frac{1}{2}(v_i - c_2 + \beta) < v_i - (\bar{b}' + 1) + \beta \Leftrightarrow \bar{b}' < c_2 - 2$. Hence, as long as there is a tie of both buyers at a current market bid $\bar{b}' < c_2 - 2$, each buyer B_i has an incentive to overbid $\bar{b}'$. Thus, a combination of strategies, that leads to a trade at a current market bid $\bar{b}' < c_2 - 2$ held by both buyers in I', cannot be an equilibrium of the game, because each buyer B_i has an incentive to overbid $\bar{b}'$ and accept a trade with S_1 at $\bar{b}' + 1$.

Trading at $p < c_2 - 1$: Special case $p = c_2 - 2$

The argument above also implies that, if $\bar{b}' = c_2 - 2$, B_i receives the same expected payoff, no matter whether he overbids $\bar{b}'$ by bidding $c_2 - 1$ or remains in the tie at $c_2 - 2$. Since B_i achieves an equal expected payoff in both cases, the original strategy combination that realizes a play, on which B_i remains in a tie at $c_2 - 2$, cannot be an impatience equilibrium of $\bar{G}$. This is so, because replacing B_i's original strategy to bid $c_2 - 2$ with the alternative strategy to bid $c_2 - 1$ in the original strategy combination results in a play, on which B_i submits a better offer, but achieves the same payoff. Hence, a quasi-pure subgame perfect equilibrium, that results in a tie of both buyers at the current market bid $\bar{b}' = c_2 - 2$, cannot be an impatience equilibrium of $\bar{G}$.

Trading at $p = c_2$ cannot be the outcome of an impatience equilibrium

Summarizing, we have shown that neither a trade at a price $p > c_2$ nor at a price $p \leq c_2 - 2$ can be the equilibrium outcome of an impatience equilibrium of the game. Thus, the only candidates left for equilibrium outcomes are a trade at the price $c_2 - 1$ or a trade at the price c_2. The only thing left to show is that the latter type of outcome cannot be the outcome of an impatience equilibrium.

Trading at $p = c_2$: Condition (A): if a trade occurs at $p = c_2$ in equilibrium, B_1 and B_2 must tie at the market bid $p = c_2$

First, let us show that a quasi-pure subgame perfect equilibrium, which leads to a trade at c_2 in the first market phase, can only be an impatience equilibrium of the game, if both B_1 and B_2 tie at c_2 in the first market phase. Assume X is a quasi-pure subgame perfect equilibrium of the game, which leads to a trade at c_2 in the first market phase and in which B_i, with $i = 1$ or 2, alone bids c_2 in some decision stage d and holds the current market bid at c_2 alone in the decision stages following d. X is not an impatience equilibrium, because B_j's, with $j = 1$ or 2 and $j \neq i$, strategy x_{Bj} in X could be replaced by a different strategy $\bar{x}_{Bj}$, which prescribes submitting c_2 in d. Since X is a quasi-pure subgame perfect equilibrium of the game, it follows that B_j cannot achieve a higher payoff by deviating from his strategy in this way. Thus, the alternative strategy combination $\bar{X}$, with $\bar{X} = (\bar{x}_{B1}, x_{B2}, x_{S1}, x_{S2})$, if $j = 1$, and $\bar{X} = (\bar{x}_{B1}, x_{B2}, x_{S1}, x_{S2})$, if $j = 2$, leads to a payoff for B_j that is equal to his equilibrium payoff in X, namely $v_j - c_2 + \beta$. (Remember that, if B_j does not trade in the first market phase at c_2, he trades in a second market phase subgame at c_2.) But, B_j submits a better offer on the play realized by $\bar{X}$ than on the play realized by X. Thus, X cannot be an impatience equilibrium of the game, if a trade at c_2 occurs in the first market phase, but only one buyer holds the current market bid at c_2.

Trading at $p = c_2$: Condition (B): if a trade occurs at $p = c_2$ in equilibrium, S_1 and S_2 must both immediately accept p

Let us again assume that X is a quasi-pure subgame perfect equilibrium of the game, which leads to a trade at c_2 in the first market phase. From the above argument, we know that X can only be an impatience equilibrium of the game, if both B_1 and B_2 tie at c_2 in the first market phase. Now, we show that X can only be an impatience equilibrium of the game, if both sellers accept a trade at c_2 in the first market phase. S_1 must accept c_2 in a quasi-pure subgame perfect equilibrium, because in any second market phase she trades at the lower price c_1. S_2 is indifferent between trading in the first market phase and in a second market phase subgame. But, she must accept c_2 in an impatience equilibrium. This is true, because, if S_2's strategy x_{S2} in X does not prescribe accepting c_2 in the first market phase, her best offer p on the play realized by X in the first market phase is greater than c_2, i.e. $p > c_2$. S_2, however, has an alternative strategy $\bar{x}_{S2}$, that prescribes submitting the better offer c_2 in the first market phase. The strategy combination $\bar{X} = (x_{B1}, x_{B2}, x_{S1}, \bar{x}_{S2})$ leads to a payoff for S_2 that is equal to her payoff in X, namely β. (Remember that we are assuming that both buyers jointly hold the current market bid c_2 in the first market phase.) Thus, X cannot be an impatience equilibrium of the game, if a trade at c_2 occurs in the first market phase and both buyers hold the current market bid at c_2, but only S_1 accepts a trade at c_2.

Trading at $p = c_2$: Conditions (A) and (B) cannot simultaneously be true in equilibrium

So far, we have established that there are two necessary condition for an impatience equilibrium X of the game, which leads to a trade at c_2 in the first market phase: (A) Both buyers must jointly hold the current market bid c_2. (B) Both sellers must accept a trade at c_2 in the first market phase. Now, we show that, if both conditions (A) and (B) are fulfilled, X is not a quasi-pure subgame perfect equilibrium of the game, because each buyer has an incentive to deviate from his strategy in X. If both sellers accept c_2 in the first market phase, each buyer B_i, with $i = 1 \dots 2$, trades at

c_2 and receives the payoff $v_i - c_2 + \beta$, since both buyers have tied at c_2. If one buyer B_i, however, refrains from bidding, instead of bidding c_2, he trades with the seller, who did not trade in the first market phase, in some second market phase. Since B_i reaches either $G_2(c_1, v_i)$ or $G_2(c_2, v_i)$, each with probability $\frac{1}{2}$, the expected price B_i pays in a second market phase is $\frac{1}{2}c_1 + \frac{1}{2}c_2$. This price is smaller than the price he pays by tieing at c_2 with B_j, with $j = 1$ or 2 and $j \neq i$, in the first market phase, since $\frac{1}{2}c_1 + \frac{1}{2}c_2 < c_2$. Thus, each buyer B_i has an incentive to deviate from his strategy in X and not tie with the other buyer at c_2. Thus, if condition (B) holds, condition (A) cannot hold in equilibrium. Since both conditions are necessary for X to be an impatience equilibrium, it follows that a strategy combination that leads to a trade at c_2 in the first market phase, cannot be an impatience equilibrium X of the game. Hence, all impatience equilibria of the game must lead to a trade at the price $c_2 - 1$.   ∎

III.E. M-Trade Markets

A general m-trade market contains a number m of non-extra-marginal traders on each market side. Additionally, some extra-marginal traders can be present. The cases with $m = 1$, i.e. 1-trade markets with and without extra-marginal traders, have been examined separately in the sections above. In this section, we generalize those results to any case with $m \geq 1$. To be able to establish these results, we use an induction over the maximum number of possible trades in a simplified alternating double auction market game and use the results for the 1-trade markets, that are 1-maximum-trade markets, as anchor in the proofs. Thus, even though the proofs are general, we basically examine those cases with more than one non-extra-marginal trader on each market side, i.e. with $m \geq 2$.

If an m-trade market, with $m \geq 2$, contains no extra-marginal traders in the first market phase, any following market phase that can be reached is also a market without extra-marginal traders. But, in the general case, with extra-marginal traders present in the first market phase, the following market phases may or may not contain extra-marginal traders. What type of following market phase is reached, depends on the play in the first market phase. Since we are interested in impatience equilibria of the game, all possible following markets must be considered. An important feature of double auction markets, however, is that no market phase that follows a number of trades in the first market phase can have a greater maximum number of possible trades than the original market. This feature is central to the inductions, that we use in the proofs of the propositions in this section.

As usual, we index the buyers in the order from the highest to the lowest redemption value and the sellers from the lowest to the highest unit cost. The traders are denoted by B_1, ..., B_m, B_{m+1}, ..., B_{nB}, and S_1, ..., S_m, S_{m+1}, ..., S_{nS}, where B_m and S_m are the marginal traders and B_{m+1} and S_{m+1} are the best extra-marginal traders. To simplify notation, we let the x denote the index m + 1, i.e. the best extra-marginal buyer is denoted by B_x ($= B_{m+1}$) and the best extra-marginal seller is denoted by S_x ($= S_{m+1}$). The player valuations are indexed correspondingly.

In Proposition E.1 an impatience equilibrium of the game is specified. This establishes the existence of such equilibria of the game. The equilibrium in Proposition E.1 is not unique, but Proposition E.2 shows that the equilibrium outcome is unique, i.e. that all impatience equilibria of the game lead to the same outcome.

To be able to specify a unique impatience equilibrium outcome for m-trade markets, we need the following technical assumption:

Assumption A2

Let w and w' be any two player valuations in a simplified alternating double auction market game. In all cases, except in the case in which the two valuations are the marginal valuations v_m and c_m, w and w' are more than two money units apart, i.e. $|w - w'| > 2$, where w and w' $\in W = \{V \cup C\}$, w $\neq$ w', and either w $\notin$ $\{v_m, c_m\}$ or w' $\notin$ $\{v_m, c_m\}$. For v_m and c_m the relationship $v_m \geq c_m$ is true.

This assumption is an extension of Assumption A1 used in the case of 1-trade markets with extra-marginal traders on both market sides. There, it was sufficient that the two extra-marginal player valuations were more than two money units apart. Now, however, we need an extension to all those pairs of player valuations, that could become the two extra-marginal valuations in some second market phase subgame. The assumption is actually much stricter than necessary, but simplifies the proofs in this form.

The case of the two marginal valuations, however, was explicitly excluded in Assumption A2 to allow for markets in which the marginal valuations are equal. Such markets are typically used in experiments. It should be noted that allowing for $v_m - c_m \leq 2$ does not cause any difficulties. The original market in such cases must have v_m and c_m as the two innermost valuations, since all other valuations are assumed to be more than two money units away from both v_m and c_m. In any second market phase subgame that can emerge, either v_m and c_m will remain the two innermost valuations or a combination of innermost valuations will be present that has a distance of more than two money units. Even if $v_m = c_m$ is true, the case we need for the induction proof is a 1-trade market without extra-marginal traders (Proposition C.1.2) for which Assumption A1 was not necessary. Additionally, a 1-trade market with extra-marginal trader on only one market side could emerge as a second market phase subgame, in which one of the two valuations is marginal and one extra-marginal. But, for the propositions concerning these markets we did not need the Assumption A1 either.

Just as in the case of Assumption A1, all indifferences that are omitted with Assumption A2 arise from the fact, that we have an environment with a discrete money unit. As in the other case, however, Assumption A2 is merely technical, because the assumed property can be achieved in any given game by a simple redefinition of the smallest money unit. Given that the smallest money unit can be chosen arbitrarily, it suffices to have the property $w \neq w'$.

Proposition E.1

Let G be an m-trade simplified alternating double auction market game. Let Assumption A2 be true in G. An impatience equilibrium of G exists with the following properties:

(1)   If the best extra-marginal player valuation on the opening market side exists and is in the market clearing price range:

   (1.1)   A total of m trades occur in equilibrium.

   (1.2)   All m trades occur at a price which lies just outside the range of feasible prices for the best extra-marginal trader of the opening market side (i.e. exactly one money unit greater than the best extra-marginal buyer's redemption value or exactly one money unit smaller than the best extra-marginal seller's unit cost).

(2)   If the best extra-marginal player valuation on the opening market side does not exist or is not in the market clearing price range:

   (2.1)   A total of m trades occur in equilibrium.

   (2.2)   The first m - 1 trades occur at a price which lies just outside the range of feasible prices for the marginal trader of the second moving market side (i.e. exactly one money unit greater than the marginal buyer's redemption value or exactly one money unit smaller than the marginal seller's unit cost).

   (2.3)   The last trade occurs at a price equal to the player valuation of the marginal trader on the second moving market side.

Proof of Proposition E.1

For the entire proof, let us assume the buyers open the market and the sellers are on the second moving market side. The opposite case can be proven analogously. The following proof only concerns m-trade markets with m $\geq$ 2, since the proofs for the 1-trade markets are given with the propositions C.1.2, C.2.2, and C.3.2.

The truncated game $\bar{G}$ and the induction assumption

Since m $\geq$ 2, more than one trade is possible. Thus, certain plays of the game G may lead to a second market phase. A r-trade market, with r $\leq$ m, begins with any such second market phase. Without loss of generality, we will assume that the original market is a μ-maximum-trade market, where $\mu \geq$ m. Since the r-trade market, that follows in the second market phase, must be a ν-maximum-trade market, with $\nu < \mu$, we can use an induction over the maximum number of possible trades left in the market for the proof. To be able to do so, we begin by assuming that the impatience equilibrium outcome of any ν-maximum-trade second market phase subgame $G_2(.,.)$ of G has the properties that the Proposition E.1 specifies. Thus and since we assume that the buyers are on the opening market side, any second market phase subgame that can evolve in G is of one of the following types with the corresponding outcome:

(i) $G_2(c_j, v_i)$, where the innermost traders, S_j and B_i, are the marginal traders. The first r - 1 trades occur at the price c_j - 1 and the last trade occurs at the price c_j.

(ii) $G_2(v_i, c_j)$, where the innermost traders, B_i and S_j, are the best extra-marginal traders. All trades occur at the price v_i + 1.

(iii) $G_2(c_j, c_{j+1})$, where the innermost traders, S_j and S_{j+1}, are the marginal seller and the best extra-marginal seller, correspondingly. The first r - 1 trades occur at the price c_j - 1 and the last trade occurs at the price c_j.

(iv) $G_2(v_i, v_{i-1})$, where the innermost traders, B_i and B_{i-1}, are the best extra-marginal buyer and the marginal buyer, correspondingly. All trades occur at the price v_i + 1.

Summarizing, we have: In any second market phase subgame $G_2(v_i,.)$, in which the lower bound of the market clearing price range is equal to the redemption value v_i of a buyer B_i, all trades occur at the price v_i + 1. In any second market phase subgame $G_2(c_j,.)$, in which the lower bound of the market clearing price range is equal to the unit cost c_j of a seller S_j, the first $r - 1$ trades occur at the price $c_j - 1$ and the last trade at the price c_j.

Given the induction assumption above, we can analyze the truncated game $\bar{G}$ instead of the original game G, where $\bar{G}$ is constructed in the following manner: $\bar{G}$ is identical to G, except that every second market phase subgame $G_2(.,.)$ of G is replaced by an end node in $\bar{G}$, which has the equilibrium payoffs of the corresponding $G_2(.,.)$. Obviously, a quasi-pure subgame perfect equilibrium of $\bar{G}$ is expanded to a quasi-pure equilibrium of G, if the strategies are expanded to cover the second market phase subgames recursively, in the way described below. Thus, we will first analyze the truncated game $\bar{G}$ for the case (1) of the proposition. Then we do the same for case (2). Finally, we use the results of the propositions C.1.1 and C.2.1 to induce the impatience equilibrium of a 2-maximum-trade market. This is possible, because the 1-trade markets considered in those propositions are 1-maximum-trade markets. Then we use the result for 2-maximum-trade markets to induce the result for 3-maximum-trade markets and so on. Obviously, by showing that the proposition holds for an arbitrary μ-maximum-trade market, we also show that it holds for an arbitrary m-trade market.

Impatience equilibrium X of the truncated game $\bar{G}$ in case (1)

At least one extra-marginal buyer exists in case (1). Additionally, the best extra-marginal buyer B_x's redemption value v_x is in the market clearing price range, i.e. $c_m < v_x$.

Construction of an impatience equilibrium X of the truncated game $\bar{G}$

Consider a strategy combination X, that contains the $n_B + n_S$ strategies x_{Bi}, with $i = 1 \dots n_B$, and x_{Sj}, with $j = 1 \dots n_S$, of the traders with the following properties:

a] In the very first decision stage d of $\bar{G}$, all x_{Bi}, with $i = 1 \dots m$, prescribe submitting $v_x + 1$, i.e. all non-extra-marginal buyers submit bids equal to the extra-marginal buyer B_x's redemption value plus one. All x_{By}, with $y = x \dots n_B$, prescribe submitting v_y, i.e. each extra-marginal buyer submits a bid equal to his own redemption value. Let d' denote the decision stage that immediately follows these choices. Let I' and $\bar{G}'$ denote the game segment and the subgame, correspondingly, that begin with d'.

b] All x_{Bi} induce the enforcement segment strategies $x'_{Bi}(I')$ on the game segment I', i.e. all buyers refrain from submitting in all decision stages in I'.

c] All x_{Sj} induce the acceptance segment strategies $x'_{Sj}(I')$ on the game segment I', i.e. each of the sellers immediately accepts the current market bid in those decision stages in I', in which the *no-loss* rule allows accepting, submits her best ask in those decision stages in I', in which the *ask-bid-spread-reduction* rule and the *no-crossing* rule allow the submission, and otherwise, refrains from submitting a new ask.

d] In each other subgame of the game, except $\bar{G}'$, the combination of subgame strategies induced by X represents some arbitrary subgame perfect equilibrium of the subgame.

We start by splitting up the truncated game $\bar{G}$ into a number of subgames after the very first decision stage. This first decision stage is not a single decision node, but the combination of the decision nodes of all B_i. Thus, each of the subgames that follow the first decision stage begins with a different combination of first bids of all B_i (one of them begins with all buyers refraining to submit a first offer). Each of these subgames has at least one subgame perfect equilibrium in behavior strategies. For all those subgames that begin with a different combination of opening bids than specified in a], we assume with d] that X induces some arbitrary subgame perfect equilibrium. Thus, to prove that X is a subgame perfect equilibrium of $\bar{G}$, we only have to show that the combination of the segment strategies $X'(I')$, that contains $x'_{Bi}(I')$, with $i = 1 \dots n_B$, and $x'_{Sj}(I')$, with $j = 1 \dots n_S$, is a subgame perfect equilibrium of $\bar{G}'$ and that none of the B_i has a reason to deviate from submitting the bids specified in a] in the very first decision stage d of the game.

Subgame perfectness in the main subgame $\bar{G}'$

First note that any subgame of $\bar{G}'$ is a game segment of $\bar{G}'$, because the truncated game $\bar{G}$ has no second market phases. This is also true for the largest game segment in $\bar{G}'$, namely, I', which is identical to $\bar{G}'$. For this reason, the segment strategies contained in $X'(I')$ are complete strategies for $\bar{G}'$ and $X'(I')$ can qualify as an equilibrium of $\bar{G}'$. To show that $X'(I')$ is a subgame perfect equilibrium of $\bar{G}'$, it suffices to show that: No game segment I'' exists in $\bar{G}'$, in which one of the traders has an incentive to deviate from the game segment strategy combination induced by $X'(I')$ on I''.

Subgame perfectness in the main subgame $\bar{G}'$:
Possible second market phase subgames

Let I'' be some arbitrary game segment of $\bar{G}'$ beginning with the decision stage d''. Let $\tilde{b}''$ be the current market bid in d''. Due to the *ask-bid-spread-reduction* rule $\tilde{b}''$ cannot be smaller than the market bid $v_x + 1$ in the first decision stage d' of $\bar{G}'$. Thus, $\tilde{b}''$ must be in the market price range of d''. Since $\tilde{b}''$ is in the market price range of d'' and the buyers are on the opening market side and all buyers refrain from submitting new bids in all their decision stages in I', it follows that the sellers face the credibly enforced price $\tilde{b}''$ in I''.

Since $\tilde{b}'' \geq v_x + 1$ and since $\tilde{b}''$ is credibly enforced, it follows that: First, some trade must occur at the price $\tilde{b}''$, due to Lemma 4. Second, no extra-marginal buyer can trade in I'', due to the *no-loss* rule. Thus, at least one trade occurs at $\tilde{b}''$, but no trade that occurs involves an extra-marginal buyer, and a second market phase subgame can begin. (Note, if all non-extra-marginal traders on both sides trade in the first market phase, a second market phase does not begin.) In any second market subgame that can begin after I'', the lower bound of the market clearing price range will be either the redemption value v_y, with $y \geq x$, of an extra-marginal buyer or the unit cost c_h, with $h \leq m$, of an non-extra-marginal seller. This is true, because of the following: The lower bound of the market clearing price range in the first market phase is the redemption value v_x of the best extra-marginal buyer. If only non-extra-marginal sellers trade, the lower bound of the market clearing price range in any reached $G_2(.,.)$ remains v_x. If one or more extra-marginal sellers trade, the intra-marginal part of the supply curve shifts less to the left than that part of the demand curve, because the number of trading non-extra-marginal sellers is smaller than number of trading non-extra-marginal buyers. This means that either some originally non-extra-marginal sellers could become extra-marginal in $G_2(.,.)$ or that some originally extra-marginal buyers could become non-extra-marginal in $G_2(.,.)$. In any case, however, the lower bound of the market clearing price range in any reached $G_2(.,.)$ is either determined by the redemption value v_y of an originally extra-marginal buyer B_y, with $y \geq x$, or by the unit cost c_h of an originally non-extra-marginal seller S_h, with $h \leq m$. Hence, after one or more trades at $\bar{b}''$ in the first market phase, any second market phase subgame that can emerge is of one of the following two types: either $G_2(v_y,.)$, with $y \geq x$, or $G_2(c_h,.)$, with $h \leq m$.

Subgame perfectness in the main subgame $\bar{G}'$: Intra-marginal sellers

In any second market phase subgame of the type $G_2(v_y,.)$, with $y \geq x$, any intra-marginal seller S_j, with $j < m$, trades at the price $v_y + 1 \leq v_x + 1$. In any second market phase subgame of the type $G_2(c_h,.)$, with $h \leq m$, three situation can arise for S_j: (i) If S_j is the marginal seller in $G_2(c_h,.)$, i.e. if $j = h$, she trades at the price $c_h = c_j \leq c_m < v_x + 1$. (Remember that we are examining the first part of Proposition E.1, in which $v_x > c_m$.) (ii) If S_j is an intra-marginal seller in $G_2(c_h,.)$, i.e. if $j < h$, she trades at the price $c_h - 1 \leq c_m - 1 < v_x + 1$. (iii) If S_j is an extra-marginal seller in $G_2(c_h,.)$, i.e. if $j > h$, she does not trade at all. Thus, in any second market phase subgame, in which S_j can trade at all, she trades at some price $p \leq v_x + 1$. Because $\bar{b}'' \geq v_x + 1 \geq p$ is true anywhere in $\bar{G}'$, S_j has no incentive to wait for a second market phase subgame. Finally, because $\bar{b}''$ is credibly enforced in I'', S_j maximizes her payoff by accepting $\bar{b}''$ immediately. If only she can accept a trade at $\bar{b}''$, it is irrelevant for her payoff, whether she accepts $\bar{b}''$ immediately or later, as long as she accepts before market period termination. If some other seller S_h, with $h \neq j$, can also accept a trade at $\bar{b}''$, S_j maximizes her payoff by accepting $\bar{b}''$ immediately, since S_h is using an acceptance segment strategy in I'. Thus, the acceptance segment strategy $x'_{Sj}(I')$, with $j = 1 \dots (m - 1)$, maximizes S_j's payoff in any game segment I'', given the other segment strategies.

Subgame perfectness in the main subgame $\bar{G}'$: The marginal seller

Note that this is also true for the marginal seller S_m, i.e. S_m has no incentive to wait for a second market phase subgame either. In any second market phase subgame of the type $G_2(v_y,.)$, with $y \geq x$, S_m trades at the price $v_y + 1 \leq v_x + 1$. In any second market phase subgame of the type $G_2(c_h,.)$, with $h \leq m$, S_m can only reach two situations: (i) Such $G_2(c_h,.)$, with $h \doteq m$, in which she is the marginal seller and trades at $c_m < v_x + 1$. (ii) Such $G_2(c_h,.)$, with $h < m$, in which she is an extra-marginal seller and does not trade at all. Thus, in any second market phase subgame, in which S_m can trade at all, she trades at some price $p \leq v_x + 1$. Hence, S_m - even in the best case - is indifferent between trading in the first market phase at a price $\bar{b}'' \geq v_x + 1$ and trading in some second market phase subgame at $p \leq v_x + 1$. Since $\bar{b}''$ is credibly enforced in I'' and since all other S_h, with $h \neq m$, are using an acceptance segment strategy in I', S_m maximizes her payoff by also accepting $\bar{b}''$ immediately. Thus, the acceptance segment strategy $x'_{Sm}(I')$ maximizes S_m's payoff in any game segment I'', given the other segment strategies.

Subgame perfectness in the main subgame $\bar{G}'$: Extra-marginal sellers

After having established that no non-extra-marginal seller has a reason to deviate from her acceptance segment strategy, let us turn to the extra-marginal sellers. The extra-marginal sellers cannot trade in any second market phase subgame, since they remain extra-marginal both in any $G_2(v_y,.)$, with $y \geq x$, and in any $G_2(c_h,.)$, with $h \leq m$. Since an extra-marginal seller can only trade in the first market phase, it follows by Corollary 2, that in any market side best reply combination of the sellers each extra-marginal seller, who makes no loss by trading at $\bar{b}''$, must accept a trade at the enforced price $\bar{b}''$. If one of the extra-marginal sellers delays the acceptance, she is certain not to trade at all, since the other sellers accept $\bar{b}''$ immediately. If an extra-marginal seller cannot accept $\bar{b}''$, because of the *no-loss* rule, all her segment strategies in I'' lead to a payoff of 0, given the strategies of the other traders. Thus, by submitting her best ask, this extra-marginal seller achieves a no smaller payoff in I'' than with any other segment strategy. Hence, the acceptance segment strategy of each extra-marginal seller maximizes her payoff in any game segment I'' of G', given the segment strategies of the others.

Subgame perfectness in the main subgame $\bar{G}'$: Non-extra-marginal buyers

Since all S_j, who can do so, accept any price that is credibly enforced by the buyers in I'' and since all other B_h, with $h \neq i$, refrain from submitting new bids in I'', B_i maximizes his payoff in I'' by also enforcing the current market bid $\bar{b}''$ in I''. To see why, let us compare B_i's payoff from enforcement to his payoff from overbidding. Remember that I'' is an arbitrary game segment in $\bar{G}'$, since we must confirm that $X'(I')$ induces equilibria in all possible subgames of $\bar{G}'$, i.e. we have to consider any possible constellation in I''.

Because all non-extra-marginal buyers submit the bid $v_x + 1$ in the decision stage d, before $\bar{G}'$, it follows that: First, the current market bid $\bar{b}''$ cannot be smaller than $v_x + 1$ anywhere in $\bar{G}'$, due to the *ask-bid-spread-reduction* rule. Second, if $\bar{b}'' = v_x + 1$, then all m non-extra-marginal buyers hold $\bar{b}''$. Considering these two conditions and the acceptance segment strategies of the sellers, the following constellations can emerge in I'' from the point of view of a non-extra-marginal buyer B_i, with $i \leq m$:

(i) $\bar{b}'' = v_x + 1$: All non-extra-marginal buyers jointly hold $\bar{b}''$. All non-extra-marginal sellers S_j, with $j \leq m$, accept $\bar{b}''$. No extra-marginal seller S_y, with $y = x \ldots n_S$, can accept a trade at $v_x + 1 < c_y$, due to the *no-loss* rule. By using an enforcement segment strategy, B_i trades at the price $v_x + 1$ in I'' and achieves the payoff $(v_i - (v_x + 1) + \beta)$. Alternatively, B_i could choose to overbid $\bar{b}''$ with $\bar{b}'' + 1$ and receive a payoff $v_i - (v_x + 2) + \beta$. Evidently, B_i has no incentive to overbid, since the price he pays by enforcement is smaller than the price he pays by overbidding.

(ii) $\bar{b}'' > v_x + 1$: A number k of non-extra-marginal buyers, with $1 \leq k \leq m$, jointly hold $\bar{b}''$. (Extra-marginal buyers cannot hold $\bar{b}'' > v_x + 1$.) All sellers S_j, with $c_j \leq \bar{b}''$, accept $\bar{b}''$. Note that at least m sellers (i.e. all non-extra-marginal sellers) accept a trade at $\bar{b}''$.

(iia) If B_i is in the group of buyers who hold $\bar{b}''$, he trades in I'' and receives a payoff $(v_i - \bar{b}'' + \beta)$. If B_i overbids $\bar{b}''$ with $\bar{b}'' + 1$, he receives a payoff $v_i - (\bar{b}'' + 1) + \beta$. Obviously, B_i has no incentive to overbid.

(iib) If B_i is not in the group of buyers who hold $\bar{b}''$, he either trades in some second market phase subgame $G_2(v_y,.)$, with $y \geq x$, at the price $v_y + 1 \leq v_x + 1 < \bar{b}''$ or trades in some second market phase subgame $G_2(c_h,.)$, with $h \leq m$, at a price $p \leq c_h \leq c_m < v_x < \bar{b}''$. (Remember that any second market phase subgame that can be reached must either be of the type $G_2(v_y,.)$, with $y \geq x$, or $G_2(c_h,.)$, with $h \leq m$, as proven further up. Remember also that $c_m < v_x$, since we are examining the first part of Proposition E.1.) Thus, no matter which type of second market phase subgame B_i reaches, the price he pays in $G_2(.,.)$ is smaller than the price he pays by overbidding. Obviously, B_i again has no incentive to overbid $\bar{b}''$ with $\bar{b}'' + 1$.

Hence, overbidding leads to a lower payoff for the non-extra-marginal buyers in all cases. (Note that B_i could overbid $\bar{b}''$ with more than one money unit in any of these cases, but that would only lead to even lower payoffs. Note also that, since $\bar{b}'' > v_x$, none of the extra-marginal buyers can trade in I''. This ensures that every non-extra-marginal buyer is able to trade in any second market phase subgame that can be reached.) Thus, the enforcement segment strategy $x'_{Bi}(I')$ maximizes a non-extra-marginal buyer B_i's payoff in any game segment I'' of $\bar{G}'$, given the segment strategies of the others.

Subgame perfectness in the main subgame $\bar{G}'$: Extra-marginal buyers

Since $\bar{b}'' > v_x$ anywhere in $\bar{G}'$, the extra-marginal buyers do not have a choice but to refrain from submitting bids in $\bar{G}'$, due to the *ask-bid-spread-reduction* rule in combination with the *no-loss* rule. Thus, the enforcement segment strategy maximizes an extra-marginal buyer's payoff in any game segment I'' of $\bar{G}'$, given the segment strategies of the others.

Hence, each seller maximizes her payoff with her acceptance segment strategy $x'_{Sj}(I')$ and each buyer maximizes his payoff with his enforcement segment strategy $x'_{Bi}(I')$ in any game segment I'' of $\bar{G}'$, given the segment strategies of the other traders. Thus, for any possible I'' in $\bar{G}'$, no trader has an incentive to deviate from his strategy contained in $X'(I')$, and it follows that $X'(I')$ is a subgame perfect equilibrium of $\bar{G}'$.

Optimality of choices in the first decision stage d

Now we have to prove that no buyer has a reason to deviate from his choice - specified in a] - in the very first decision stage d of the truncated game $\bar{G}$. As shown above - given the strategies of the others -, if a non-extra-marginal buyer B_i submits $v_x + 1$ in d, then he trades at the price $v_x + 1$. This leads to an expected payoff $v_i - (v_x + 1) + \beta$. If he deviates by submitting a bid greater than $v_x + 1$, he earns less, since - even in the very best case - he will trade at price $v_x + 2$ and receive $v_i - (v_x + 2) + \beta$. If B_i deviates by submitting a bid smaller than $v_x + 1$ (or deviates by refraining to submit a bid), one of three outcomes are possible, since all other B_h, with $h \neq i$ and $h \leq m$, bid $v_x + 1$ in d: (i) B_i does not trade at all and receives 0. (ii) B_i does not trade in the first market phase. Instead he trades in some second market phase subgame $G_2(v_x,.)$ at the price $v_x + 1$ and receives a payoff of $v_i - (v_x + 1) + \beta$. (Note that any second market phase subgame that can be reached is of the type $G_2(v_x,.)$, because only non-extra-marginal buyers and sellers can trade in the first market phase at $v_x + 1$ and because v_x is the lower bound of the market clearing price range in the first market phase.) (iii) B_i overbids the opening bid $v_x + 1$ of the other buyers in some later decision stage of the first market phase and trades alone in the first market phase. Since B_i must bid at least $v_x + 2$ to overbid the other buyers, his payoff can at maximum be $v_i - (v_x + 2) + \beta$.

Clearly, B_i's payoff in the case (i) is smaller than his payoff from bidding $v_x + 1$ in d. In the cases (ii) and (iii) B_i's payoff is also smaller than his payoff from bidding $v_x + 1$ in d, as already explained above. Thus, no deviation from the opening bid $v_x + 1$ can lead to a higher payoff for a non-extra-marginal B_i, with $i \leq m$, given the other strategies. Extra-marginal buyers do not have an incentive to deviate from their opening bids either, since - given the strategies of the others - they cannot trade in the game, no matter which opening bid they choose. Since no buyer has an incentive to deviate from his choice in d and since $X'(I')$ is a subgame perfect equilibrium of $\bar{G}'$, X is a subgame perfect equilibrium of the truncated game $\bar{G}$.

Quasi-pureness and impatience

X is a quasi-pure subgame perfect equilibrium of the $\bar{G}$, because the local strategies of the buyers in d are pure and the segment strategies of all traders in I' are pure segment strategies. But, X is also an impatience equilibrium of the game. This is so, because each non-extra-marginal buyer B_i's best offer must be $v_x + 1$ in the first market phase on the play realized by any strategy combination $\bar{X} = (x_{B1}, ..., x_{Bi-1},$ $\bar{x}_{Bi}, x_{Bi+1}, ..., x_{BnB}, x_{S1}, ..., x_{SnS})$ which leads to a payoff equal to B_i's equilibrium payoff in X; otherwise B_i, with $i = 1 ... m$, would trade at a different expected price or not at all in the first market phase. But, since x_{Bi} prescribes submitting $v_x + 1$ in the very first decision stage, no $\bar{x}_{Bi}$ can specify submitting $v_x + 1$ any earlier. Similarly, every alternative strategy of each of the non-extra-marginal sellers S_j, with $j = 1 ... m$, with the same payoff as x_{Sj}, must prescribe submitting $v_x + 1$, but cannot prescribe doing so any earlier than x_{Sj} does. Lastly, no alternative strategy of an extra-marginal buyer B_k, with $k = x ... n_B$, or seller S_y, with $y = x ... n_S$, can lead to a trade in the first market phase, given the other strategies. Thus, B_k and S_y must submit their best offers, v_k and c_y, correspondingly, as soon as possible in an impatience equilibrium. This is also true in X.

Impatience equilibrium X of the truncated game $\bar{G}$ in Case (2)

In case (2) either no extra-marginal buyer exists or the best extra-marginal buyer B_x's redemption value v_x is not in the market clearing price range.

Construction of an impatience equilibrium X of the truncated game $\bar{G}$

Consider a strategy combination X, that contains the $n_B + n_S$ strategies x_{Bi}, with $i = 1 \ldots n_B$, and x_{Sj}, with $j = 1 \ldots n_S$, of the traders with the following properties:

a] In the very first decision stage d of $\bar{G}$, all x_{Bi}, with $i = 1 \ldots m$, prescribe submitting $c_m - 1$, i.e. all non-extra-marginal buyers submit bids equal to the marginal seller S_m's unit cost minus one. All x_{By}, with $y = x \ldots n_B$, prescribe submitting v_y, i.e. each extra-marginal buyer submits a bid equal to his own redemption value. Let d′ denote the decision stage that immediately follows these choices. Let I′ and $\bar{G}'$ denote the game segment and the subgame, correspondingly, that begin with d′.

b] All x_{Bi} induce the enforcement segment strategies $x'_{Bi}(I')$ on the game segment I′, i.e. all buyers refrain from submitting in all decision stages in I′.

c] All x_{Sj} induce the acceptance segment strategies $x'_{Sj}(I')$ on the game segment I′, i.e. each of the sellers immediately accepts the current market bid in those decision stages in I′, in which the *no-loss* rule allows accepting, submits her best ask in those decision stages in I′, in which the *ask-bid-spread-reduction* rule and the *no-crossing* rule allow the submission, and otherwise, refrains from submitting a new ask.

d] In each other subgame of the game, except $\bar{G}'$, the combination of subgame strategies induced by X represents some arbitrary subgame perfect equilibrium of the subgame.

Again, we start by splitting up the truncated game $\bar{G}$ into a number of subgames after the very first decision stage. This first decision stage is not a single decision node, but the combination of the decision nodes of all B_i. Thus, each of the subgames that follow the first decision stage begins with a different combination of first bids of all B_i (one of them begins with all buyers refraining to submit a first offer). Each of these subgames has at least one subgame perfect equilibrium in behavior strategies. For all those subgames that begin with a different combination of opening bids than specified in a], we assume with d] that X induces some arbitrary subgame perfect equilibrium. Thus, to prove that X is a subgame perfect equilibrium of $\bar{G}$, we only have to show that the combination of the segment strategies $X'(I')$, that contains $x'_{Bi}(I')$, with $i = 1 \ldots n_B$, and $x'_{Sj}(I')$, with $j = 1 \ldots n_S$, is a subgame perfect equilibrium of $\bar{G}'$ and that none of the B_i has a reason to deviate from submitting the bids specified in a] in the very first decision stage d of the game.

Subgame perfectness in the main subgame $\bar{G}'$

First note that any subgame of $\bar{G}'$ is a game segment of $\bar{G}'$, because the truncated game $\bar{G}$ has no second market phases. This is also true for the largest game segment in $\bar{G}'$, namely, I', which is identical to $\bar{G}'$. For this reason, the segment strategies contained in $X'(I')$ are complete strategies for $\bar{G}'$ and $X'(I')$ can qualify as an equilibrium of $\bar{G}'$. To show that $X'(I')$ is a subgame perfect equilibrium of $\bar{G}'$, it suffices to show that: No game segment I'' exists in $\bar{G}'$, in which one of the traders has an incentive to deviate from the game segment strategy combination induced by $X'(I')$ on I''.

Subgame perfectness in the main subgame $\bar{G}'$:
Possible second market phase subgames

Let I'' be some arbitrary game segment of $\bar{G}'$ beginning with the decision stage d''. Let $\bar{b}''$ be the current market bid in d''. Due to the *ask-bid-spread-reduction* rule $\bar{b}''$ cannot be smaller than the market bid $c_m - 1$ in the first decision stage d' of $\bar{G}'$. Thus, $\bar{b}''$ must be in the market price range of d''. Since $\bar{b}''$ is in the market price range of d'' and the buyers are on the opening market side and all buyers refrain from submitting new bids in all their decision stages in I', it follows that the sellers face the credibly enforced price $\bar{b}''$ in I''.

Since $\bar{b}'' \geq c_m - 1$ and since $\bar{b}''$ is credibly enforced, it follows that: First, some trade must occur at the price $\bar{b}''$, due to Lemma 4. Second, no extra-marginal buyer can trade in I'', due to the *no-loss* rule. (Remember that, if extra-marginal buyers exist, their redemption values are all smaller than $c_m - 1$, since we are examining the second part of Proposition E.1, in which $v_x < c_m$.) Thus, at least one trade occurs at $\bar{b}''$, but no trade that occurs involves an extra-marginal buyer, and a second market phase subgame can begin. (Note, if all non-extra-marginal traders on both sides trade in the first market phase, a second market phase does not begin.) In any second market subgame that can begin after I'', the lower bound of the market clearing price range will be either the redemption value v_y, with $y \geq x$, of an extra-marginal buyer or the unit cost c_h, with $h \leq m$, of an non-extra-marginal seller. This is true, because of the following: The lower bound of the market clearing price range in the first market phase is the unit cost c_m of the marginal seller S_m. If only intra-marginal sellers trade, the lower bound of the market clearing price range in any reached $G_2(.,.)$ remains c_m. If the marginal seller trades, the lower bound of the market clearing price range in any reached $G_2(.,.)$ changes either to the redemption value v_x of the best extra-marginal buyer B_x or to the unit cost c_h of some intra-marginal seller S_h, with $h < m$. If one or more extra-marginal sellers trade, the intra-marginal part of the supply curve shifts less to the left than that part of the demand curve, because the number of trading non-extra-marginal sellers is smaller than number of trading non-extra-marginal buyers. This means that either some originally non-extra-marginal sellers could become extra-marginal in $G_2(.,.)$ or that some originally extra-marginal buyers could become non-extra-marginal in $G_2(.,.)$. In any case, however, the lower bound of the market clearing price range in $G_2(.,.)$ is either determined by the redemption value v_y of an originally extra-marginal buyer B_y, with $y \geq x$, or by the unit cost c_h of an originally non-extra-marginal seller S_h, with $h \leq m$. Hence, after one or more trades at $\bar{b}''$ in the first market phase, any second market phase subgame that can emerge is of one of the following two types: either $G_2(v_y,.)$, with $y \geq x$, or $G_2(c_h,.)$, with $h \leq m$.

Subgame perfectness in the main subgame $\bar{G}'$: Intra-marginal sellers

In any second market phase subgame of the type $G_2(v_y,.)$, with $y \geq x$, any intra-marginal seller S_j, with $j < m$, trades at the price $v_y + 1 \leq v_x + 1 < c_m - 1$. This is true, because $v_x < c_m$, since we are examining the second part of Proposition E.1 and because it follows by Assumption A2 that $v_x + 2 < c_m$. In any second market phase subgame of the type $G_2(c_h,.)$, with $h \leq m$, three situation can arise for S_j: (i) If S_j is the marginal seller in $G_2(c_h,.)$, i.e. if $j = h$, she trades at the price $c_h = c_j < c_m - 1$. (Remember that, due to Assumption A2, $c_j + 2 < c_m$.) (ii) If S_j is an intra-marginal seller in $G_2(c_h,.)$, i.e. if $j < h$, she trades at the price $c_h - 1 \leq c_m - 1$. (iii) If S_j is an extra-marginal seller in $G_2(c_h,.)$, i.e. if $j > h$, she does not trade at all. Thus, in any second market phase subgame, in which S_j can trade at all, she trades at some price $p \leq c_m - 1$. Because $\bar{b}'' \geq c_m - 1 \geq p$ is true anywhere in $\bar{G}'$, S_j has no incentive to wait for a second market phase subgame. Finally, because $\bar{b}''$ is credibly enforced in I'', S_j maximizes her payoff by accepting $\bar{b}''$ immediately. If only she can accept a trade at $\bar{b}''$, it is irrelevant for her payoff, whether she accepts $\bar{b}''$ immediately or later, as long as she accepts before market period termination. If some other seller S_h, with $h \neq j$, can also accept a trade at $\bar{b}''$, S_j maximizes her payoff by accepting $\bar{b}''$ immediately, since S_h is using an acceptance segment strategy in I'. Thus, the acceptance segment strategy $x'_{Sj}(I')$, with $j = 1 \ldots (m - 1)$ maximizes S_j's payoff in any game segment I'', given the other segment strategies.

Subgame perfectness in the main subgame $\bar{G}'$: The marginal seller

Note that this is also true for the marginal seller S_m, i.e. S_m has no incentive to wait for a second market phase subgame either. In any second market phase subgame of the type $G_2(v_y,.)$, with $y \geq x$, S_m does not trade at all, due to the *no-loss* rule and since $v_y + 1 \leq v_x + 1 < c_m - 1 < c_m$. This is true, because $v_x < c_m$, since we are examining the second part of Proposition E.1 and because it follows by Assumption A2 that $v_x + 2 < c_m$. In any second market phase subgame of the type $G_2(c_h,.)$, with $h \leq m$, S_m can only reach two situations: (i) Such $G_2(c_h,.)$, with

$h = m$, in which she is the marginal seller and trades at the price c_m. (ii) Such $G_2(c_h,.)$, with $h < m$, in which she is an extra-marginal seller and does not trade at all. Thus, in any second market phase subgame, in which S_m can trade at all, she trades at some price $p = c_m$. Hence, if $\bar{b}'' > c_m$, then S_m strictly prefers trading in I'' to trading in a $G_2(c_m,.)$. But, even if $\bar{b}'' = c_m$, S_m is indifferent between trading in I'' and trading in any $G_2(c_m,.)$. Finally, if $\bar{b}'' < c_m$, then S_m has no choice, but to trade in a $G_2(c_m,.)$. Thus, if $\bar{b}'' > c_m - 1$, S_m maximizes her payoff by accepting $\bar{b}''$ immediately, since $\bar{b}''$ is credibly enforced in I'' and since all other S_h, with $h \neq m$, are using an acceptance segment strategy in I'. If $\bar{b}'' = c_m - 1$, S_m achieves a no smaller payoff by submitting her best ask in the first market phase than with any other segment strategy. Hence, the acceptance segment strategy $x'_{Sm}(I')$ maximizes S_m's payoff in any game segment I'', given the other segment strategies.

Subgame perfectness in the main subgame $\bar{G}'$: Extra-marginal sellers

After having established that no non-extra-marginal seller has a reason to deviate from her acceptance segment strategy, let us turn to the extra-marginal sellers. The extra-marginal sellers cannot trade in any second market phase subgame, since they remain extra-marginal both in any $G_2(v_y,.)$, with $y \geq x$, and in any $G_2(c_h,.)$, with $h \leq m$. Since an extra-marginal seller can only trade in the first market phase, it follows by Corollary 2, that in any market side best reply combination of the sellers each extra-marginal seller, who makes no loss by trading at $\bar{b}''$, must accept a trade at the enforced price $\bar{b}''$. If one of the extra-marginal sellers delays the acceptance, she is certain not to trade at all, since the other sellers accept $\bar{b}''$ immediately. If an extra-marginal seller cannot accept $\bar{b}''$, because of the *no-loss* rule, all her segment strategies in I'' lead to a payoff of 0, given the strategies of the other traders. Thus, by submitting her best ask, this extra-marginal seller achieves a no smaller payoff in I'' than with any other segment strategy. Hence, the acceptance segment strategy of each extra-marginal seller maximizes her payoff in any game segment I'' of G', given the segment strategies of the others.

Subgame perfectness in the main subgame $\bar{G}'$: Non-extra-marginal buyers

Since all S_j, who can do so, accept any price that is credibly enforced by the buyers in I'' and since all other B_h, with $h \neq i$, refrain from submitting new bids in I'', B_i maximizes his payoff in I'' by also enforcing the current market bid $\bar{b}''$ in I''. To see why, let us compare B_i's payoff from enforcement to his payoff from overbidding. Remember that I'' is an arbitrary game segment in $\bar{G}'$, since we must confirm that $X'(I')$ induces equilibria in all possible subgames of $\bar{G}'$, i.e. we have to consider any possible constellation in I''.

Because all non-extra-marginal buyers submit the bid $c_m - 1$ in the decision stage d, before $\bar{G}'$, it follows that: First, the current market bid $\bar{b}''$ cannot be smaller than $c_m - 1$ anywhere in $\bar{G}'$, due to the *ask-bid-spread-reduction* rule. Second, if $\bar{b}'' = c_m - 1$, then all m non-extra-marginal buyers hold $\bar{b}''$. Considering these two conditions and the acceptance segment strategies of the sellers, the following constellations can emerge in I'' from the point of view of a non-extra-marginal buyer B_i, with $i \leq m$:

(i) $\bar{b}'' = c_m - 1$: All m non-extra-marginal buyers jointly hold $\bar{b}''$. All (m - 1) intra-marginal sellers S_j, with $j < m$, accept $\bar{b}''$. No other seller, including S_m, can accept a trade at $c_m - 1$, due to the *no-loss* rule. By using an enforcement segment strategy, B_i trades at the price $c_m - 1$ in I'' with a probability of $\frac{m-1}{m}$. With a probability of $\frac{1}{m}$ B_i reaches some second market phase subgame of the type $G_2(c_m,.)$ and trades at the price c_m. Thus, B_i's expected payoff from enforcing $\bar{b}''$ is $\frac{m-1}{m} (v_i - \bar{b}'' + \beta) + \frac{1}{m} (v_i - c_m + \beta)$. Alternatively, B_i can overbid $\bar{b}''$ with $\bar{b}'' + 1$, in which case he receives a payoff $v_i - (\bar{b}'' + 1) + \beta$ by trading in the first market phase. Replacing $\bar{b}''$ with $c_m - 1$ in both terms, it becomes evident that B_i has no incentive to overbid, since the expected price he pays by enforcement is $c_m - \frac{m-1}{m}$ and, thus, is smaller than the price he pays to overbid, for all $m \geq 2$.

(ii) $\bar{b}'' \geq c_m$: A number k of non-extra-marginal buyers, with $1 \leq k \leq m$, jointly hold $\bar{b}''$. (Extra-marginal buyers cannot hold $\bar{b}'' \geq c_m > v_x$.) All sellers S_j, with $c_j \leq \bar{b}''$, accept $\bar{b}''$. Note that at least m sellers accept a trade at $\bar{b}''$, since all non-extra-marginal sellers can accept a trade at $\bar{b}'' \geq c_m \geq c_j$, for all $j \leq m$.

(iia) If B_i is in the group of buyers who hold $\bar{b}''$, he trades in I'' and receives a payoff $(v_i - \bar{b}'' + \beta)$. If B_i overbids $\bar{b}''$ with $\bar{b}'' + 1$, he receives a payoff $v_i - (\bar{b}'' + 1) + \beta$. Obviously, B_i has no incentive to overbid.

(iib) If B_i is not in the group of buyers who hold $\bar{b}''$, he either trades in some second market phase subgame $G_2(v_y,.)$, with $y \geq x$, at the price $v_y + 1 \leq v_x + 1 < c_m \leq \bar{b}''$ or trades in some second market phase subgame $G_2(c_h,.)$, with $h \leq m$, at the expected price $c_h - \frac{(m-k)-1}{(m-k)} \leq c_m - \frac{(m-k)-1}{(m-k)} < c_m \leq \bar{b}''$. (Remember that any second market phase subgame that can be reached must either be of the type $G_2(v_y,.)$, with $y \geq x$, or $G_2(c_h,.)$, with $h \leq m$, as proven further up. Remember also that $v_x + 2 < c_m$, since we are examining the second part of Proposition E.1, where $v_x < c_m$ and due to Assumption A2.) Thus, no matter which type of second market phase subgame B_i reaches, the price he pays in $G_2(.,.)$ is smaller than the price he pays by overbidding. Obviously, B_i again has no incentive to overbid $\bar{b}''$ with $\bar{b}'' + 1$.

Hence, overbidding leads to a lower payoff for the non-extra-marginal buyers in all cases. (Note that B_i could overbid $\bar{b}''$ with more than one money unit in any of these cases, but that would only lead to even lower payoffs. Note also that, since $\bar{b}'' \geq c_m > v_x$, none of the extra-marginal buyers can trade in I''. This ensures that every non-extra-marginal buyer will be able to trade in any second market phase subgame.) Thus, the enforcement segment strategy $x'_{Bi}(I')$ maximizes a non-extra-marginal buyer B_i's payoff in any game segment I'' of $\bar{G}'$, given the segment strategies of the others.

Subgame perfectness in the main subgame $\bar{G}'$: Extra-marginal buyers

Since $\bar{b}'' \geq c_m > v_x$ anywhere in $\bar{G}'$, the extra-marginal buyers do not have a choice but to refrain from submitting bids in $\bar{G}'$, due to the *ask-bid-spread-reduction* rule in combination with the *no-loss* rule. Thus, the enforcement segment strategy maximizes an extra-marginal buyer's payoff in any game segment I'' of $\bar{G}'$, given the segment strategies of the others.

Hence, each seller maximizes her payoff with her acceptance segment strategy $x'_{Sj}(I')$ and each buyer maximizes his payoff with his enforcement segment strategy $x'_{Bi}(I')$ in any game segment I'' of $\bar{G}'$, given the segment strategies of the other traders. Thus, for any possible I'' in $\bar{G}'$, no trader has an incentive to deviate from his strategy contained in $X'(I')$, and it follows that $X'(I')$ is a subgame perfect equilibrium of $\bar{G}'$.

Optimality of choices in the first decision stage d

Now we have to prove that no buyer has a reason to deviate from his choice specified in a] in the very first decision stage d of the truncated game $\bar{G}$. As shown above - given the strategies of the others -, if a non-extra-marginal buyer B_i submits c_m - 1 in d, he either trades at the price c_m - 1 in the first market phase or trades at the price c_m in some second market phase. Thus, the bid c_m - 1 in d leads to an expected payoff $v_i - (c_m - \frac{m-1}{m}) + \beta$. If he deviates by submitting a bid greater than c_m - 1, he earns less, since - even in the very best case - he trades at the price c_m and receives $v_i - c_m + \beta$. If B_i deviates by submitting a bid smaller than c_m - 1 (or deviates by refraining to submit a bid), one of three outcomes are possible, since all other B_h, with h $\neq$ i and h $\leq$ m, bid c_m - 1 in d: (i) B_i does not trade at all and receives a payoff of 0. (ii) B_i does not trade in the first market phase, but trades in some second market phase subgame $G_2(c_m,.)$ with the marginal seller S_m at c_m and receives a payoff of $v_i - c_m + \beta$. (iii) B_i overbids the opening bid c_m - 1 of the other buyers in some later decision stage of the first market phase and trades alone in the first market phase. Since B_i must bid at least c_m to overbid the other buyers, his payoff can at maximum be $v_i - c_m + \beta$.

Clearly, B_i's payoff in the case (i) is smaller than his payoff from bidding c_m - 1 in d. In the cases (ii) and (iii) B_i's payoff is also smaller than his payoff from bidding c_m - 1 in d, as already explained above. Thus, no deviation from the opening bid c_m - 1 can lead to a higher payoff for a non-extra-marginal buyer B_i, with i $\leq$ m, given the other strategies. Extra-marginal buyers do not have an incentive to deviate from their opening bids either, since - given the strategies of the others - they cannot trade in the game, no matter which opening bid they choose. Since no buyer has an incentive to deviate from his choice in d and since $X'(I')$ is a subgame perfect equilibrium of $\bar{G}'$, X is a subgame perfect equilibrium of the truncated game $\bar{G}$.

Quasi-pureness and impatience

X is a quasi-pure subgame perfect equilibrium of the $\bar{G}$, because the local strategies of the buyers in d are pure and the segment strategies of all traders in I' are pure segment strategies. But, X is also an impatience equilibrium of the game. This is so, because each non-extra-marginal buyer B_i's best offer must be $c_m - 1$ in the first market phase on the play realized by any strategy combination $\bar{X} = (x_{B1}, ..., x_{Bi-1}, \bar{x}_{Bi}, x_{Bi+1}, ..., x_{BnB}, x_{S1}, ..., x_{SnS})$, which leads to a payoff equal to B_i's equilibrium payoff in X; otherwise B_i, with $i = 1 ... m$, would trade at a different expected price or not at all in the first market phase. But, since x_{Bi} prescribes submitting $c_m - 1$ in the very first decision stage, no $\bar{x}_{Bi}$ can specify submitting $c_m - 1$ any earlier. Similarly, every alternative strategy of each of the intra-marginal sellers S_j, with $j = 1 ... (m - 1)$, with the same payoff as x_{Sj}, must prescribe submitting $c_m - 1$, but cannot prescribe doing so any earlier than x_{Sj} does. Additionally, no alternative strategy of the marginal seller S_m in the first market phase can lead to a trade in the first market phase, given the other strategies. Thus, S_m must submit her best offer c_m as soon as possible in an impatience equilibrium. This is true in X. Lastly, no alternative strategy of an extra-marginal buyer B_k, with $k = x ... n_B$, or seller S_y, with $y = x ... n_S$, can lead to a trade in the first market phase, given the other strategies. Thus, B_k and S_y must submit their best offers, v_k and c_y, correspondingly, as soon as possible in an impatience equilibrium. This is also true in X.

Induction over the market size

With this we have proven the following induction step: if Proposition E.1 holds for all ν-maximum-trade markets, where $\nu < \mu$, it also holds for any μ-maximum-trade market. It follows by the propositions C.1.1 and C.2.1 that Proposition E.1 holds for all 1-maximum-trade markets. Thus, we can conclude that Proposition E.1 also holds for all 2-maximum-trade markets. Therefore, the result also holds for all 3-maximum-trade markets and so forth. Since the result holds for an arbitrary μ-maximum-trade market, it also holds for an arbitrary m-trade market. ∎

Proposition E.2

Let G be an m-trade simplified alternating double auction market game. Let Assumption A2 be true in G. Any combination of pure strategies that is an impatience equilibrium of G - on the equilibrium path - leads to the following outcome:

(1)   If the best extra-marginal player valuation on the opening market side exists and is in the market clearing price range:

    (1.1)   A total of m trades occur in equilibrium.

    (1.2)   All m trades occur at a price which lies just outside the range of feasible prices for the best extra-marginal trader of the opening market side (i.e. exactly one money unit greater than the best extra-marginal buyer's redemption value or exactly one money unit smaller than the best extra-marginal seller's unit cost).

(2)   If the best extra-marginal player valuation on the opening market side does not exist or is not in the market clearing price range:

    (2.1)   A total of m trades occur in equilibrium.

    (2.2)   The first m - 1 trades occur at a price which lies just outside the range of feasible prices for the marginal trader of the second moving market side (i.e. exactly one money unit greater than the marginal buyer's redemption value or exactly one money unit smaller than the marginal seller's unit cost).

    (2.3)   The last trade occurs at a price equal to the player valuation of the marginal trader on the second moving market side.

Proof of Proposition E.2

For the entire proof, we assume the buyers open the market. The other case can be proven analogously. The following proof only concerns m-trade markets with $m \geq 2$, since the proofs for the 1-trade markets are given with the propositions C.1.2, C.2.2, and C.3.2. We must only show that strategy combinations leading to other outcomes cannot be impatience equilibria.

The truncated game $\bar{G}$ and the induction assumption

Since $m \geq 2$ trades are possible, certain plays of the game G lead to a second market phase. Which second market phase is reached, depends on the history of play. But, no matter which second market phase subgame is reached, it is always a r-trade market, where $r \leq m$. Without loss of generality, we will assume that the original market is a μ-maximum-trade market, where $\mu \geq m$. Since the r-trade market, that follows in the second market phase, must be a ν-maximum-trade market, with $\nu < \mu$, we can use an induction over the maximum number of possible trades left in the market for the proof. To be able to do so, we begin by assuming that the impatience equilibrium outcome of any ν-maximum-trade second market phase subgame $G_2(.,.)$ of G has the properties that the Proposition E.2 specifies. Thus and since we assume that the buyers are on the opening market side, any second market phase subgame that can evolve in G is of one of the following types with the corresponding outcome:

(i) $G_2(c_j,v_i)$, where the innermost traders, S_j and B_i, are the marginal traders. The first $r - 1$ trades occur at the price $c_j - 1$ and the last trade occurs at the price c_j.

(ii) $G_2(v_i,c_j)$, where the innermost traders, B_i and S_j, are the best extra-marginal traders. All trades occur at the price $v_i + 1$.

(iii) $G_2(c_j,c_{j+1})$, where the innermost traders, S_j and S_{j+1}, are the marginal seller and the best extra-marginal seller, correspondingly. The first $r - 1$ trades occur at the price $c_j - 1$ and the last trade occurs at the price c_j.

(iv) $G_2(v_i,v_{i-1})$, where the innermost traders, B_i and B_{i-1}, are the best extra-marginal buyer and the marginal buyer, correspondingly. All trades occur at the price $v_i + 1$.

Summarizing, we have: In any second market phase subgame $G_2(v_i,.)$, in which the lower bound of the market clearing price range is equal to the redemption value v_i of a buyer B_i, all trades occur at the price $v_i + 1$. In any second market phase subgame $G_2(c_j,.)$, in which the lower bound of the market clearing price range is equal to the unit cost c_j of a seller S_j, the first $r - 1$ trades occur at the price $c_j - 1$ and the last trade at the price c_j.

Hence - as in the Proof of Proposition E.1 -, instead of analyzing the original game G, we analyze the truncated game $\bar{G}$, in which the second market phase subgames of G are replaced with end nodes with the corresponding impatience equilibrium payoffs of the r-trade second market phase subgames. Thus, we will first analyze the truncated game $\bar{G}$ for the case (1) of the proposition. Then we do the same for case (2). Finally, we use the results of the propositions C.1.2 and C.2.2 to induce the impatience equilibrium of a 2-maximum-trade market. This is possible, because the 1-trade markets considered in those propositions are 1-maximum-trade markets. Then we use the result for 2-maximum-trade markets to induce the result for 3-maximum-trade markets and so on. Obviously, by showing that the proposition holds for an arbitrary μ-maximum-trade market, we also show that it holds for an arbitrary m-trade market.

Impatience equilibrium X of the truncated game $\bar{G}$ in Case (1)

By Lemma 3 we know that market period termination without a trade cannot be an equilibrium outcome. Thus, the only other possible outcome of $\bar{G}$ that must be checked is that a trade at some other price than that specified in Proposition E.2 occurs in the first market phase. At least one extra-marginal buyer exists in case (1). Additionally, the best extra-marginal buyer B_x's redemption value v_x is in the market clearing price range, i.e. $c_m < v_x$. Since the buyers open the market, the only trading price in the first market phase, which is compatible with the proposition, is the price $v_x + 1$.

A trade occurs at $p > v_x + 1$

Let us first check the cases in which a trade occurs at a price $p > v_x + 1$ in the first market phase. If a trade occurs at the price p, the current market bid $\tilde{b}'$ in some decision stage d' must have been equal to p. Let I' denote the game segment that begins with the decision stage d'. In I' some seller must have accepted a trade at the price $\tilde{b}' > v_x + 1$.

A trade occurs at $p > v_x + 1$: Possible second market phase subgames

Before we proceed, let us first establish that only second market phase subgames either of the type $G_2(v_y,.)$, with $y \geq x$, or of the type $G_2(c_h,.)$, with $h \leq m$, could have been reached after I', i.e. after any number of trades at $p > v_x + 1$ in the first market phase. The lower bound of the market clearing price range in the first market phase is the redemption value v_x of the best extra-marginal buyer, since we are examining the first part of Proposition E.2, in which $c_m < v_x$ is true. Thus, if trades occur at a price $p > v_x + 1$ in the first market phase, only non-extra-marginal buyers can have been involved, since extra-marginal buyers cannot trade $p > v_x$, due to the *no-loss* rule. If only non-extra-marginal sellers trade, the lower bound of the market clearing price range in any reached $G_2(.,.)$ remains v_x. If one or more extra-marginal sellers trade, the intra-marginal part of the supply curve shifts less to the left than that part of the demand curve, because the number of trading non-extra-marginal sellers is smaller than number of trading non-extra-marginal buyers. This means that either some originally non-extra-marginal sellers could become extra-marginal in $G_2(.,.)$ or that some originally extra-marginal buyers could become non-extra-marginal in $G_2(.,.)$. In any case, however, the lower bound of the market clearing price range in $G_2(.,.)$ is either determined by the redemption value v_y of an originally extra-marginal buyer B_y, with $y \geq x$, or by the unit cost c_h of an originally non-extra-marginal seller S_h, with $h \leq m$. Hence, after one or more trades at $p > v_x + 1$ in the first market phase, any second market phase subgame that can emerge is of one of the following two types: either $G_2(v_y,.)$, with $y \geq x$, or $G_2(c_h,.)$, with $h \leq m$.

In the impatience equilibrium of any such second market phase subgame, that follows after one or more trades at $p > v_x + 1$ in the first market phase, the expected equilibrium price p_2 is smaller than or equal to $v_x + 1$, i.e. $p_2 \leq v_x + 1$. Let us show why.

If a $G_2(v_y,.)$, with $y \geq x$, is reached, then $p_2 = v_y + 1$. Here $p_2 \leq v_x + 1$, because $p_2 = v_y + 1 \leq v_x + 1$, for all $y \geq x$.

If a $G_2(c_h,.)$, with $h \leq m$, is reached, in which r non-extra-marginal traders are present, two different cases must be considered:

(i)   For any non-extra-marginal buyer B_i, with $i \leq m$, the expected price in a $G_2(c_h,.)$, with $h \leq m$, is $p_2 = c_h - \frac{(r-1)}{r}$, since he trades at the price $c_h - 1$ with probability $\frac{(r-1)}{r}$ and trades at the price c_h with probability $\frac{1}{r}$. Here, $p_2 < v_x + 1$, because $c_m < v_x$ and, thus, $p_2 = c_h - \frac{(r-1)}{r} \leq c_h \leq c_m < v_x + 1$, for all $h \leq m$.

(ii)   For any non-extra-marginal seller S_j, with $j \leq m$, the expected price in a $G_2(c_h,.)$, with $h \leq m$, is $p_2 = c_h - 1$, if S_j is intra-marginal in $G_2(c_h,.)$, i.e. $j < h$. The expected price is $p_2 = c_h$, if S_j is marginal in $G_2(c_h,.)$, i.e. $j = h$. Finally, if S_j is extra-marginal in $G_2(c_h,.)$, i.e. $j > h$, S_j cannot trade in $G_2(c_h,.)$. Thus, if S_j can trade at all in $G_2(c_h,.)$, $p_2 < v_x + 1$, because $c_m < v_x$ and, thus, $p_2 \leq c_h \leq c_m < v_x + 1$, for all $h \leq m$.

A trade occurs at $p > v_x + 1$: All non-extra-marginal sellers accept p

Thus, if a trade at a price greater than $v_x + 1$ occurred in the first market phase, all non-extra-marginal sellers must have accepted the current market bid $\bar{b}' > v_x + 1$ in the game segment I' that begins with the decision stage d'. This is so, because each non-extra-marginal seller S_j, with $j = 1 \ldots m$, achieves a greater payoff by trading at $\bar{b}' > v_x + 1$ in I' than by trading at a lower price

$p_2 \leq v_x + 1 < \bar{b}'$ in any $G_2(.,.)$. Even if S_j only has a $\frac{1}{m}$ chance of trading at $\bar{b}' > v_x + 1$ in I', her expected payoff from accepting a trade at $\bar{b}'$ in I' is greater than the expected payoff from trading in any second market phase subgame.

A trade occurs at $p > v_x + 1$: All non-extra-marginal buyers have an incentive to refrain from submitting p

If $k > 1$ buyers together held the market bid $\bar{b}' > v_x + 1$, there must have been some decision stage d'' before d', in which they submitted $\bar{b}'$, but each of them had an incentive to deviate from his strategy. This is so, because each could have achieved a higher expected payoff by simply refraining to bid, instead of bidding $\bar{b}'$ in d''. Without loss of generality, let us assume B_i refrains from bidding $\bar{b}'$ in d''. Since all non-extra-marginal sellers accept $\bar{b}' > v_x + 1$, held by the other $(k - 1)$ buyers, only $(k - 1)$ of the sellers trade at the price $\bar{b}'$. Then B_i reaches some second market phase subgame $G_2(.,.)$, in which he trades at a lower expected price $p_2 \leq v_x + 1 < \bar{b}'$. Thus, if $k > 1$ buyers tie at some $\bar{b}' > v_x + 1$ in d'', each of them has an incentive to deviate from his strategy.

If only one buyer B_i held the market bid $\bar{b}' > v_x + 1$ in d'', this buyer also had an incentive to deviate from his strategy. This is so, because B_i could have chosen to enforce some other price $p^* < \bar{b}'$ and $p^* \geq v_x + 1$ in I'. Because of the argument above, the other buyers would not have overbid p^*. Clearly, a trade at any $p^* < \bar{b}'$ leads to a higher payoff for B_i than $\bar{b}'$. Thus, if a strategy combination leads to a current market bid $\bar{b}' > v_x + 1$, it cannot be a quasi-pure subgame perfect equilibrium of the game, since - no matter whether a single buyer or many buyers together hold $\bar{b}'$ - at least one of the buyers has an incentive to deviate from his strategy.

A trade occurs at $p < v_x + 1$

Now, consider some price $p < v_x + 1$. If a trade occurs at the price p, the current market bid $\bar{b}'$ in some decision stage d' must have been equal to p. Let I' denote the game segment that begins with d'. In I' some seller must have accepted a trade at the price $\bar{b}' < v_x + 1$.

A trade occurs at $p < v_x + 1$: Possible second market phase subgames

Before we proceed, let us first establish that only second market phase subgames either of the type $G_2(v_h,.)$, with $h \leq x$, or of the type $G_2(c_y,.)$, with $y \geq x$, could have been reached after I', i.e. after any number of trades at $\bar{b}' < v_x + 1$ in the first market phase. The lower bound of the market clearing price range in the first market phase is the redemption value v_x of the best extra-marginal buyer, since we are examining the first part of Proposition E.2, in which $c_m < v_x$ is true. If a trade occurs at a price $p < v_x + 1$ in the first market phase, only non-extra-marginal sellers can have been involved, since extra-marginal sellers cannot trade at $p < v_x + 1 < c_x$, due to the *no-loss* rule and since $c_x > v_x + 2$ by Assumption A2. If only non-extra-marginal buyers trade, the lower bound of the market clearing price range in any reached $G_2(.,.)$ remains v_x. If one or more extra-marginal buyers trade, the intra-marginal part of the demand curve shifts less to the left than that part of the supply curve, because the number of trading non-extra-marginal buyers is smaller than number of trading non-extra-marginal sellers. This means that either some originally non-extra-marginal buyers could become extra-marginal in $G_2(.,.)$ or that some originally extra-marginal sellers could become non-extra-marginal in $G_2(.,.)$. In any case, however, the lower bound of the market clearing price range in $G_2(.,.)$ is either determined by the redemption value v_h of an originally non-extra-marginal buyer B_h, with $h \leq m$, or by the redemption value v_x of the originally best extra-marginal buyer B_x, or by the unit cost c_y of an originally extra-marginal seller S_y, with $y \geq x$. Hence, after one or more trades at $p < v_x + 1$ in the first market phase, any second market phase subgame that can emerge is of one of the following two types: either $G_2(v_h,.)$, with $h \leq x$, or $G_2(c_y,.)$, with $y \geq x$.

A trade occurs at $p < v_x + 1$: Only one non-extra-marginal seller accepts p

In the impatience equilibrium of any such second market phase subgame, that follows after one or more trades at $p < v_x + 1$ in the first market phase, the expected equilibrium price p_2, at which an non-extra-marginal seller S_j, with $j \leq m$, trades, is greater or equals to $v_x + 1$, i.e. $p_2 \geq v_x + 1$. Let us show why.

If a $G_2(v_h,.)$, with $h \leq x$, is reached, then $p_2 = v_h + 1$. Here $p_2 \geq v_x + 1$, because $p_2 = v_h + 1 \geq v_x + 1$, for all $h \leq x$.

If a $G_2(c_y,.)$, with $y \geq x$, is reached, then $p_2 = c_y - 1$, since $j < y$. Here $p_2 > v_x + 1$, because $p_2 = c_y - 1 \geq c_x - 1 > v_x + 1$, for all $y \geq x$. (Remember that $c_x < v_x + 2$, due to the Assumption A2.)

Thus, if more than one non-extra-marginal seller accepted a trade at $\bar{b}' < v_x + 1$ in I', this cannot have been an equilibrium. This is so, because in any second market phase subgame, that can be reached after I', these sellers trade at a higher price $p_2 > v_x + 1 > \bar{b}'$. Hence, all non-extra-marginal sellers prefer waiting for a second market phase to trading at a price $\bar{b}' < v_x + 1$ in I'. With this argument, however, we cannot rule out that an equilibrium exists in which a trade at $\bar{b}' < v_x + 1$ occurs: if $\bar{b}'$ is a credibly enforced price, it follows by Lemma 4, that at least one seller must accept a trade at the enforced price in order to prevent market period termination. If, however, one seller chooses to accept $\bar{b}'$, all others are better off waiting for a second market phase subgame. Thus, if a trade at $\bar{b}' < v_x + 1$ occurred in I', only one seller S_j accepted a trade at this price.

Obviously, some decision stage d'' must have existed, after $\bar{b}'$ was first submitted, but before S_j accepted $\bar{b}'$, in which S_j could have submitted an ask $\bar{a}'' = \bar{b}' + 1$. Let I'' denote the game segment and $\bar{G}''$ denote the subgame that begins with d'', i.e. with S_j's submission of $\bar{b}' + 1$. If any one of the buyers accepts a trade at $\bar{b}' + 1$ in $\bar{G}''$, S_j receives a higher payoff than on the original play that led to a trade at $\bar{b}'$, because she trades at a higher price. Thus, if any of the buyers accepts $\bar{b}' + 1$ in $\bar{G}''$, S_j has an incentive to deviate from her original strategy. Hence, to prove that a quasi-pure subgame perfect equilibrium cannot lead to a price $p < v_x + 1$, it is sufficient to show that at least one of the buyers prefers to trade with S_j at $\bar{b}' + 1$ in $\bar{G}''$, i.e. overbids any current market bid $\bar{b}' < v_x + 1$. It suffices to show that S_j will be able to trade in $\bar{G}''$, since, for any strategy combination that realizes a play that leads to a trade at a price $\bar{b}' < v_x + 1$, an alternative strategy combination exists, in which S_j's strategy choice is altered in such a way, that the realized play reaches the subgame $\bar{G}''$, in which S_j achieves a greater payoff.

A trade occurs at $p < v_x + 1$
Reduced number of possible second market phase subgames

Since only one seller S_j, with $j \le m$, accepts a trade at $\bar{b}' < v_x + 1$ in I', the number of reachable second market phase subgames is reduced. If S_j trades with a non-extra-marginal buyer, a second market phase subgame of the type $G_2(v_x,.)$ is reached. If S_j trades with an extra-marginal buyer, a second market phase subgame either of the type $G_2(v_m,.)$ or of the type $G_2(c_x,.)$ is reached. This is so, because, after S_j trades with an extra-marginal buyer, $(m - 1)$ originally non-extra-marginal sellers face m originally non-extra-marginal buyers. If $v_m < c_x$ (or c_x does not exist), the originally marginal buyer B_m becomes the best extra-marginal buyer in $G_2(v_m,.)$ and his redemption value v_m is the lower bound of the market clearing price range in $G_2(v_m,.)$. Thus, any $G_2(v_m,.)$ is a $(m - 1)$-trade market. But, if $v_m > c_x$, the originally extra-marginal seller S_x becomes marginal in $G_2(c_x,.)$ and her unit cost c_x is the lower bound of the market clearing price range $G_2(c_x,.)$. Thus, any $G_2(c_x,.)$ is an m-trade market. Hence, only three types of second market phase subgames can follow the game segment I', in which a trade occurs at the price $\bar{b}' < v_x + 1$:

(i)   If a non-extra-marginal buyer trades in I', a $(m - 1)$-trade market of the type $G_2(v_x,.)$ emerges, in which only non-extra-marginal buyers trade at $v_x + 1$.

(ii)   If an extra-marginal buyer trades in I' and $v_m < c_x$, a $(m - 1)$-trade market of the type $G_2(v_m,.)$ emerges, in which only intra-marginal buyers trade at $v_m + 1$.

(iii)   If an extra-marginal buyer trades in I' and $v_m > c_x$, an m-trade market of the type $G_2(c_x,.)$ emerges, in which only non-extra-marginal buyers trade at an expected price of $c_x - \frac{(m-1)}{m}$, (at $c_x - 1$ with probability $\frac{(m-1)}{m}$ and at c_x with probability $\frac{1}{m}$).

Since $v_x + 1 < v_m + 1$, a non-extra-marginal buyer trades at a lower expected price in a $G_2(v_x,.)$ than in a $G_2(v_m,.)$. Similarly, since $v_x + 2 < c_x$, by Assumption A2, it follows that $v_x + 1 < c_x - 1 < c_x - \frac{(m-1)}{m}$. Thus, a non-extra-marginal buyer trades at a lower expected price in a $G_2(v_x,.)$ than in a $G_2(c_x,.)$. The important result for the following payoff comparisons is: $v_x + 1$ is the lowest price, at which a non-extra-marginal buyer B_i, with $i \leq m$, can trade at in any second market phase subgame $G_2(.,.)$.

Since $\tilde{b}' < v_x + 1 \leq v_x$, all non-extra-marginal buyers and - at least - the best extra-marginal buyer could have held the market bid $\tilde{b}'$ in d'. Thus, if $k \geq 1$ buyers held $\tilde{b}'$ in d', three situations are possible in d'': (i) The best extra-marginal buyer B_x is in the group of buyers holding the current market bid $\tilde{b}'$, but some non-extra-marginal buyer B_i, with $i \leq m$, exists, who is not in that group. (ii) The best extra-marginal buyer B_x is in the group of the buyers holding the current market bid $\tilde{b}'$ and all non-extra-marginal buyers are also in the group. (iii) The best extra-marginal buyer B_x is not in the group of the buyers holding the current market bid $\tilde{b}'$.

A trade occurs at $p < v_x + 1$
Case (i): non-extra-marginal buyer overbids, if B_x is in the group of buyers holding $\tilde{b}' = p < v_x + 1$, but some non-extra-marginal buyer is not

Let us first examine (i), i.e. B_x holds $\tilde{b}'$, but some non-extra-marginal buyer B_i, with $i \leq m$, does not hold $\tilde{b}'$ in d''. In this case, B_i has an incentive to overbid $\tilde{b}'$ with $\tilde{b}' + 1$, i.e. to accept S_j's offer. If B_i accepts S_j's offer, he trades at the price $\tilde{b}' + 1 \leq v_x + 1$ in I''. In contrast, if B_i does not overbid $\tilde{b}'$, only one trade occurs in the first market phase at the price $\tilde{b}'$, since only one seller accepts a trade at $\tilde{b}' < v_x + 1$. With the probability $\frac{(k-1)}{k}$, where $k \geq 1$ is the number of buyers holding $\tilde{b}'$ in d', this trade involves some non-extra-marginal buyer. In this case, B_i reaches some second market phase of type $G_2(v_x,.)$ and trades at the price $v_x + 1$. With probability $\frac{1}{k}$ the trade in the first market phase involves the best extra-marginal buyer B_x. In this case, B_i reaches some second market phase either of type $G_2(v_m,.)$ or of the type $G_2(c_x,.)$. In either case B_i trades at a higher price p_2 in any $G_2(v_m,.)$ or $G_2(c_x,.)$ than in any $G_2(v_x,.)$, since - as shown above - either

$p_2 = v_m + 1 > v_x + 1$ or $p_2 = c_x - \frac{(m-1)}{m} > v_x + 1$. The expected price, at which B_i trades by not overbidding is therefore greater than the price he pays by overbidding $\tilde{b}'$ with $\tilde{b}' + 1$, since $\frac{(k-1)}{k} (v_x + 1) + \frac{1}{k} p_2 > v_x + 1 \geq \tilde{b}' + 1$. Hence, any strategy combination that leads to a trade at a current market bid $\tilde{b}' < v_x + 1$, held by the best extra-marginal buyer and some $k < m$ non-extra-marginal buyers in I'', cannot be an equilibrium of the game, because each non-extra-marginal buyer B_i not holding the current market bid $\tilde{b}'$ has an incentive to overbid $\tilde{b}'$ and to accept a trade with S_j at $\tilde{b}' + 1$.

A trade occurs at $p < v_x + 1$
Case (ii): non-extra-marginal buyer overbids, if B_x and all non-extra-marginal buyers are in the group of buyers holding $\tilde{b}' = p < v_x + 1$

Now, let us examine case (ii), i.e. B_x and all non-extra-marginal buyers are in the group of buyers holding the current market bid $\tilde{b}'$ in d''. If one of the non-extra-marginal buyers B_i, with $i \leq m$, breaks the tie at $\tilde{b}'$ by accepting S_j's offer to trade at $\tilde{b}' + 1$ in I'', he pays $\tilde{b}' + 1 \leq v_x + 1$ and receives $v_i - (\tilde{b}' + 1) + \beta$. If B_i, however, remains in the tie with the other buyers at $\tilde{b}'$, he trades at the price $\tilde{b}'$ in the first market phase with probability $\frac{1}{k}$, since only one seller will accept a trade at $\tilde{b}' < v_x + 1$. With the probability $\frac{(k-2)}{k}$ some other non-extra-marginal buyer trades at $\tilde{b}'$ in the first market phase. In this case, B_i reaches some second market phase of type $G_2(v_x,.)$ and trades at the price $v_x + 1$. Finally, with probability $\frac{1}{k}$ the trade in the first market phase involves the best extra-marginal buyer B_x. In this case, B_i reaches some second market phase either of type $G_2(v_m,.)$ or of the type $G_2(c_x,.)$. In either case B_i trades at a higher price p_2 in any $G_2(v_m,.)$ or $G_2(c_x,.)$ than in any $G_2(v_x,.)$, since either $p_2 = v_m + 1 > v_x + 1$ or $p_2 = c_x - \frac{(m-1)}{m} > v_x + 1$, as shown above. Therefore, the expected price, at which B_i at trades, if he does not overbid $\tilde{b}'$, is greater than $\tilde{b}' + 1$, the price he trades at, if he overbids $\tilde{b}'$ with $\tilde{b}' + 1$. This is true, because $\frac{1}{k} (v_x + 1) + \frac{(k-2)}{k} (v_x + 1) + \frac{1}{k} p_2 > v_x + 1 \geq \tilde{b}' + 1$. Hence, B_i has an incentive to overbid $\tilde{b}'$ with $\tilde{b}' + 1$ as long as $\tilde{b}' < v_x + 1$. Thus, a combination of strategies, that leads to a trade at a current market bid $\bar{b}' < v_x + 1$, that is held by the best extra-marginal buyer B_x together with all non-extra-marginal buyers B_i, with $i \le m$, in I', cannot be an equilibrium of the game, because each B_i has an incentive to overbid $\bar{b}'$ and accept a trade with S_j at $\bar{b}' + 1$.

A trade occurs at $p < v_x + 1$
Case (iii): B_x overbids, if B_x is not in the group holding $\bar{b}' = p < v_x + 1$

Now, let us examine case (iii), i.e. B_x is not in the group of $k \ge 1$ buyers holding the current market bid $\bar{b}'$ in d''. As long as $\bar{b}' < v_x$, the best extra-marginal buyer B_x can overbid $\bar{b}'$ and accept S_j's offer to trade at $\bar{b}' + 1$ in I''. By doing so, he pays $\bar{b}' + 1 \le v_x$ and receives $v_x - (\bar{b}' + 1) + \beta$. If B_x, however, does not overbid $\bar{b}'$, he does not trade in the first market phase and cannot trade in any second market phase subgame that is reached. This is so, because only non-intra-marginal buyers can trade in any reachable $G_2(.,.)$, as shown above. Hence, B_x has an incentive to overbid $\bar{b}'$ with $\bar{b}' + 1$ as long as $\bar{b}' < v_x$. Thus, a combination of strategies, that leads to a trade at a current market bid $\bar{b}' < v_x$, that is not held by the best extra-marginal buyer B_x, but by some non-extra-marginal buyers in I', cannot be an equilibrium of the game, because B_x has an incentive to overbid $\bar{b}'$ and accept a trade with S_j at $\bar{b}' + 1$.

A trade occurs at $p < v_2 + 1$
Special case (iii) where $p = v_x$ and B_x is not in the group of buyers holding $\bar{b}' = p = v_2$: cannot be an impatience equilibrium

Note, however, that the argument above does not apply to the case $\bar{b}' = v_x$, because B_x cannot overbid v_x with $v_x + 1$, due to the *no-loss* rule. Thus, if B_x is not in the group of buyers holding the current market bid $\bar{b}' = v_x$ in d', he can neither trade in the first market phase, by overbidding $\bar{b}'$ with $\bar{b}' + 1$, nor can he trade in any second market phase subgame reached after I'. This implies that, if $\bar{b}' = v_x$ and B_x is not in the group of buyers holding the current market bid $\bar{b}' = v_x$ in d', B_x receives a payoff of 0. But, we can show that this situation cannot arise in an impatience equilibrium of the game.

Assume X is a quasi-pure subgame perfect equilibrium of the game, which leads to a trade involving some non-extra-marginal buyer B_i, with $i \leq m$, and some non-extra-marginal seller S_j, with $j \leq m$, at the price v_x in the first market phase. From the discussion of the cases (i), (ii), and (iii) above, it follows that this can only be the case, if on the play realized by X, B_x does not hold current market bid at v_x together with the non-extra-marginal buyers, who submitted v_x in some decision stage d. X is not an impatience equilibrium, because B_x's strategy x_{Bx} in X could be replaced by a different strategy $\bar{x}_{Bx}$, which prescribes submitting v_x in d, instead of submitting some lower bid or not submitting at all. Since X is a quasi-pure subgame perfect equilibrium of the game, it follows that B_x does not achieve a higher payoff by deviating from his strategy in this way. The play realized by the strategy combination $\bar{X} = (x_{B1}, ..., x_{Bm}, \bar{x}_{Bx}, x_{Bx+1}, ..., x_{BnB}, x_{S1}, ..., x_{SnS})$ leads to a payoff for B_x that is equal to his equilibrium payoff in X, namely 0. But, B_x submits a better offer in the first market phase on the play realized by $\bar{X}$ than on the play realized by X. Thus, X cannot be an impatience equilibrium of the game. Hence, a combination of strategies, that leads to a trade at a market bid $\bar{b}' < v_x + 1$, that is not held by the best extra-marginal buyer B_x cannot be an impatience equilibrium.

Summarizing, we have shown that neither a trade at some price $p > v_x + 1$ nor at some price $p < v_x + 1$ in the first market phase can be the equilibrium outcome of an impatience equilibrium of the game. Thus, the only candidate left for equilibrium outcome are trades at the price $v_x + 1$ in the first market phase.

Impatience equilibrium X of the truncated game $\bar{G}$ in Case (2)

In case (2) either no extra-marginal buyer exists or the best extra-marginal buyer's redemption value v_x is not in the market clearing price range, i.e. $c_m > v_x$. Since the buyers open the market, the only trading price in the first market phase, which is compatible with the proposition, is the price $c_m - 1$.

A trade occurs at $p > c_m - 1$

Let us first check the cases in which a trade occurs at a price $p > c_m - 1$, i.e. $p \geq c_m$ in the first market phase. If a trade occurs at the price p, the current market bid $\bar{b}'$ in some decision stage d' must have been equal to p. Let I' denote the game segment that begins with the decision stage d'. In I' some seller accepted a trade at the price $\bar{b}' \geq c_m$.

A trade occurs at $p > c_m - 1$: Possible second market phase subgames

Before we proceed, let us first establish that only second market phase subgames either of the type $G_2(v_y, .)$, with $y \geq x$, or of the type $G_2(c_h, .)$, with $h \leq m$, could have been reached after I', i.e. after any number of trades at $p \geq c_m$ in the first market phase. The lower bound of the market clearing price range in the first market phase is the unit cost c_m of the marginal seller, since we are examining the second part of Proposition E.2, in which $c_m > v_x$ is true. Thus, if trades occur at a price $p \geq c_m$ in the first market phase, only non-extra-marginal buyers can have been involved, since extra-marginal buyers cannot trade $p \geq c_m > v_x$, due to the *no-loss* rule. If only intra-marginal sellers trade, the lower bound of the market clearing price range in any reached $G_2(.,.)$ remains c_m. If the marginal seller trades, the lower bound of the market clearing price range in any reached $G_2(.,.)$ changes either to the redemption value v_x of the best extra-marginal buyer or to the unit cost c_h of some intra-marginal seller S_h, with $h < m$. (Note that this is also true, if $c_m = v_m$.) If one or more extra-marginal sellers trade, the intra-marginal part of the supply curve shifts less to the left than that part of the demand curve, because the number of trading non-extra-marginal sellers is smaller than number of trading non-extra-marginal buyers. This means that either some originally non-extra-marginal sellers could become extra-marginal in $G_2(.,.)$ or that some originally extra-marginal buyers could become non-extra-marginal in $G_2(.,.)$. In any case, however, the lower bound of the market clearing price range in $G_2(.,.)$ is either determined by the redemption value v_y of an originally extra-marginal buyer B_y, with $y \geq x$, or by the unit cost c_h of an originally non-extra-marginal seller S_h, with $h \leq m$. Hence, after one or more trades at $p \geq c_m$ in the first market phase, any second market phase subgame that can emerge is of one of the following two types: either $G_2(v_y,.)$, with $y \geq x$, or $G_2(c_h,.)$, with $h \leq m$.

In the impatience equilibrium of any such second market phase subgame, that follows after one or more trades at $p \geq c_m$ in the first market phase, the expected equilibrium price p_2 is smaller than or equal to c_m, i.e. $p_2 \leq c_m$. Let us show why.

If a $G_2(v_y,.)$, with $y \geq x$, is reached, then $p_2 = v_y + 1$. Here $p_2 < c_m$, because we are examining the second part of Proposition E.2, in which $c_m > v_x$ is true. Thus, it follows that $p_2 = v_y + 1 \leq v_x + 1 < c_m$, for all $y \geq x$. (Remember that $c_m > v_x + 2$, due to Assumption A2.)

If a $G_2(c_h,.)$, with $h < m$, is reached, in which r non-extra-marginal traders are present, three different cases must be considered:

(i) For any non-extra-marginal buyer B_i, with $i \leq m$, the expected price in a $G_2(c_h,.)$, with $h < m$, is $p_2 = c_h - \frac{(r-1)}{r}$, since he trades at $c_h - 1$ with probability $\frac{(r-1)}{r}$ and trades at c_h with probability $\frac{1}{r}$. Here, $p_2 < c_m$, because $p_2 = c_h - \frac{(r-1)}{r} \leq c_h < c_m$, for all $h < m$.

(ii) For any intra-marginal seller S_j, with $j < m$, the expected price in a $G_2(c_h,.)$, with $h < m$, is $p_2 = c_h - 1$, if S_j is intra-marginal in $G_2(c_h,.)$, i.e. $j < h$. The expected price is $p_2 = c_h$, if S_j is marginal in $G_2(c_h,.)$, i.e. $j = h$. In either case, $p_2 < c_m$, because $p_2 \leq c_h < c_m$, for all $h < m$.

(iii) The marginal seller S_m cannot trade in a $G_2(c_h,.)$, with $h < m$, since $c_h < c_m$.

If a $G_2(c_h,.)$, with $h = m$, i.e. a $G_2(c_m,.)$, is reached, in which r non-extra-marginal traders are present, again three different cases must be considered:

(i) For any non-extra-marginal buyer B_i, with $i \leq m$, the expected price in a $G_2(c_m,.)$ is $p_2 = c_m - \frac{(r-1)}{r}$, since he trades at $c_m - 1$ with probability $\frac{(r-1)}{r}$ and trades at c_m with probability $\frac{1}{r}$. Here, $p_2 \leq c_m$, because $p_2 = c_m - \frac{(r-1)}{r} \leq c_m$.

(ii)   For any intra-marginal seller S_j, with $j < m$, the expected price in a $G_2(c_m,.)$ is $p_2 = c_m - 1$. Here, $p_2 < c_m$, because $p_2 = c_m - 1 < c_m$.

(iii)   For the marginal seller S_m, the expected price in a $G_2(c_m,.)$ is $p_2 = c_m$.

A trade occurs at $p > c_m$: All non-extra-marginal sellers accept p

Let us first examine the case that a trade at a price greater than c_m occurred in the first market phase, i.e. that in the game segment I' that begins with the decision stage d' the current market bid is $\bar{b}' > c_m$. In this case, all non-extra-marginal sellers must accept a trade at $\bar{b}'$. This is so, because each non-extra-marginal seller S_j, with $j = 1 \dots m$, achieves a greater payoff by trading at $\bar{b}' > c_m$ in I' than by trading at a lower price $p_2 \le c_m < \bar{b}'$ in any second market phase subgame $G_2(.,.)$. Even if S_j only has a $\frac{1}{m}$ chance of trading at $\bar{b}' > c_m$ in I', her expected payoff from accepting a trade at $\bar{b}'$ in I' is greater than the expected payoff from trading in any second market phase subgame.

A trade occurs at $p > c_m$
All non-extra-marginal buyers have an incentive to refrain from submitting p

If $k > 1$ buyers together held the market bid $\bar{b}' > c_m$, there must have been some decision stage d'' before d', in which they submitted $\bar{b}'$, but each of them had an incentive to deviate from his strategy. This is so, because each could have achieved a higher expected payoff by simply refraining to bid, instead of bidding $\bar{b}'$ in d''. Without loss of generality, let us assume B_i refrains from bidding $\bar{b}'$ in d''. Since all non-extra-marginal sellers accept $\bar{b}' > c_m$, held by the other $(k - 1)$ buyers, only $(k - 1)$ of the sellers trade at the price $\bar{b}'$. Then B_i reaches some second market phase subgame $G_2(.,.)$, in which he trades at a lower expected price $p_2 \le c_m < \bar{b}'$. Thus, if $k > 1$ buyers tie at $\bar{b}' > c_m$ in d'', each of them has an incentive to deviate from his strategy.

If only one buyer B_i held the market bid $\bar{b}' > c_m$ in d'', this buyer also had an incentive to deviate from his strategy. This is so, because B_i could have chosen to enforce some other price $p^* < \bar{b}'$ and $p^* \ge c_m$ in I'. Because of the argument above, the other buyers would not have overbid p*. Evidently, any $p* < \tilde{b}'$ leads to a higher payoff for B_i than $\tilde{b}'$. Thus, if a strategy combination leads to a current market bid $\tilde{b}' > c_m$, it cannot be a quasi-pure subgame perfect equilibrium of the game, since - no matter whether a single buyer alone or a number of buyers together hold $\tilde{b}'$ - at least one of the buyers has an incentive to deviate from his strategy.

A trade occurs at $p = c_m$: Condition (A) for an impatience equilibrium:
All intra-marginal sellers must have accepted p

Now, let us examine the case that a trade at a price equal to c_m occurred in the first market phase, i.e. that in the game segment I' that begins with the decision stage d' the current market bid is $\tilde{b}' = c_m$. In this case, all intra-marginal sellers must accept a trade at $\tilde{b}'$. This is so, because each intra-marginal seller S_j, with $j < m$, achieves a greater payoff by trading at $\tilde{b}' = c_m$ in I' than by trading at a lower price $p_2 < c_m = \tilde{b}'$ in any second market phase subgame $G_2(.,.)$, as shown above. Even if S_j only has a $\frac{1}{m}$ chance of trading at $\tilde{b}' = c_m$ in I', her expected payoff from accepting a trade at $\tilde{b}'$ in I' is greater than the expected payoff from trading in any second market phase subgame.

A trade occurs at $p = c_m$: Condition (B) for an impatience equilibrium:
All non-extra-marginal buyers must have tied at p

Given that all (m - 1) intra-marginal sellers accept a trade in the first market phase at $\tilde{b}' = c_m$, all m non-extra-marginal buyers together must hold the current market bid $\tilde{b}'$ in any quasi-pure subgame perfect equilibrium. To see why, consider any case in which only k, with $k < m$, non-extra-marginal buyers hold $\tilde{b}'$ in d'. Assume B_i is in the group of k buyers holding $\tilde{b}'$ in d'. Being in this group, B_i trades at $\tilde{b}' = c_m$ in the first market phase. If B_i, however, does not tie with the others at $\tilde{b}' = c_m$ in d', he does not trade in the first market phase, in which (k - 1) trades occur at c_m. Instead, B_i reaches a second market phase subgame of type $G_2(c_m,.)$, in which he trades at $c_m - \frac{(r-1)}{r}$. Since only intra-marginal sellers and only non-extra-marginal buyers trade in the first market phase, any second market phase that can be reached after I' is of type $G_2(c_{m'},.)$. In this $G_2(c_{m'},.)$, $r = m - (k - 1)$ non-extra-marginal traders are present on each market side. Since $k < m$ and since we are examining m-trade markets, with $m \geq 2$, it follows that $r > 1$ in $G_2(c_{m'},.)$. Thus, the expected price B_i trades at in $G_2(c_{m'},.)$ is smaller than c_m, the price he trades at, if he is in a tie in the first market phase, since $c_m - \frac{(r-1)}{r} < c_m$, for all $r > 1$. If, however, $k = m$, it follows that $r = 1$ and that B_i is indifferent between trading in the first market phase and trading in some second market phase subgame. Hence, if a quasi-pure subgame perfect equilibrium X of the game exists, in which trades at c_m occur in the first market phase, then: (B) All m non-extra-marginal buyers must have tied in some decision stage d' at the current market bid $\tilde{b}' = c_m$. (A) All $(m - 1)$ intra-marginal sellers must have accepted $\tilde{b}' = c_m$ in the game segment I' starting with d'.

A trade occurs at $p = c_m$: Condition (C) for an impatience equilibrium: The marginal seller must also have accepted p

Now, we show that, if a quasi-pure subgame perfect equilibrium X of the game exists, X can only be an impatience equilibrium of the game, if - in addition to the conditions (A) and (B) - the following condition (C) is fulfilled: (C) The marginal seller S_m accepts a trade at $\tilde{b}' = c_m$ in I', i.e. in the first market phase. Given the conditions (A) and (B), $(m - 1)$ trades occur in the first market phase involving only non-extra-marginal traders. If S_m does not accept a trade at $\tilde{b}' = c_m$ in the first market phase, any second market phase subgame that can be reached is of the type $G_2(c_{m'},.)$ with S_m as the only non-extra-marginal seller and exactly one non-extra-marginal buyer present. In this $G_2(c_{m'},.)$, S_m trades at the price c_m. Thus, the payoff S_m achieves in $G_2(c_{m'},.)$ is equal to the payoff of accepting $\tilde{b}' = c_m$ in the first market phase. But, in an impatience equilibrium of the game, S_m must accept c_m in the first market phase. This is true, because, if S_m's strategy x_{Sm} in X does not prescribe accepting c_m in the first market phase, her best offer in the first market phase must be greater than c_m (or she does not submit an ask in the first market phase) on the play realized by X. S_m, however, has an alternative strategy $\bar{x}_{Sm}$, that prescribes submitting the better offer c_m in the first market phase. The play realized by the strategy combination $\bar{X} = (x_{B1}, ..., x_{BnB}, x_{S1}, ..., x_{Sm-1}, \bar{x}_{Sm}, x_{Sx}, ..., x_{SnS})$ leads to a payoff for S_m that is equal to her payoff in X, since all m non-extra-marginal buyers jointly hold the current market bid c_m in the first market phase, due to condition (A). Thus, X can only be a quasi-pure subgame perfect equilibrium of the game, if conditions (A) and (B) hold, and can only be an impatience equilibrium of the game, if condition (C) also holds, i.e. if S_m accepts a trade at c_m in the first market phase.

A trade at $p = c_m$ cannot be the outcome of an impatience equilibrium: Conditions (A), (B), and (C) cannot simultaneously be true in equilibrium

Finally, we now show that, if all three conditions (A), (B), and (C) hold for some strategy combination X, X cannot be a quasi-pure subgame perfect equilibrium of the game, because each buyer has an incentive to deviate from his strategy in X. If all three conditions (A), (B), and (C) hold, every non-extra-marginal buyer B_i, with $i \leq m$, trades at c_m in the first market phase. If the buyer B_i, however, refrains from bidding, instead of bidding c_m, he trades in some second market phase with that non-extra-marginal seller, who did not trade in the first market phase. Which second market phase subgame is reached, depends on the random draw, which determines which (m - 1) of the non-extra-marginal sellers trade with the (m - 1) non-extra-marginal buyers - excluding B_i. Thus, with probability $\frac{1}{m}$, B_i reaches a $G_2(c_m,.)$, in which the marginal seller S_m is present and trades at c_m. With probability $\frac{(m-1)}{m}$, however, B_i reaches a $G_2(.,.)$, in which the marginal seller S_m is not present. In any of these second market phase subgames, B_i trades at some price $p_2 < c_m$, as shown above. Thus, the expected price B_i trades at in any second market phase subgame, if he does not tie with the other (m - 1) buyers at $\bar{b}' = c_m$ in the first market phase, is smaller than c_m, because $\frac{1}{m} c_m + \frac{(m-1)}{m} p_2 < \frac{1}{m} c_m + \frac{(m-1)}{m} c_m = c_m$. Each buyer B_i, therefore, has an incentive to deviate from his strategy in X and not to tie with the other buyers at $\bar{b}' = c_m$ in d'.

Hence, we have shown that all three conditions (A), (B), and (C) are necessary for some strategy combination X to be an impatience equilibrium of the game. But, if all three conditions hold, X cannot be a quasi-pure subgame perfect equilibrium of the game. It therefore follows that, if a strategy combination X leads to a trade at the price c_m in the first market phase of the game, X cannot be an impatience equilibrium of the game.

A trade occurs at $p < c_m - 1$

Now, consider some price $p < c_m - 1$. If a trade occurs at the price p, the current market bid $\bar{b}'$ in some decision stage d' must have been equal to p. Let I' denote the game segment that begins with d'. In I' some seller must have accepted a trade at the price $\bar{b}' < c_m - 1$.

A trade occurs at $p < c_m - 1$: Possible second market phase subgames

Before we proceed, let us first establish that only second market phase subgames either of the type $G_2(v_h,.)$, with $h \leq m$, or of the type $G_2(c_y,.)$, with $y \geq m$, could have been reached after I', i.e. after any number of trades at $p < c_m - 1$ in the first market phase. The lower bound of the market clearing price range in the first market phase is the unit cost c_m of the marginal seller, since we are examining the second part of Proposition E.2, in which $c_m > v_x$ is true. If a trade occurs at a price $p < c_m - 1$ in the first market phase, only intra-marginal sellers can have been involved, due to the *no-loss* rule. If only non-extra-marginal buyers trade, the lower bound of the market clearing price range in any reached $G_2(.,.)$ remains c_m. (Note that this is also true if $v_m = c_m$.) If one or more extra-marginal buyers trade, the intra-marginal part of the demand curve shifts less to the left than that part of the supply curve, because the number of trading non-extra-marginal buyers is smaller than number of trading non-extra-marginal sellers. This means that either some originally non-extra-marginal buyers could become extra-marginal in $G_2(.,.)$ or that some originally extra-marginal sellers could become non-extra-marginal in $G_2(.,.)$. In any case, however, the lower bound of the market clearing price range in $G_2(.,.)$ is either determined by the redemption value v_h of an originally non-extra-marginal buyer B_h, with $h \leq m$, or by the unit cost c_y of an originally marginal or extra-marginal seller S_y, with $y \geq m$. Hence, after one or more trades at $p < c_m - 1$ in the first market phase, any second market phase subgame that can emerge is of one of the following two types: either $G_2(v_h, .)$, with $h \leq m$, or $G_2(c_y, .)$, with $y \geq m$. In the impatience equilibrium of any such second market phase subgame, that follows after one or more trades at $p < c_m - 1$ in the first market phase, the expected equilibrium price p_2, at which an intra-marginal seller S_j, with $j < m$, trades, is greater or equals to $c_m - 1$, i.e. $p_2 \geq c_m - 1$. Let us show why.

If a $G_2(v_h, .)$, with $h \leq m$, is reached, then $p_2 = v_h + 1$. Here $p_2 > c_m - 1$, because $v_m \geq c_m$ and, thus, $p_2 = v_h + 1 \geq v_m + 1 > c_m - 1$, for all $h \leq m$.

If a $G_2(c_y, .)$, with $y \geq m$, is reached, then $p_2 = c_y - 1$, since $j < m$. Here $p_2 \geq c_m - 1$, because $p_2 = c_y - 1 \geq c_m - 1$, for all $y \geq m$.

A trade occurs at $p < c_m - 1$: Only one non-extra-marginal seller accepts p

Thus, it follows that if more than one seller accepted a trade at $\bar{b}' < c_m - 1$ in I', this cannot have been an equilibrium. This is so, because only intra-marginal sellers can trade at $\bar{b}' < c_m - 1$ in I' and because in any second market phase subgame, that can be reached after I', the intra-marginal sellers trade at a higher price $p_2 \geq c_m - 1 > \bar{b}'$. Hence, all intra-marginal sellers prefer waiting for a second market phase to trading at a price $\bar{b}' < c_m - 1$ in I'. With this argument, however, we cannot rule out that an equilibrium exists in which a trade at $\bar{b}' < c_m - 1$ occurs: if $\bar{b}'$ is a credibly enforced price, it follows by Lemma 4, that at least one seller must accept a trade at the enforced price in order to prevent market period termination. If, however, one seller chooses to accept $\bar{b}'$, all others are better off waiting for a second market phase subgame. Thus, if a trade at $\bar{b}' < c_m - 1$ occurred in I', only one seller S_j, with $j < m$, accepted a trade at this price.

Obviously, some decision stage d'' must have existed, after $\bar{b}'$ was first submitted, but before S_j accepted $\bar{b}'$, in which S_j had the alternative to submit an ask $\bar{a}'' = \bar{b}' + 1$. Let I'' denote the game segment and $\bar{G}''$ denote the subgame that begins with d'', i.e. with S_j's submission of $\bar{b}' + 1$. If any one of the buyers accepts a trade at $\bar{b}' + 1$ in $\bar{G}''$, S_j receives a higher payoff than on the original play that led to a trade at $\bar{b}'$, because she trades at a higher price. Thus, if any of the buyers accepts $\bar{b}' + 1$ in $\bar{G}''$, S_j has an incentive to deviate from her original strategy. Hence, to prove that a quasi-pure subgame perfect equilibrium cannot lead to a price $p < c_m - 1$, it is sufficient to show that at least one of the buyers prefers to trade with S_j at $\bar{b}' + 1$ in $\bar{G}''$, i.e. overbids any current market bid $\bar{b}' < c_m - 1$. Showing that S_j will be able to trade in $\bar{G}''$ is sufficient, because, for any strategy combination that realizes a play that leads to a trade at a price $\bar{b}' < c_m - 1$, an alternative strategy combination exists, in which S_j's strategy choice is altered in such a way, that the realized play reaches the subgame $\bar{G}''$, in which S_j achieves a greater payoff.

A trade occurs at $p < c_m - 1$
Reduced number of possible second market phase subgames

Since only one seller S_j, with $j < m$, accepts a trade at $\bar{b}' < c_m - 1$ in I', the number of reachable second market phase subgames is reduced. If S_j trades with a non-extra-marginal buyer, a second market phase subgame of the type $G_2(c_m,.)$ is reached. If S_j trades with an extra-marginal buyer, a second market phase subgame either of the type $G_2(v_m,.)$ or of the type $G_2(c_x,.)$ is reached. This is so, because, after S_j trades with an extra-marginal buyer, $(m - 1)$ originally non-extra-marginal sellers face m originally non-extra-marginal buyers. If $v_m < c_x$ (or c_x does not exist), the originally marginal buyer B_m becomes the best extra-marginal buyer in $G_2(v_m,.)$ and his redemption value v_m is the lower bound of the market clearing price range in $G_2(v_m,.)$. Thus, any $G_2(v_m,.)$ is a $(m - 1)$-trade market. But, if $v_m > c_x$, the originally extra-marginal seller S_x becomes marginal in $G_2(c_x,.)$ and her unit cost c_x is the lower bound of the market clearing price range $G_2(c_x,.)$. Thus, any $G_2(c_x,.)$ is an m-trade market. Hence, only three types of second market phase subgames can follow the game segment I', in which a trade occurs at the price $\bar{b}' < c_m - 1$:

(i)   a $(m - 1)$-trade market of the type $G_2(c_m, .)$, in which a non-extra-marginal buyer trades at an expected price of $c_m - \frac{(m-2)}{(m-1)}$, since he trades at $c_m - 1$ with probability $\frac{(m-2)}{(m-1)}$ and at c_m with probability $\frac{1}{(m-1)}$,

(ii)   an m-trade market of the type $G_2(c_x, .)$, in which a non-extra-marginal buyer trades at an expected price of $c_x - \frac{(m-1)}{m}$, since he trades at $c_x - 1$ with probability $\frac{(m-1)}{m}$ and at c_x with probability $\frac{1}{m}$, and

(iii)   a $(m - 1)$-trade market of the type $G_2(v_m, .)$, in which a non-extra-marginal buyer trades at an expected price of $v_m + 1$.

Since $c_m - \frac{(m-2)}{(m-1)} \leq c_m < c_x - 1 < c_x - \frac{(m-1)}{m}$, it follows that a non-extra-marginal buyer trades at a lower expected price in a $G_2(c_m, .)$ than in a $G_2(c_x, .)$. (Remember that, by Assumption A2, $c_x > c_m + 2$.) Similarly, since $c_m - \frac{(m-2)}{(m-1)} \leq c_m \leq v_m < v_m + 1$, a non-extra-marginal buyer trades at a lower expected price in a $G_2(c_m, .)$ than in a $G_2(v_m, .)$. The important result for the following payoff comparisons is: $c_m - \frac{(m-2)}{(m-1)}$ is the lowest price, at which a non-extra-marginal buyer B_i, with $i \leq m$, can trade at in any second market phase subgame $G_2(., .)$.

Since $\bar{b}' < c_m - 1 < v_m$, because $v_m \geq c_m$, there are at least two non-extra-marginal buyers, who could have held the market bid $\bar{b}'$ in d'. Thus, if $k \geq 1$ buyers held $\bar{b}'$ in d', two situations are possible in d'': either (i) some non-extra-marginal buyer B_i, with $i \leq m$, exists, who is not in the group of the buyers holding the current market bid $\bar{b}'$ or (ii) all non-extra-marginal buyers are in the group holding $\bar{b}'$.

A trade occurs at $p < c_m - 1$
Case (i): non-extra-marginal buyer overbids, if he is not in the group of buyers holding $\bar{b}' = p < c_m - 1$

Let us first examine (i), i.e. some non-extra-marginal buyer B_i does not hold $\bar{b}'$ in d''. In this case, B_i has an incentive to overbid $\bar{b}'$ with $\bar{b}' + 1$, i.e. to accept S_j's offer. If B_i accepts S_j's offer, he trades at the price $\bar{b}' + 1 \leq c_m - 1$ in I''. In contrast, if B_i does not overbid, B_i can trade at the minimum expected price

$c_m - \frac{(m-2)}{(m-1)}$ in some $G_2(.,.)$. Therefore, as long as $\bar{b}' < c_m - 1$, B_i pays a lower price by overbidding in I'' than by trading in a $G_2(.,.)$, since $\bar{b}' + 1 \leq c_m - 1 <$ $c_m - \frac{(m-2)}{(m-1)}$, for all $\bar{b}' < c_m - 1$. (Remember that we are examining m-trade markets, with $m \geq 2$.) Hence, any strategy combination that leads to a trade at any current market bid $\bar{b}' < c_m - 1$, held by $k < m$ buyers in I'', cannot be an equilibrium of the game, because a non-extra-marginal buyer not holding the current market bid $\bar{b}'$ has an incentive to overbid $\bar{b}'$ and to accept a trade with S_j at $\bar{b}' + 1$.

A trade occurs at $p < c_m - 1$

Case (ii): non-extra-marginal buyer overbids, if all non-extra-marginal buyers are in the group of buyers holding $\bar{b}' = p < c_m - 1$

Now, let us examine case (ii), i.e. all non-extra-marginal buyers are in the group of $k \geq m$ buyers holding the current market bid $\bar{b}'$ in d''. If one of the non-extra-marginal buyers B_i, with $i \leq m$, breaks the tie at $\bar{b}'$ by accepting S_j's offer to trade at $\bar{b}' + 1$ in I'', he pays $\bar{b}' + 1 \leq c_m - 1$ and receives $v_i - (\bar{b}' + 1) + \beta$. If B_i, however, remains in the tie with the other buyers at $\bar{b}'$, then he trades at the price $\bar{b}'$ with probability $\frac{1}{k}$, since only one seller accepts a trade at $\bar{b}' < c_m - 1$. Because $k \geq m$, B_i's maximum expected payoff of trading in the first market phase is $\frac{1}{m}(v_i - \bar{b}' + \beta)$. As shown above, the lowest expected price B_i must pay in any following second market phase subgame $G_2(.,.)$ is $c_m - \frac{(m-2)}{(m-1)}$. Thus, B_i's maximum expected payoff of remaining in a tie with the other buyers, at the price $\bar{b}'$, in the first market phase, is $\frac{1}{m}(v_i - \bar{b}' + \beta) + \frac{(m-1)}{m}(v_i - (c_m - \frac{(m-2)}{(m-1)}) + \beta) =$ $v_i - (\frac{1}{m}\bar{b}' + \frac{(m-1)}{m}c_m - \frac{(m-2)}{m}) + \beta$. Hence, overbidding leads to a greater payoff for B_i than remaining in the tie at $\bar{b}'$, as long as $v_i - (\bar{b}' + 1) + \beta >$ $v_i - (\frac{1}{m}\bar{b}' + \frac{(m-1)}{m}c_m - \frac{(m-2)}{m}) + \beta$. This is equivalent to $\bar{b}' + 1 <$ $\frac{1}{m}\bar{b}' + \frac{(m-1)}{m}c_m - \frac{(m-2)}{m} \Leftrightarrow m\bar{b}' + m < \bar{b}' + (m-1)c_m - m + 2 \Leftrightarrow (m-1)\bar{b}'$ $< (m-1)c_m - 2(m-1) \Leftrightarrow \bar{b}' < c_m - 2$. Hence, as long as there is a tie of all m non-extra-marginal buyers B_i at a current market bid $\bar{b}' < c_m - 2$, each B_i has an incentive to overbid $\bar{b}'$. Thus, a combination of strategies, that leads to a trade at a current market bid $\bar{b}' < c_m - 2$ held by all non-extra-marginal buyers in I', cannot be an equilibrium of the game, because each B_i has an incentive to overbid $\bar{b}'$ and accept a trade at $\bar{b}' + 1$.

A Trade at $p = c_m - 2$ cannot be the outcome of an impatience equilibrium

Note that the argument above also implies that, if $\bar{b}' = c_m - 2$, B_i - at maximum - receives the same expected payoff by remaining in the tie at $c_m - 2$ as he does by overbidding $\bar{b}' = c_m - 2$ with $\bar{b}' + 1 = c_m - 1$. However, even if B_i achieves an equal expected payoff in both cases, the original strategy combination that realizes a play, on which B_i remains in a tie at $c_m - 2$, cannot be an impatience equilibrium of $\bar{G}$. This is so, because replacing B_i's original strategy to bid $c_m - 2$ with the alternative strategy to bid $c_m - 1$ in the original strategy combination results in a play, on which B_i submits a better offer, but achieves the same expected payoff. Hence, a quasi-pure subgame perfect equilibrium, that realizes a play on which all non-extra-marginal buyers tie at a current market bid $\bar{b}' = c_m - 2$, cannot be an impatience equilibrium of $\bar{G}$.

Summarizing, we have shown that neither a trade at some price $p \geq c_m$ nor at some price $p \leq c_m - 2$ in the first market phase can be the equilibrium outcome of an impatience equilibrium of the game. Thus, the only candidate left for equilibrium outcome are trades at the price $c_m - 1$ in the first market phase.

Induction over the market size

With this we have proven the following induction step: if Proposition E.2 holds for all ν-maximum-trade markets, where $\nu < \mu$, it also holds for any μ-maximum-trade market. It follows by the propositions C.1.2 and C.2.2 that Proposition E.2 holds for all 1-maximum-trade markets. Thus, we can conclude that Proposition E.2 also holds for all 2-maximum-trade markets. Therefore, the result also holds for all 3-maximum-trade markets and so forth. Since the result holds for an arbitrary μ-maximum-trade market, it also holds for an arbitrary m-trade market. ∎

III.F. Résumé and Discussion

The game theoretic analysis of the simplified alternating double auction market game is of an extremely complicated nature. This is mainly so, because of the dynamic structure of the game. Although the game is finite, the dynamics of submitting and re-submitting offers creates an enormous strategy space for each player, since strategies can be conditioned on any of the numerous possible histories of play. Even with a small number of players (e.g. two players on each market side) and a small number of possible alternative offers (e.g. the prices 0 to 9 and the alternative to refrain from submitting), the game tree of a simplified alternating double auction market game is far beyond the scope of any practical graphical representation. Additionally, the termination rule creates the problem, that the plays of the game can be of very different lengths, when the length of a play is measured in offer cycles from the start.

To be able to cope with the difficulties of the analysis arising from the considerable size of the game, we introduced the concept of *game segments* and *market phases*. The former split the game into segments starting at some arbitrary decision stage and ending either with a contract or with market period termination. The latter are defined as maximal segments in the sense that they are not proper substructures of larger segments, i.e. the segment of the game between two adjacent contracts (or between the start of the game and the first contract, or between the previous contract and the end of the game).

Using these concepts we defined *quasi-pure subgame perfect equilibria*, in which the strategy of each trader induces a local pure strategy in the first decision stage of every market phase and a pure segment strategy in the game segment that follows the realized play of the first decision stage in every market phase. Thus, a quasi-pure subgame perfect equilibrium has the property that all local strategies on the play realized in any market phase subgame are pure local strategies. This concept is very helpful in the complicated context of the simplified alternating double auction market game, because it allows us to specify only those parts of the strategies, which are actually used on or *close to* the equilibrium path of each of the market phase subgames. Hence, the strategies are specified and pure in some parts and generic in other parts. In the proofs, we demonstrated that, no matter which subgame equilibria the generic parts of the strategies induce in the corresponding subgames, these subgames are not reached on the equilibrium path of the entire game.

The first main result of the analysis is that quasi-pure subgame perfect equilibria exist in any simplified alternating double auction market game (Proposition E.1), if every two of the player valuations in the market - except the marginal valuations - are more than two money units apart (Assumption A2). Since we did not exclude the case of equal marginal valuations, the result also holds for typical experimental markets, with a unique competitive equilibrium price. (Note, however, that the general existence of quasi-pure subgame perfect equilibria, does not establish the general existence of pure subgame perfect equilibria, since the set of pure subgame perfect equilibria is a subset of the set of quasi-pure subgame perfect equilibria.) The analysis also showed that different types of quasi-pure subgame perfect equilibria can exist in a simplified alternating double auction market game.

One type of equilibrium was shown to exist in every market in which extra-marginal traders are present on the opening market side and the player valuation of the best extra-marginal player on the market opening side is in the market clearing price range. Let us call this type of quasi-pure subgame perfect equilibria *type 1A equilibria* in the following. In a type 1A equilibrium, the non-extra-marginal traders on the market opening side submit offers in the first decision stage of every market phase, that are one money unit outside the range of feasible offers for the best extra-marginal trader on their own market side, i.e. one money unit greater than the redemption value of the best extra-marginal buyer or one money unit smaller than the unit cost of the best extra-marginal seller. Each extra-marginal trader on the opening market side submits his own player valuation in the first decision stage of every market phase. After the first offer, the traders on the opening market side refrain from submitting new offers in the remaining part of each market phase. All non-extra-marginal traders on the second moving market side immediately accept a trade at the offered price. Each extra-marginal trader on the second moving market side submits his own player valuation.

Thus, a type 1A equilibrium leads to as many trades as non-extra-marginal traders are present on each market side. The trades involve only non-extra-marginal traders, with all trades at a price just outside the range of feasible offers of the best extra-marginal trader on the opening market side. The outcome of such type 1A equilibria is in line with traditional competitive market analysis: Only non-extra-marginal traders trade. All non-extra-marginal traders trade. All trades occur at the same price within the market clearing price range.

For the case of 1-trade markets with extra-marginal traders on the opening market side, Proposition C.2.1 also establishes the existence of a second type of quasi-pure subgame perfect equilibrium: a *type 1B equilibrium*. In this type 1B equilibrium, the marginal trader on the opening market side submits the player valuation of the best extra-marginal trader on the opening market side in the first decision stage of the game. His strategy, however, also specifies submitting a better offer (i.e. overbidding the first bid or underasking the first ask), if a tie with the best extra-marginal trader should occur at the first offer. Due to this *threat*, the best extra-marginal trader knows that he cannot trade in the market, no matter which offer he submits in the first decision stage. On the one hand, if he submits an offer other than his own player valuation, he knows that he will not be able to overbid or underask the marginal trader's first offer later. On the other hand, if he submits his player valuation, he knows that his offer will be overbid or underasked by the marginal trader in the next offer cycle. Thus, since he knows he will not be able to trade, the best extra-marginal trader on the opening market side chooses not submit his own player valuation in the first decision stage. Hence, the marginal trader, by using such a *threat* strategy, trades at a price equal to the player valuation of the best extra-marginal trader, instead of at a price just outside the feasible range of offers for the best extra-marginal trader.

Note that the outcome of such type 1B equilibria is also in line with traditional competitive market analysis: Only non-extra-marginal traders trade. All non-extra-marginal traders trade. All trades occur at the same price within the market clearing price range (more precisely: exactly on the boundary of the market clearing price range). Note, however, that type 1B equilibria are very sensitive to the *patience* of the best extra-marginal trader on the opening market side. He must choose not to submit his best offer early, even though he is indifferent between submitting this offer early and not doing so, since he does not trade in any case. In other words, if the best extra-marginal trader on the opening market side does decide to submit his best offer in the first decision stage, a type 1B equilibrium breaks down.

In every market in which the best extra-marginal valuation on the opening market side either does not exist or is not in the market clearing price range another type of equilibrium was shown to exist. Let us call this type of quasi-pure subgame perfect equilibria *type 2A equilibria* in the following. In a type 2A equilibrium, all non-extra-marginal traders on the opening market side submit an offer p in the first decision stage of every market phase. If intra-marginal traders are present in the market, p is one money unit outside the range of feasible offers for the marginal trader on the other market side, i.e. one money unit greater than the redemption value of the marginal buyer of the second moving market side or one money unit smaller than the unit cost of the marginal seller of the second moving side. If no intra-marginal traders are present (i.e. only marginal and extra-marginal traders are present), p is equal to the player valuation of the marginal trader on the other market side. Each extra-marginal trader on the opening market side submits his own player valuation in the first decision stage of every market phase. After the first offer, the traders on the opening market side refrain from submitting new offers in the remaining part of each market phase. All intra-marginal traders on the second moving market side immediately accept a trade at the offered price. Each extra-marginal trader on the second moving market side submits his own player valuation.

Thus, a type 2A equilibrium leads to as many trades as non-extra-marginal traders are present on each market side. All trades involve only non-extra-marginal traders, with most of the trades - all, except the last one - at a price just outside the range of feasible offers of the marginal trader on the second moving market side. The last trade occurs at a price equal to the player valuation of the marginal trader on the second moving market side. The outcome of such type 2A equilibria is close to, but not completely in line with traditional competitive market analysis: Although only non-extra-marginal traders trade and all non-extra-marginal traders trade, the trades occur at two different prices, most of which are just outside the market clearing price range. Only the last trade occurs at a price within the market clearing price range (more precisely: exactly on the boundary of the market clearing price range).

For the case of 2-trade markets without extra-marginal traders, Proposition D.1 also establishes the existence of a two other types of quasi-pure subgame perfect equilib-rium: a *type 2B equilibrium* and a *type 2C equilibrium*. The type 2B equilibrium is very similar to the type 2A equilibrium. The only difference is that the first trades occur at a price that lies two money units outside the range of feasible offers for the marginal trader of the second moving market side. The type 2B equilibria are very sensitive to the *patience* of the non-extra-marginal traders on the opening market side. Each of these traders has a ½ chance of trading in the first market phase and a ½ chance of trading in the second market phase. If either of them chooses not to wait, even though he is indifferent between trading early and trading later, a type 2B equilibrium breaks down.

The type 2C equilibrium is quite different. Here, both non-extra-marginal traders on the opening market side submit the player valuation of the marginal trader on the second moving market side in the first decision stage of the game. This equilibrium hinges on the fact that only the intra-marginal trader on the second moving market side accepts a trade in the first market phase. In contrast, the marginal trader on the second market side waits to trade in the second market phase. This trade in the second market phase, however, is at the same price as the trade in the first market phase, i.e. at a price equal to the player valuation of the marginal trader on the second moving market side.

The outcome of such type 2C equilibria is in line with traditional competitive market analysis: Only non-extra-marginal traders trade. All non-extra-marginal traders trade. The trades occur at a price within the market clearing price range (more precisely: exactly on the boundary of the market clearing price range). Note, however, that type 2C equilibria are very sensitive to the *patience* of the marginal trader on the second moving market side, who must choose to wait, even though he is indifferent between trading early and trading later. In other words, if the marginal trader on the second moving market side does decide to accept the trade in the first market phase, a type 2C equilibrium breaks down.

Thus, the examination of the quasi-pure subgame perfect equilibria of the simplified alternating double auction market game uncovered only two types of equilibria that are robust concerning the *patience* of indifferent traders, namely type 1A and type 2A equilibria. It therefore seemed worthwhile to restrict our attention to only those quasi-pure subgame perfect equilibria that fulfill a specific *impatience* criterium. Thus, we introduced the concept of *impatience equilibria*. These are quasi-pure subgame perfect equilibria in which none of the players' strategies can be replaced by an alternative strategy that leads to the same expected payoff, but specifies the submission of a better or equal offer in an earlier stage.

The second main result of the analysis is that the equilibrium outcome of all impatience equilibria of a simplified alternating double auction market game is unique. The outcome of all impatience equilibria of any market, in which an extra-marginal trader is present on the opening market side and his valuation is in the market clearing price range, always corresponds to the outcome of the type 1A equilibria.

The outcome of all impatience equilibria of any market, in which an extra-marginal trader is not present on the opening market side or is present, but his valuation is not in the market clearing price range, always corresponds to the outcome of the type 2A equilibria (Proposition E.2).

Hence, the game theoretic analysis of the simplified alternating double auction market game leads to the following main results, if the set of player valuations in the game has the property, that any two valuations - except the marginal valuations - are more than two money units apart (check figure III.1 in section III.C.1 for a graphical display of the equilibrium prices in each of the possible cases):

Result 1:   Every simplified alternating double auction market game has at least one impatience equilibrium.

Result 2:   All impatience equilibria of a simplified alternating double auction market game lead to market efficiency, i.e. exactly as many trades occur as non-extra-marginal traders are present and only non-extra-marginal traders trade.

Result 3:   All trades in any impatience equilibria of a simplified alternating double auction market game occur at prices which are either one money unit inside, one money unit outside, or exactly on that bound of the market clearing price range that favors the opening market side. The relevant case is uniquely determined by the innermost valuation at that bound of the market clearing price range.

Unlike the prediction by competitive market clearing, the prediction by an impatience equilibrium is always a point prediction, no matter which size the market clearing price range has. In other words, where competitive market clearing only predicts a range of possible trading prices, the impatience equilibria predict a specific price, either just inside or just outside the bound of the market clearing price range.

The price predicted by the game theoretic analysis, however, is always biased in favor of the opening market side. This means that the price predicted is close to the lower bound of the market clearing price range, if buyers are on the opening market side, but close the upper bound, if sellers are on the opening market side. As explained above, however, this bias does not lead to extreme outcomes relative to the entire market price range, but only relative to the market clearing price range. In other words, the part of the market rent outside the market clearing price range is distributed amongst the non-extra-marginal buyers and sellers in (almost) the same way as in the competitive market clearing. Only the distribution of that part of the market rent, which lies within the market clearing price range is biased in favor of the opening market side. As mentioned before, this bias is not in conflict with the competitive market clearing, which does not predict a specific distribution of this part of the market rent.

An important implication of the result of the game theoretic analysis is that a minimal amount of competition in simplified alternating double auction market game is sufficient to induce game theoretic equilibria that are very close to traditional, competitive market clearing. Even in a case with only three traders, in which the extra-marginal trader is on the market opening side, the one trade that is possible occurs within the market clearing price range. Thus, it seems that non-cooperative strategic behavior is the clue to the astonishing observation that practically all double auction markets examined in experiments tend to efficient market clearing at competitive prices, even when very few traders are present. Here, the game theoretic analysis seems to shed at least some light into the phenomenon that SMITH (1982) calls a "scientific mystery".

A direct comparison of the game theoretic results and the experimental observations, however, is subject to two serious objections: First, the simplified alternating double auction market game, as examined, only allows for one valuation per player, whereas in many experimental setups (including the one we report later), each subject was assigned a number of valuations. Second, the simplified alternating double auction market game, as examined, is a game of almost perfect information, whereas in all experiments (in all those we know of), subjects had only private information on their own valuations. Without giving a proof, we argue that - although both objections are well founded - a comparison of the theoretic and the experimental results, nevertheless, gives valuable insight into the problem.

The multiplicity of player valuations can indeed change the character of equilibrium outcomes in a simplified alternating double auction market game. Imagine, for example, that in a 1-trade market with extra-marginal traders on the opening market side, the roles of the marginal trader and of some extra-marginal trader are played by the same player T. If T does not play the role of the best extra-marginal trader, the equilibrium outcome remains as in the above analysis, i.e. a trade occurs at a price within the market clearing price range, but just outside the range of feasible offers of the best extra-marginal trader. This is evidently true, because T faces competition by the best extra-marginal trader, just as he would, if he only played the role of the marginal trader. However, if T plays the role of the best extra-marginal trader along with the role of the marginal trader, the equilibrium of the game changes: now, a trade occurs at a price well outside the market clearing price range, namely either at the player valuation of the marginal trader on the second moving market side or just outside the range of feasible prices for the second best extra-marginal trader on the opening market side. This is true, since T can simply dispose of his second unit, i.e. *create* a different market, in which the best extra-marginal player valuation is missing. In this way, T can change one of the bounds of the market clearing price range and can gain from not competing with himself.

From the discussion above, it seems clear that players with multiple valuations can only influence the outcome of the market, if they control the innermost valuation. But, even when all extra-marginal valuations of the opening market side are withheld, the market clearing price range does not necessarily expand to the size of the entire market price range. This is so, because, when no extra-marginal valuations are present on one market side, the bound of the market clearing price range is determined by the marginal valuation on the other market side. Thus, in settings, like the ones we use in our experiment, in which the marginal valuation of each market side is close to the best extra-marginal valuation of the other market side, the effect of the withdrawal of the extra-marginal valuations is small, relative to the entire market price range.

With this we do not assert that the multiplicity of valuations has little or no effect on the game theoretic results, especially since it is quite possible that a number of complicated, history dependent equilibria could emerge in that case. However, given the above argument and the limitations to subjects capabilities to calculate such equilibria in the short duration of an experiment, it seems plausible to believe that the strategic influence of the multiplicity of valuations is not of great significance in an experimental setting. Thus, we will - perhaps somewhat too boldly - compare the experimental results of markets, in which subjects are assigned multiple values, to the theoretic results for markets of the same size, implicitly assuming that subjects tend to play the role of each trader they represent independently of the roles of the other traders they represent.

The second objection, concerning the complete information in the theoretic analysis compared to incomplete information in the experimental setting, presents a long standing debate in the literature. The basic question is whether a static experimental repetition of a game of incomplete information leads to an equilibrium of the game itself or to the equilibrium of a corresponding game with complete information. Meanwhile, there seems to be a general - but yet unproven - understanding amongst many scholars, that the latter is the case. (For an extensive discussion of the problem, see FRIEDMAN 1993.) In fact, SELTEN (1995) even argues that subjects who are given less information are more likely to tend to the complete information equilibrium than subjects who are fully informed. The intuition behind this argument is that fully informed subjects can more easily discover cooperative (collusive) possibilities in the game.

JORDAN (1991) considers players who repeatedly play the Baysian equilibrium of a normal-form game with incomplete information. After each play the players update their beliefs according to Bayes' Rule. JORDAN proves that - under certain circumstances - such a process converges to an equilibrium of the complete information game. In fact, the players considered do not even have to be completely informed on the parameters of the game, when the complete information game equilibrium is reached. It seems sufficient for them to accumulate *just enough* information to be able to play their part in that equilibrium. COX, SHACHAT, and WALKER (1997) present some experimental evidence for JORDAN's assertion that the complete information equilibrium is *learned* in static repetitions of the corresponding incomplete information game. Since in both papers, however, only simple normal form games are examined, the results must still be seen as preliminary rather than full-fledged. We, nevertheless, join this group of scientists in the conjecture that the equilibria of the complete information game can be adequate predictors of the experimental outcomes, if a sufficient number of static repetitions of the incomplete information game are conducted. Hence, in the following chapters, we compare the experimental results of static repetitions of an alternating double auction market game with an incomplete information setting to the baseline provided by the results from the game theoretic analysis of the simplified alternating double auction market game, that were presented in this chapter.

IV. An Experimental Investigation of the Alternating Double Auction Market

IV.A. Introduction

Double auction markets have fascinated experimental economics researchers ever since Vernon SMITH reported his experiments in 1962. The fascination lies in the fact, that these markets exhibit an extremely high efficiency and a remarkable convergence to the market clearing price, although the underlying games are exceptionally complex. However, neither the abundant experimental nor the scarce game theoretic research on double auction markets, existing to date, has yet entirely clarified the process of price formation in these markets. The main goal of the investigation presented here, is to find regularities in the behavior of traders in experimental discriminatory price double auction markets, which might help explain the process of price formation.

The process of price formation in experimental double auction markets has been previously studied by a number of researchers (EASLEY and LEDYARD 1983 and 1993, CASON and FRIEDMAN 1993 and 1994, GJERSTAD and DICKHAUT 1997). In their very thorough studies, CASON and FRIEDMAN (1993 and 1994) obtain mixed results. They basically compare predictions of the theoretic approaches by WILSON (1987) and FRIEDMAN (1991) to the zero-intelligence trader approach by GODE and SUNDER (1993). They come to the conclusion that "none of the current models adequately explains price formation in double auction markets." (See CASON and FRIEDMAN 1994, p. 19.)

In our experimental investigation of the alternating double auction market, we take a similar path. The outcomes of the experimental sessions are compared to all approaches mentioned above, including zero-intelligence trader market simulations, that are used as a benchmark for the evaluation of the data. The advantage of employing the alternative double auction market institution for our experiment is that we have an additional benchmark: the result of the game theoretic analysis, presented in the previous chapter.

In light of the examined experimental data, we come to the conclusion that the price formation in the markets is best described by the following two approaches: First, the simple rules postulated by the *learning direction theory* (SELTEN and STOECKER 1986, SELTEN and BUCHTA 1998) explain the tendency of contract prices to converge to the market clearing price. Second, the *anchor price hypothesis*, presented in the last section of this chapter, enables us to predict the direction of price convergence and the structure of price paths, i.e. the sequence of contract prices within a market period. The prediction is simply based on the location of the *anchor price*, the first contract price in the first market period of a session. The *anchor price* itself is predicted to be greater than the market clearing price p^*, if buyers open the first period's market, and smaller than p^*, if sellers open the first period's market.

The rest of the chapter is organized in the following manner. In the next section the experimental setup is described in detail. In section VI.3. the results of the experiment are discussed and evaluated. That section also contains detailed descriptions of the benchmarks used and the theories compared. Finally, in the last section, the results of the investigation are summarized and the two main conclusions on the process of price formation are presented.

IV.B. The Experimental Setup

All sessions of the experiment took place in the computerized laboratory of the economics department of the University of Bonn, Germany (*Laboratorium für experimentelle Wirtschaftsforschung*). A mixed subject pool was recruited, mainly consisting of undergraduate students of law and economics. The subjects signed up one or two weeks before each session took place. At sign up time they were only informed that they would take part in an experiment in the laboratory and would be paid in accordance with their performance in the experiment.

Each session began with a forty-five minute verbal introduction, in which the instruction handouts were read aloud and explained. Typical computer screen shots were shown. Examples for action sequences were given. It was stressed that all screen shots and examples contained values clearly out of the range of the experimental session. (The handouts and screen shots are contained in the appendix.) The subjects were encouraged to ask questions. The session was started only after all questions had been answered and all subjects had agreed to proceed.

A total of 86 subjects were divided into ten independent subject groups: one group of twelve, one group of ten, and eight groups of eight. Each group took part in a first session. Eight of the ten groups took part in a second session. The division line between the first and second sessions of a single subject group was drawn by a short break. The subjects were told that after the break a new market with completely different parameters would begin. They were also told, however, that they would remain on the same market side they were on during the first session. They were neither allowed to leave the cubicles nor to communicate during the break. Since only subjects who had never before taken part in a market experiment were allowed to sign up, the first session with each subject group is a session with inexperienced subjects. If a group took part in a second session, that is considered to be a session with experienced subjects.

An overview of all sessions is given in the tables IV.1 and IV.2. Each session name starts with the market type, M1 or M2. The number of participating subjects follows in hexadecimal representation (A = 10 and C = 12). The following number represents is a counter. The names end with a 0 or a 1, depending on the subjects' experience level. Thus, the session M2A11 was the first session with 10 experienced participants in market type M2. Sessions with names differing only in the last digit (experience) were run with exactly the same subject group.

Table IV.1 - Experimental parameters for the sessions with inexperienced subjects.

Session		M1810	M1820	M1830	M1840	M1C10	M2810	M2820	M2830	M2840	M2A10
Number of participants	buyers	4	4	4	4	6	4	4	4	4	5
	sellers	4	4	4	4	6	4	4	4	4	5
	total	8	8	8	8	12	8	8	8	8	10
Experienced group		no	no	no	no	no	no	no	no	no	no
Number of market periods		5	5	5	5	4	5	5	5	5	5
Number of no-offer-cycles		2	2	2	2	3	3	2	2	2	3
Decision time limit in minutes	for the very first decision	5	5	5	5	5	5	5	5	5	5
	for all other first period decisions	2	2	2	2	2	2	2	2	2	2
	for all other decisions	1	1	1	1	1	1	1	1	1	1
Exchange rate in DM per point		.1	.15	.15	.15	.1	.05	.15	.15	.15	.1
Trading commission	in points	2	2	2	2	2	3	2	2	2	5
	in DM	.2	.3	.3	.3	.2	.15	.3	.3	.3	.5
Period bonus in DM	for buyers	1.3	.9	.9	.9		.35	.15	.15	.15	.1
	for sellers	.1	.15	.15	.15		.35	.9	.9	.9	.7
Opening market side	period 1	sellers	sellers	buyers	buyers	buyers	buyers	sellers	buyers	sellers	buyers
	period 2	sellers	buyers	buyers	sellers	buyers	buyers	sellers	sellers	sellers	sellers
	period 3	buyers	sellers	buyers	buyers	buyers	buyers	buyers	buyers	buyers	sellers
	period 4	buyers	sellers	sellers	buyers	sellers	buyers	buyers	buyers	sellers	sellers
	period 5	buyers	sellers	buyers	buyers		sellers	buyers	sellers	buyers	buyers
Demand & Supply	market type	M1	M1	M1	M1	M1	M2	M2	M2	M2	M2
	demand function	281-2q	410-2q	913-2q	231-2q	791-2q	393-6q	781-6q	323-6q	878-6q	369-3q
	supply function	231+3q	360+3q	863+3q	181+3q	721+3q	323+4q	711+4q	253+4q	808+4q	324+2q
	buyers' rent	90	90	90	90	182	126	126	126	126	108
	sellers' rent	135	135	135	135	273	84	84	84	84	72
	ratio of buyers' to sellers' rent	2:3	2:3	2:3	2:3	2:3	3:2	3:2	3:2	3:2	3:2
	market price range	45	45	45	45	65	60	60	60	60	40

Table IV.2 - Experimental parameters for the sessions with experienced subjects.

Session		M1811	M1821	M1831	M1841	M2811	M2821	M2841	M2A11
Number of participants	buyers	4	4	4	4	4	4	4	5
	sellers	4	4	4	4	4	4	4	5
	Total	8	8	8	8	8	8	8	10
Experienced group		yes	yes	yes	yes	yes	yes	yes	yes
Number of market periods		5	5	5	5	5	5	5	5
Number of no-offer-cycles		2	2	2	2	3	2	2	3
Decision time limit in minutes	for the very first decision	5	5	5	5	5	5	5	5
	for all other first period decisions	2	2	2	2	2	2	2	2
	for all other decisions	1	1	1	1	1	1	1	1
Exchange rate in DM per point		.1	.15	.15	.15	.05	.15	.15	.1
Trading commission	in points	2	2	2	2	3	2	2	5
	in DM	.2	.3	.3	.3	.15	.3	.3	.5
Period bonus in DM	for buyers	1.3	.9	.9	.9	.35	.15	.15	.1
	for sellers	.1	.15	.15	.15	.35	.9	.9	.7
Opening market side	period 1	buyers	sellers	buyers	sellers	buyers	sellers	buyers	sellers
	period 2	buyers	buyers	sellers	sellers	buyers	buyers	sellers	sellers
	period 3	sellers	sellers	sellers	sellers	buyers	sellers	buyers	buyers
	period 4	buyers	sellers	buyers	buyers	buyers	sellers	sellers	sellers
	period 5	buyers	buyers	buyers	buyers	buyers	sellers	sellers	sellers
Demand & Supply	market type	M1	M1	M1	M1	M2	M2	M2	M2
	demand function	849-2q	756-2q	456-2q	648-2q	807-6q	458-6q	601-6q	824-3q
	supply function	799+3q	710+3q	406+3q	598+3q	737+4q	388+4q	531+4q	779+2q
	buyers' rent	90	90	90	90	126	126	126	108
	sellers' rent	135	135	135	135	84	84	84	72
	ratio of buyers' to sellers' rent	2:3	2:3	2:3	2:3	3:2	3:2	3:2	3:2
	market price range	45	45	45	45	60	60	60	40

Two different market types, M1 and M2, were used in the experiment. The demand and supply curves of both markets are step functions with constant step sizes. In the market M1 buyers receive two-fifths and sellers three-fifths of the total rent in competitive equilibrium, i.e. the slope (step size) of the demand curve is two-thirds of the slope of the supply curve. In the market M2 buyers receive three-fifths and sellers two-fifths of the total rent in competitive equilibrium, i.e. the slope (step size) of the demand curve is one and one-half times the slope of the supply curve.

The demand and supply curves of the market types are of the following form:

M1: $demand^{M1} = 4(n_B + 1) - 2q + p^*$ $supply^{M1} = -6(n_S + 1) + 3q + p^*$

M2: $demand^{M2} = 12(n_B - .5) - 6q + p^*$ $supply^{M2} = -8(n_S - .5) + 4q + p^*$

The first term of each function is used to normalize the markets in such a way, that the proportion of intra- to extra-marginal units (the relative size of the market clearing quantity to the total number of units in the market) remains constant over all market sizes, as long as $n_B = n_S$, i.e. as long as the number of buyers equals the number of sellers. Note that $n_B = n_S$ was true for all sessions. The second term of each function specifies the slope in quantity q. The markets are *normal* in the sense that demand falls and supply rises in price. The last term of each function is the market clearing price p^*. The demand and supply curves of every session were both *shifted* with the unique market clearing price p^* of that session, i.e. no two sessions had the same p^*. All demand and supply curves, however, were well within the *technical* price range [1...9999], with the lowest market clearing price at 211 and the highest at 893. The market clearing quantity q^* was the same for all sessions of the same market type and same number of traders. (Note that - strictly speaking - q^* is not uniquely specified in any of the examined markets, since the marginal redemption value is equal to the marginal unit cost in all markets. However, because a *trading commission* is paid to any trader who trades, q^* will always be supposed to take the larger of the two possible values, i.e. supposed to include the trade by the marginal traders.)

Every buyer in every session was assigned four redemption values and every seller four unit costs. Both redemption values and unit costs were assigned sequentially. The highest redemption price was assigned to the first buyer B_1, the second highest to the second buyer B_2, etc. Once every buyer was assigned one redemption value, the sequence started over: the $n_B + 1$ highest redemption price was assigned to B_1, the $n_B + 2$ highest to B_2, etc. The unit costs were assigned analogously to the sellers 1 to n_S. Thus, every session had a distinct hierarchy in roles: from the *best* buyer B_1 to the *worst* buyer B_{nR} and from the *best* seller S_1 to the *worst* seller S_{nS}. The figures IV.1 and IV.2 show the p*-normalized demand and supply curves for the market types M1 and M2, respectively. Since the only information given to the subjects about the distribution of redemption values and unit costs of the other traders was that these were all in the range [1...9999], the subjects knew neither of the existence of such an hierarchy, nor of their relative position.

Each session's market was run five times, i.e. five market periods with identical values. (Session M1C10 was an exception with only four periods.) The only difference between one period and the next was that in the subsequent period the entire offer and trade history of all preceding periods was readily accessible on screen. If an experienced session followed the first session of a subject group, however, the history of the completed first (inexperienced) session was not available in the second (experienced) session.

In each first session of every subject group, the subjects were randomly assigned a market side (buyer or seller) and a role. If a second session followed the first session, the subjects were systematically re-ordered in opposite order. Thus, not only the face value of the market clearing price was changed (by shifting the demand and supply curves) from inexperienced session to experienced session, but also the position of the equilibrium relative to every trader's redemption values or unit costs.

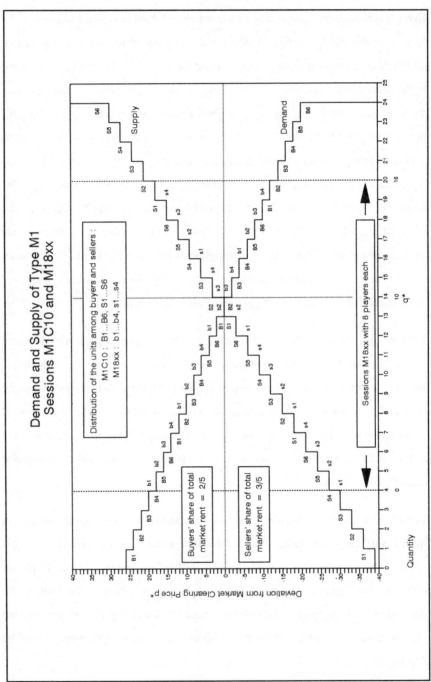

Figure VI.1 - Display of demand and supply in the market type M1 sessions.

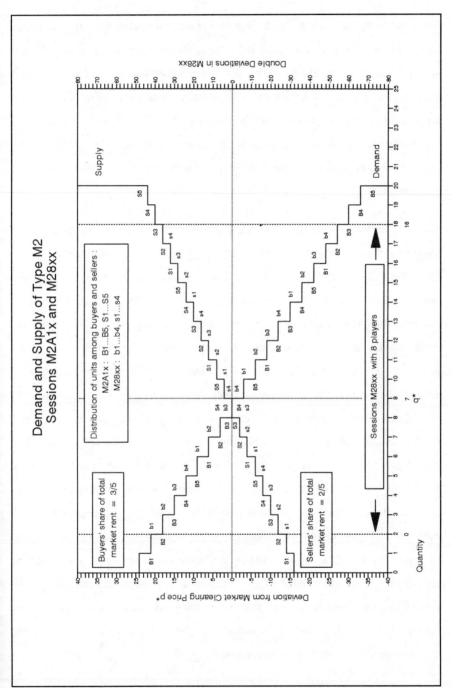

Figure VI.2 - Display of demand and supply in the market type M2 sessions.

According to the rules of the alternating double auction market, each market period ended after a number z of adjacent offer cycles without new offers. This number z was set to three (z = 3) in the first six sessions, but was reduced to two cycles without new offers (z = 2) in the later sessions. The extra *alert* cycle seemed superfluous, since the subjects of the early sessions had proven very attentive concerning the period ending condition. Reducing z shortened the duration of the sessions, without eliminating the possibility of signalling rigid commitment to the last submitted offer.

Although the time the subjects had for a decision was limited by the experimental software and the number of possible cycles in a period are inherently limited in the game, the first session that was run, M1C10, consumed much more time than was expected. One reason for the extremely long duration was that subjects, who wanted to refrain from submitting an offer, had realized that they could do so by simply waiting for the decision time to elapse. In contrast, those who wanted to submit did so in only a few seconds. Thus, in every bidding or asking round there were a number of extremely quick submissions followed by a long time span of inactivity, until the decision time had finally elapsed. To avoid putting the subjects under stricter decision time limitations, which might have seriously affected the quality of the subjects' decisions, another way of speeding up the experiment was introduced for all sessions after the first session. A *period bonus* was paid to every subject after each completed period. The period bonus was also used to compensate the weaker market side, depending on the parameters of the session. For this reason, only its existence, but not its value was publicly made known in the verbal introduction at session start. The exact value of the small monetary incentive was only shown privately on the terminal screens each time a period ended. It was stressed that the period bonus had no connection whatsoever to the profits from trading and would not compensate potential losses in the market.

Unlike the period bonus, the *trading commission* was equal for both market sides and its value publicly announced before each session. The trading commission was paid to both parties of any contract made. Like in many auction experiments, the commission was paid to ensure marginal trade at p* would occur. Depending on the session, commissions of DM .2 to DM .5 were paid. In most sessions the trading commission was smaller than 1% of the total market rent. The subjects were clearly instructed that the commissions cannot be used to loosen the *no-loss* condition, by balancing potential losses of an offer.

The profits from trading were transformed to German Marks at a rate of DM .05 to DM .15 per point, depending on the session. In all sessions the exchange rates were calculated in such a manner, that potential payoffs to the subjects would - on average - exceed the standard hourly wage of student helpers. Immediately after each session, the subjects were paid separately and anonymously.

IV.C. The Results of the Experimental Investigation

In this section the results of the experimental investigation of the alternating double auction market are presented and evaluated. For an informative evaluation of the results of the experiment, it seems necessary to first present a number of bench-marks that allow the appraisal of different aspects of the outcomes. This is done in the first subsection of this section. In the subsections thereafter, the experimental results are presented from a number of different perspectives, such as efficiency, price convergence, etc., that can all be subsumed in the term *market features*. Instead of comparing all market features to one benchmark after another, each feature is evaluated using the appropriate benchmarks in the subsection in which it is presented. The evaluation of individual behavior, whenever of interest, is also embedded into the different subsections on the examined market features. The individual behavior is a central theme of the subsection on the boundedly rational approaches to traders' behavior in double auction markets. A general overview of the results is presented in the last section of this chapter.

The volume of data collected in the experiment is so large that the inclusion of the entire data set on paper did not seem practical. (On request, the entire data set can be obtained on diskettes from the author.) Instead, summary results of the data of each experimental session is included in the Appendix VII.B. The figures VII.B.1 to VII.B.18 in that appendix display the price paths, i.e. the sequence and prices of the contracts, of every period of every session. The panels VII.B.1 to VII.B.18 in that appendix contain the summary data of every period of every session and the corresponding all-period averages. If not specifically noted otherwise, the all-period averages of the considered parameters are examined in the following subsections. The parameters and the terms used to refer to them will be described in the course of the investigation.

IV.C.1. Benchmarks for Evaluating the Experimental Results

IV.C.1.a. The Walrasian Competitive Equilibrium

In spite of all new approaches to the analysis of double auction markets, the Walrasian market clearing price and quantity have remained the most commonly used benchmarks for the evaluation of market outcomes in general and will also be the primary benchmarks in the analysis of our experimental data. It seems that the most important reason for the survival of the market clearing price and quantity as chief benchmarks is the general result obtained in many experimental examinations of the double auction market: Experimental double auction markets tend to be highly efficient, with trading quantity very close to the market clearing quantity and with prices converging towards market clearing price, after repeated runs of the market. The method of experimentation was first applied to oral double auction markets as early as 1962 (SMITH 1962). The evidence that has been gathered ever since, in the host of experiments with many different forms of the double auction market, has - all in all - established the robustness of the early results. (See PLOTT 1982, FRIEDMAN 1993, and KAGEL 1995 for overviews on the experimental work.) Thus, using the market clearing price and quantity as benchmarks for the data analysis means more than just comparing results to Walrasian theory, it also means comparing outcomes to the well established findings of the experimental work on double auction markets.

In the following, two other useful benchmarks for the evaluation of the experimental outcomes are discussed, which represent the two extremes with regard to the extent of rationality demanded from the trading agents. The game theoretic approaches, that are presented first, are the one extreme, assuming that fully rational traders interact strategically and decide optimally in the markets. The other extreme is the purely stochastic approach with *zero-intelligence* traders, who are assumed to make completely random, non-strategic, and non-optimal choices.

IV.C.1.b. Game Theoretic Approaches

The trading agents in the non-cooperative game theoretic approaches are not Walrasian *price takers*, but act strategically in the game delineated by the rules of the market. These are *hyper-rational* players who can evaluate the strategic situation with no time delay and calculate the mutually optimal moves. The (hyper-) rationality assumption finds its climax in the common knowledge of rationality, which guarantees that each player knows that all other players are rational and that they all know that all others are rational, and so forth.

Apart from the fact that the (hyper-) rationality assumption of game theory has been critizied on logical grounds (see, for example, BINMORE 1987), the application of game theoretic analyses to experimentally examined double auction market games faces a great number of difficulties, especially due to the complexity of such games. This, probably, is the reason why game theoretic examinations of double auction market games are so rare. We will concentrate on the following two examinations, which seem to be the most extensive: The application of game theoretic analysis to a double auction game with imperfect information by WILSON (1987) and the analysis of the alternating double auction market game, as a game with almost perfect information, that is presented in the previous chapter.

Next to other problems which arise when comparing game theoretic predictions to the experimental outcomes, the major difficulty seems to be that the assumptions needed by any game theoretic analysis of double auction markets are hardly verifiable in an experimental setting. This not only includes the assumption of the risk-neutrality of players' preferences during the entire game and the assumption that each player can only trade a single unit, which are common to both approaches. It also includes the assumption of common knowledge on market parameters, which take different forms in the two considered approaches. The almost perfect information game approach, that we presented in the previous chapter, assumes common and full knowledge of the players on demand and supply curves. This is obviously not the case in the experiments. Similarly, the assumption of common knowledge concerning the distribution of the *player types* (i.e. concerning the probability distributions of randomly drawn redemption values and unit costs), that is made in the game theoretic analysis of a double auction market game with imperfect information presented by WILSON (1987), is not fulfilled in the presented experiments either. (Note that the only the discriminatory price double auction market experiment, that we know of, in which this assumption is fulfilled is the experiment reported by CASON and FRIEDMAN 1994.)

Another difficulty with WILSON's approach is that the theoretic outcome of the *endgame*, in which a single buyer or a single seller remains in the market, is not specified. This outcome, however, is necessary for the suggested construction of the equilibrium strategies. This difficulty is pointed out in the paper (see WILSON 1987, pp. 376-377) and possible resolutions are suggested (see WILSON 1987, pp. 408-410). Nevertheless, it remains unclear, exactly how the proposed solution can be derived for the game underlying the presented experiment.

In spite of the mentioned difficulties, it seems appropriate to examine at least some implications of the game theoretic approaches. One common implication of both analyses is that trading prices should be *close* to the market clearing price p*. In the approach presented in this paper, the *closeness* to p* is exactly specified: since marginal traders are the innermost traders in all experimental markets, the game theoretic prediction is that the first trades occur at p* - 1, if buyers open the market, and at p* + 1, if sellers open the market. The last trade is always predicted to occur at p*.[3] WILSON's imperfect information approach does not specify the *closeness* to p* precisely. The analysis, however, does make the prediction that prices are closer to p*, the more traders are present in the market.

Another common implication of both analyses is that the trading quantity should be equal to (in our approach) or almost equal to (in WILSON's approach) the market clearing quantity q*. Furthermore, the markets in both types of game theoretic equilibria are very efficient (100% efficient in our analysis).

Finally, the two approaches differ in their implications about the timing of trades and the order in which the players trade. In the impatience equilibrium that we constructed in the previous chapter, all but one of the trades occur immediately in the very first offer cycle. Only the last trade, that must involve the marginal trader of the second moving market side, occurs in the second offer cycle. (Note that this is impossible in the given experimental setting, since each subject plays the role of four players in the game. Since the subjects are restricted to trading one unit after the other, three offer cycles are required in the market type M1 and two are required in the market type M2 for all trades to occur as soon as possible.) Our analysis, however, does not per se refute any trading order that fulfills the necessary condition concerning the game theoretic equilibrium price. In other words, all

---

3) At this point, we should note that the technical assumption that all player valuations (except the two marginal values) are at least two money units apart does not hold in all our market parameterizations. This was due to the fact that the experiment was conducted before the theoretic analysis was completed.

experimental outcomes fulfilling the two conditions that (i) the marginal trader on the second moving market side is last to trade and that (ii) all trades - except the last trade - occur at a price just outside the range of feasible prices of that trader, can potentially be in line with the game theoretic equilibria. In case of such outcomes, the observed individual bidding and asking behavior must be examined more closely to correctly assess, whether subjects were playing equilibrium strategies.

The predictions of WILSON's analysis are different in this aspect. There trading occurs one unit after the other in equilibrium. Each trade is preceded with a delay in which only *non-serious* offers are submitted. These delays are used by the players to signal the location of their player valuations relative to the range of possible valuations. The better a trader's player valuation is, the shorter the delay which that player chooses. Even though the precise size of these delays is not easily assessable (and is dependent on the *parameterization of time*), the result has a well testable implication: at any given point in time, the next trade that occurs is always a trade between the two best traders present in the market. A number of authors (EASLEY and LEDYARD 1983 and 1993, CASON and FRIEDMAN 1993 and 1994), including WILSON himself (1987, p.411), have reported that this predicted strict order of trading is typically not observed in double auction experiments. We will later show that neither this prediction nor the prediction made by our game theoretic analysis, concerning the sequence of trades, hold for our data.

IV.C.1.c. The Zero-Intelligence Traders Approach

Markets populated with zero-intelligence trading agents were first introduced by GODE and SUNDER (1993). A zero-intelligence trading agent (or zero-intelligence trader) - in principle - is a random device that draws (i.e. *submits*) an offer from the range of feasible offer with a uniform probability. Zero-intelligence trading agents are assigned the roles of the traders in a market and restricted to the sequence of actions and the range of feasible offers of the underlying market rules. In this way a market outcome is randomly generated from the range of possible outcomes. Thus, a Monte-Carlo study with zero-intelligence traders provides a simulation distribution of market outcomes generated from random draws only.

Ever since their introduction, simulations with zero-intelligence trader markets have been used more and more frequently as benchmarks for the evaluation of market outcomes. (See, for example, CASON and FRIEDMAN 1994). The main reason for usefulness of the simulation of a market populated with zero-intelligence traders as a benchmark is that the outcome of such a market represents the most natural and plausible random draw of the complex market outcome. In this way, Monte-Carlo studies with zero-intelligence traders supply an elegant solution to a problem that is outside the scope of analytic practicability.

The panels contained in the Appendix VII.C. display the results of simulation runs of the alternating double auction market populated with zero-intelligence trader agents. In each offer cycle of a simulated market, each zero-intelligence trading agent submits an offer that is randomly drawn from the range of feasible offers with uniform probability. The range of feasible offers of such an agent is always limited by the *no-loss* rule, i.e. an offer that if accepted would lead to a loss for the agent is non-feasible. The range of offers is further limited by the technical price range, i.e. the range of positive integers in which all offers must lie. (In the experiments, this range contained all positive integers with less than 5 digits, i.e. from 1 to 9999. Note that subjects were informed on this technical price range in the instructions and could not input an offer outside this range at their terminals.)

In the reported simulations, the zero-intelligence trading agents were further bounded by the *no-crossing* and the *ask-bid-spread-reduction* rules, just as the human counterparts in the experiment. (Remember that the former rule disallows asks lower than the current market bid and bids higher than the current market ask, given that a corresponding market offer of the other market side is standing, i.e. open to acceptance. The latter rule requires any new ask to be smaller than the current market ask and any new bid to be greater than the current market bid, given that a corresponding market offer of the own market side is standing.)

Completely analogous to his human counterpart, a zero-intelligence trading agent was not allowed to submit an offer, if the lower bound of the range of his feasible offers was greater than the upper bound of that range. (This, for example, was the case for a buyer agent in an offer cycle in which the current market bid was greater than his redemption value, since any bid in this offer cycle that complied with the *ask-bid-spread-reduction* rule, was disallowed by the *no-loss* rule.) Thus, as soon as no more profitable trades were possible in a simulation market, the zero-intelligence traders quickly locked themselves into a situation in which none of them could submit a new offer. Hence, a simulated market could be terminated in finite time, by using the same market period termination criterion as in the experiment, namely that a market period ends, when a pre-specified number z of offer cycles passes without any new offers. The value of z was set to 3 in all simulations.

Each reported simulation consisted of 10000 rounds. A simulation round corresponded to an experimental session, i.e. consisted of five market periods. Since each market period with zero-intelligence trading agents - in principle - is a random and independent draw of the market outcome, there is no inter-temporal carry-over from one market period to the next, as one would expect from subjects' decisions in an experiment. Thus, the 10000 rounds of each simulation effectively correspond to 50000 independent market period observations, i.e. 10000 rounds times 5 periods each. This number seemed large enough to guarantee robust results, especially since the comparison of the preliminary 2000 round simulations to the 10000 round simulations revealed almost no difference in the values.

The reported simulation results in the panels of Appendix VII.C. are ordered in the same manner as the experimental results reported in the panels of the Appendix VII.B. In both appendices, prices and offers are normalized on the market clearing price, i.e. reported as deviation from p*. (We refer the reader to the introductions of those appendices for more details on the organization of the panels.)

The reported values are - in general - averages of the corresponding parameter over the 10000 rounds of the simulations. In some cases, however, an additional line titled *median value in the simulation* is inserted. A value reported in such a line is the median of the simulation distribution of the parameter reported in the previous line. Note that the values reported in the column *Average of Periods* are averages of the five period averages over the 10000 rounds, not averages of the corresponding parameter over 50000 rounds. In this column, in lines titled *median value in the simulation*, the median of the distribution of the five period averages is reported. This median can be decisively different than the median of the distribution of the value itself, since the distribution of the five period averages is generally more *tightly* distributed. (The notes concerning the lines containing median values are analogously true for the lines titled *lower quartile bound in the simulation* and *upper quartile bound in the simulation*. Here, the boundary values for the lower or upper quartile of the simulation distribution of the corresponding parameters are reported.)

At the end of each simulation result panel, the simulation distribution of price paths is reported. We postpone the detailed description of this data to the subsection concerning price paths, further down in this chapter.

Some of the results of the zero-intelligence trader simulations are sensitive to the location of the support of the distribution of random offers (i.e. the technical price range from which these offers are drawn). For this reason two types of simulations are reported: *baseline simulations* (contained in the panels numbered **VII.C.0.**x) and *full price range simulations* (contained in the panels numbered **VII.C.1.**x).

The *full price range simulations* allow the zero-intelligence trading agents to submit offers in the same technical price range that subjects in the experiment were faced with, i.e. buyer agents could bid prices between 0 and their redemption value and seller agents could ask prices between their unit cost and 9999. (Note that the zero-intelligence traders in these simulations - just like their human counterparts in the experiment - could only submit offers in the full price range in those offer cycles in which their offers were not limited by the *no-crossing* and the *ask-bid-spread-reduction* rule.) The panels VII.C.1.1 - VII.C.1.18 show the results of the full price range simulations for the market parameters (number of traders and demand and supply) of each of the experimental sessions. These panels are sorted as the panels in the Appendix VII.B. (and as in all following tables): the panels corresponding to the inexperienced sessions are displayed first, ordered from the market type 1 to the market type 2 sessions and from sessions with less players to those with more players. Then the panels of the experienced sessions follow in a similar manner.

Examining these panels, the reader can easily verify that the prices in all simulations are biased upwards. This bias is simply due to the fact that the zero-intelligence seller agents have a much larger support for the random draw of their offers than the buyer agents. Thus, the price bias is, obviously, also dependent on the location of the market clearing price p^*. The lower p^*, the more the prices are biased upwards. Additionally, a comparison of the panels reveals that the bias of prices is also dependent on the market type. In the M2 markets prices are biased even more upward than in the M1 markets. Lastly, the bias is also dependent on the number of traders in the market. Each of these conditions influences the ratio of the sizes of the two market sides' range of offers. The more the range of feasible asks, relative to the range of feasible bids, is increased - ceteris paribus -, the more the prices are biased upwards.

As far as the price bias is due to an asymmetric technical price range (i.e. the location of the market clearing price p^* in the technical price range), we agree with GODE and SUNDER (1997, footnote 2, page 6) that the bias causes an implausible distortion of the simulation results. In fact we believe to have some experimental underpinning for the use of a symmetric technical price range: The experimental results presented in the panels of the Appendix VII.B. show that in all but one session (M1280) subjects submitted less than 5% of their offers outside a range of p^* - 200 to p^* + 200. Moreover, there seems to be no clear tendency for sellers to submit offers outside this range more frequently than buyers, even though the technical price range was biased extremely in favor of the sellers. For these reasons, a number of *baseline simulations* were run using a symmetric technical price range of p^* - 200 to p^* + 200.

In the baseline simulations, the price bias due to the asymmetry of the experimental technical price range is obviously eliminated. The results of the baseline simulations, however, remain influenced by the slopes of the demand and supply curves and the number of traders in the market. Thus, for a comparison of the simulation results to the experimental results it is not sufficient to examine a single type of simulation. Simulations with each experimented combination of market type and player number are necessary. Four combinations were employed in the experiments: market type M1 with 8 players, market type M1 with 12 players, market type M2 with 8 players, and market type M2 with 10 players. The results of the corresponding zero-intelligence trader simulations are reported in the panels VII.C.0.1a, VII.C.0.1b, VII.C.0.2a, and VII.C.0.2b.

In addition to the simulations based on the experimental market parameters, a baseline simulation (panel VII.C.0.0) and a full price range simulation (panel VII.C.1.0) of a market with completely symmetric demand and supply curves were run. The simulations for this type M0 market are instructive, because they present a benchmark for the evaluation of the effects of the asymmetric rent distribution on the price development. In the market type M1 simulations, average prices are below the market clearing price p^*. Buyers submit a smaller number of disadvantageous offers than sellers (i.e relatively less bids greater than p^* than asks smaller than p^*). Both effects are due to the asymmetric distribution of market rent, which is in favor of the sellers in the market type M1. In the market type M2 simulations, the situation is reversed: the buyers have a relatively larger share of the market rent and, thus, average prices are above p* and buyers submit disadvantageous offers more often than sellers. That zero-intelligence trading agents have a disadvantage when they are on the market side with a larger portion of the market rent becomes completely evident by examination of the market type M0 simulation results in panel VII.C.0.0. Since the demand and supply curves are completely symmetric in M0, the conditions for an *ideal* double auction market populated with zero-intelligence traders are fulfilled: efficiency is almost at 100%, average prices are at p*, and both buyers and sellers submit the same number of offers of which the same percentage are disadvantageous. Note, however, that the *ideal* conditions are eliminated, if the bounds of the technical price range are not symmetric around p*. In this case, as demonstrated by full price range simulation of the market M0 reported in panel VII.C.1.0, the market side with the greater support for its random draws of offers has an advantage.

IV.C.2. Market Efficiency

Most of the research on auction market experiments has focused on the two central questions, whether experimental markets are efficient and whether the prices tend to move towards the market clearing price. In most cases in which the experimental market was organized as a double auction, both questions were answered positively. Inquiring whether the corresponding results also hold for the alternating double auction market seems to be a natural starting point of the analysis. Should this market prove to be different in these basic features, the question must be posed, which institutional characteristics cause the distinction. Should it prove to be similar, one can tentatively assume that further analogies with other double auction institutions exist.

For the measurement of market efficiency we employ a coefficient proposed by PLOTT and SMITH (1978). The *efficiency coefficient* is defined as:

$$\text{efficiency coefficient} = \frac{\Sigma \text{ payoffs subjects received}}{\Sigma \text{ achievable payoffs in competitive equilibrium}}$$

Since trading commissions (but not the period bonuses) are included in *received* and *achievable* payoffs, the efficiency coefficient can only be equal to one, if all intra-marginal and marginal, but none of the extra-marginal units have been traded. In this case all surplus of the market has been exploited by the traders.

Like most previous double auction market experiments and as implied by both the game theoretic and the zero-intelligence simulation benchmarks, the alternating double auction markets of this experiment prove to be highly efficient. The average efficiency coefficient for the inexperienced groups is .984 and for the experienced groups .986. This result coincides with the observation that in most markets the traded quantity is very close to the market clearing quantity: In the total of 89 markets that were run, a deviation of two units is observed only in 3 markets, a deviation of one unit in 31 markets, and no deviation in the rest of the markets. Finally, in those sessions in which early market periods are less than 95% efficient, efficiency tends to rise in later periods. (See the tables IV.3 and IV.4 and the panels in Appendix VII.B for the data.)

IV.C.3. Convergence of Prices to the Market Clearing Price

The figures VII.B.1 to VII.B.18 in the Appendix VII.B. illustrate the observed contract prices in all sessions of the experiment. A quick visual inspection reveals that the prices in the early periods of most sessions - especially of the sessions with inexperienced subjects - are not very close to the market clearing price p*. In later market periods, however, the prices tend to get closer to p*. In this subsection, we attempt to confirm this initial visual impression that prices converge to p*.

Checking for the convergence of contract prices to the market clearing price p* is a more demanding task than checking the efficiency of the market. The analysis of the dynamics of price development requires a unit of measurement, with which the distance of contract prices to market clearing p* can be evaluated. Since the alternating double auction does not induce a uniform price for all trades in a market period, a compound measure is needed, which incorporates all price information. For this reason a new measure of *nearness* of prices to p* is introduced, which can be used to evaluate observed prices in discriminatory price auctions.

The *market clearing price concordance coefficient* is a measure for the *nearness* of contract prices to the market clearing price p* in any discriminatory price auction, in which prices are limited by a *no-loss* condition. The concept of the market clearing price concordance coefficient is to compare the observed deviation to maximum possible deviation of contract prices from p*. The observed deviation is obtained by summing up the absolute deviations of all observed contract prices from p*. The maximum possible deviation, for a given number of contracts, depends on the distribution of non-extra-marginal player valuations.

Let E denote the set of all absolute deviations of non-extra-marginal player valuations (i.e. intra-marginal and marginal redemption values and unit costs) from p*:

$$E = \{\{ v_{ik} - p* \mid v_{ik} \geq p* \} \cup \{ p* - c_{jl} \mid c_{jl} \leq p* \}\}$$
$$\text{with } i = 1...n_B; \ j = 1...n_S; \ k = 1...q; \ l = 1...q.$$

Next let $\{\epsilon_h\}$ denote the sequence of all elements of E, sorted from maximum to minimum, so that $\epsilon_h \geq \epsilon_{h+1}$ holds true $\forall$ h, with ϵ_h and $\epsilon_{h+1} \in$ E and h = 1...#E. Now note that, any trade satisfying the *no-loss* condition must involve a non-extra-marginal trader at least on one market side. Thus, the greatest possible distance that the first contract price can have to p* is equal to ϵ_1, i.e. equal to the maximum of all p*-normalized non-extra-marginal player valuations. The second largest possible distance to p* is equal to ϵ_2, and so forth. Finally, assuming that m contracts are made, the sum of observed absolute deviations of the contract prices from p* cannot exceed the sum of the first m elements of $\{\epsilon_h\}$.

Table IV.3 - Equilibrium quantity and price, price convergence, and efficiency in the sessions with inexperienced subjects.

Session		M1810	M1820	M1830	M1840	M1C10	M2810	M2820	M2830	M2840	M2A10
Market clearing quantity (q*)		10	10	10	10	14	7	7	7	7	9
Deviation of traded quantities from q*	in period 1	0	0	0	-1	2	0	0	-1	-1	-1
	in period 2	-1	0	0	0	1	0	0	0	0	-1
	in period 3	-1	-1	-1	0	-1	-1	0	-1	0	-1
	in period 4	-1	-1	-1	0	0	-1	0	-1	0	0
	in period 5	1	0	0	0	.5	1	0	-1	0	0
	average	-.4	-.4	-.4	-.2	.5	-.2	0	-.8	-.2	.2
Market clearing price (p*)		261	390	893	211	763	351	739	281	836	342
Deviation of the opening price of period 1 from p*		-11	0	-21	1	14	9	-9	22	-17	-12
Deviation of average prices from p*	in period 1	-6.90	-7.20	-8.00	2.00	1.50	7.57	-6.43	6.00	-9.00	-6.75
	in period 2	-9.56	-.10	-2.10	-.60	3.47	1.00	-7.43	-1.29	-7.00	-4.40
	in period 3	-6.67	-5.22	.56	-1.10	1.31	9.17	-4.14	-2.83	-6.29	-4.10
	in period 4	-4.89	-3.22	-2.67	-1.50	-.50	7.50	-2.43	-.67	-6.43	-3.56
	in period 5	-3.18	-4.90	.60	-2.10		3.88	-.57	-1.50	-5.14	-.22
	average	-6.24	-4.13	-2.32	-.66	1.44	5.82	-4.20	-.06	-6.77	-3.81
Rank correlation coefficient of average prices to period		.80	.20	.60	-1.00	-.67	-.20	.80	-.40	.80	1.00
Significance level of RCC		.04	.41	.12	.01	.17	.41	.04	.24	.04	.01
Market clearing price concordance coefficient κ	in period 1	.58	.59	.53	.83	.75	.67	.73	.46	.64	.53
	in period 2	.48	.81	.85	.85	.78	.73	.69	.86	.71	.62
	in period 3	.64	.72	.87	.79	.92	.63	.83	.82	.71	.64
	in period 4	.73	.82	.79	.85	.91	.70	.86	.87	.73	.70
	in period 5	.77	.70	.91	.85		.72	.90	.88	.77	.69
	average	.64	.73	.79	.83	.84	.69	.80	.78	.72	.64
Rank correlation coefficient of κ to period		.80	.20	.60	.10	.67	.20	.80	.80	1.00	.80
Significance level of RCC		.04	.41	.12	.50	.17	.41	.04	.04	.01	.04
Efficiency	in period 1	.97	1.00	.99	.99	.97	1.00	1.00	.99	.94	.92
	in period 2	.92	.96	.99	.98	.98	1.00	1.00	1.00	1.00	.99
	in period 3	.92	.99	.97	1.00	.99	.99	1.00	.99	1.00	1.00
	in period 4	.98	.99	.95	1.00	.99	.99	1.00	.99	1.00	.99
	in period 5	.99	1.00	.99	1.00		.97	1.00	.97	1.00	.99
	average	.95	.99	.98	1.00	.98	.99	1.00	.99	.99	.98

Table IV.4 - Equilibrium quantity and price, price convergence, and efficiency in the sessions with experienced subjects.

Session		M1811	M1821	M1831	M1841	M2811	M2821	M2841	M2A11
Market clearing quantity (q^*)		10	10	10	10	7	7	7	9
Deviation of traded quantities from q^*	in period 1	-1	1	0	-2	0	0	-1	0
	in period 2	-1	0	-1	-2	0	-1	-1	0
	in period 3	0	0	0	0	0	0	0	0
	in period 4	1	0	-1	0	0	0	0	0
	in period 5	0	0	0	-8	0	0	0	0
	average	-.2	.2	-.4	-.8	0	-.2	-.4	0
Market clearing price (p^*)		829	740	436	628	765	416	559	797
Deviation of the opening price of period 1 from p^*		-7	0	-2	-8	5	-15	-10	-7
Deviation of average prices from p^*	in period 1	-2.44	-2.09	.70	-7.25	6.00	-.71	-2.00	-2.11
	in period 2	.00	-2.30	3.11	-6.13	2.00	1.83	-2.83	-1.67
	in period 3	.20	-3.70	1.80	-2.90	.86	-2.57	-2.00	-.22
	in period 4	.64	-4.30	2.00	-2.10	-.29	-1.71	-1.43	-.11
	in period 5	.90	-2.00	1.60	-1.50	.00	-1.86	-1.43	-.44
	average	-.14	-2.88	1.84	-3.97	1.71	-1.09	-1.94	-.91
Rank correlation coefficient of average prices to period		1.00	-.20	.00	1.00	-.80	-.40	.60	.60
Significance level of RCC		.01	.41	.59	.01	.04	.24	.12	.12
Market clearing price concordance coefficient κ	in period 1	.87	.81	.93	.62	.70	.66	.78	.81
	in period 2	.98	.87	.83	.68	.74	.81	.89	.85
	in period 3	.97	.79	.86	.81	.80	.83	.92	.94
	in period 4	.93	.75	.89	.86	.81	.83	.94	.96
	in period 5	.95	.89	.91	.87	.88	.84	.92	.97
	average	.94	.82	.89	.78	.79	.79	.89	.91
Rank correlation coefficient of κ to period		.00	.00	.20	1.00	1.00	.90	.70	1.00
Significance level of RCC		.59	.59	.41	.01	.01	.04	.10	.01
Efficiency	in period 1	.99	.96	1.00	.83	1.00	1.00	.97	1.00
	in period 2	.95	1.00	.96	.91	1.00	.99	.99	1.00
	in period 3	1.00	.99	1.00	.99	1.00	1.00	1.00	1.00
	in period 4	.99	1.00	.96	.99	1.00	1.00	1.00	1.00
	in period 5	1.00	1.00	1.00	.99	1.00	1.00	1.00	1.00
	average	.99	.99	.98	.94	1.00	1.00	.99	.99

The *market clearing price concordance coefficient* $\kappa(m)$ for the number of observed contracts m is defined as:

$$
\kappa(m) = \begin{cases} 0 & \text{for } m = 0 \text{ or } \sum_{h=1}^{m} \epsilon_h = 0 \\[2em] 1 - \dfrac{\sum_{h=1}^{m} |\, p_h - p* \,|}{\sum_{h=1}^{m} \epsilon_h} & \text{otherwise} \end{cases}
$$

where p_h denotes the h-th contract price and ϵ_h the h-th element of the descending sequence $\{\epsilon_h\}$ of absolute deviations of all non-extra-marginal player valuations from p*.

$\kappa(m)$ measures the proportion of concordance (one minus the proportion of deviation) of contract prices with p*, given the number m of observed contracts. It equals one, if all contract prices were exactly equal to p*, i.e. the prices were totally concordant with p*. It equals zero, if the observed deviations of contract prices were maximal (or if no contracts were observed or if the maximum possible deviation from p* was zero).

The main advantage of the market clearing price concordance coefficient $\kappa(m)$ in comparison to other measures of *nearness* suggested in the literature is that it is normalized on the 0-1-interval. The 0-1-normalization enables comparisons between experimental markets independent of the parameters used. For example, imagine comparing two experimental markets, of which one has more elastic and one less elastic demand and supply schedules. A nominally equal deviation from p* in both markets results in a smaller $\kappa(m)$ of the first market, since the maximum possible deviation from p* in the first market is smaller than in the second market. Due to the fact that, given the same deviation from p*, excess demand and excess supply are larger in the first than in the second market it seems plausible that the latter market is considered to be *nearer* to market clearing price than the former.

Additionally, the 0-1-normalization causes $\kappa(m)$ to be insensitive towards the face value of the market parameters. In contrast to $\kappa(m)$, the "coefficient of convergence" suggested by SMITH (1962), for example, is strongly dependent on the face value of the market clearing price. The "coefficient of convergence" is the ratio of the standard deviation of contract prices from p* to p*. Thus, if a market is simply *shifted* up, i.e. all redemption values, unit costs, and contract prices are increased by a constant, the coefficient decreases. It must be noted, however, that this coefficient (like others suggested by KETCHAM, SMITH, AND WILLIAMS 1984) was primarily used to analyze the convergence dynamics of a single market and not to compare a number of different markets.

The average $\kappa(m)$ of the inexperienced groups' sessions ranges from .64 to .84, with an over all sessions average of .75 (see table IV.3). In the sessions of the experienced groups the average $\kappa(m)$ ranges from .78 to .94, with an over all sessions average of .85 (see table IV.4). Hence, on average, the prices in the experienced markets can be considered as 10% *nearer* to the market clearing price, than those of the inexperienced markets. To test the hypothesis that experience has a positive effect on the average *nearness* of market prices to p*, we examine the eight subject groups that took part in both an inexperienced and in an experienced session. Subtracting the average experienced $\kappa(m)$ of each group from the average inexperienced $\kappa(m)$ of the same group yields seven positive and one negative difference. The binomial test rejects the null hypothesis that positive and negative differences are equally likely at the 5% level (one-sided). Thus, we can conclude that the prices in the experienced markets are significantly closer to p* than in the inexperienced markets.

The observed average $\kappa(m)$ in every case is larger than the median of the distribution of five period average $\kappa(m)$ in both the baseline and the full price range zero-intelligence trader simulations (see the panels in the Appendix VII.C.). (In fact, this is also true for all but two of the single period $\kappa(m)$. The only exceptions are the period 1 $\kappa(m)$ of the sessions M1830 and M2A10. Both of these values, however, are extremely close to the corresponding zero-intelligence trader simulation median values.) Obviously, the binomial test rejects the null hypothesis that an experimentally observed average $\kappa(m)$ is equally likely to be greater or smaller than the corresponding median of the simulation distribution of the five period average $\kappa(m)$ of the zero-intelligence trader simulation. This is significant at the .1% level (one-sided) for the ten observations with inexperienced groups and at the .5% level (one-side) for the eight observations with experienced groups.

The game theoretic analysis (for the game with almost perfect information) predicts $\kappa(m)$ values of .95 (for the markets with 8 traders) and .96 (for the markets with more than 8 traders). Note that these values are smaller than one, since the first trades in each market are predicted to occur at p* - 1 or p* + 1, depending on the market side that opens the market. Only five of the observed $\kappa(m)$ reach these benchmarks: periods 2, 3, and 5 of the session M1811 and periods 4 and 5 of the session M2A11. Both of these sessions are sessions with experienced subjects. This observation, together with the result that prices in experienced sessions are significantly closer to market clearing price, seems to confirm the hypothesis that the outcomes of the market experiment tend to converge to the game theoretic equilibrium outcome of the game with almost perfect information. (Note that outcomes may converge to equilibrium outcome, even though behavior does not converge to equilibrium behavior.) However, to check the hypothesis that a static repetition of the game also tends to shift experimental outcomes in the direction of the game theoretic equilibrium outcome, we will have to take a closer look at the development of $\kappa(m)$ from period to period in each session.

If the prices in a session of alternating double auction market experiment converge to market clearing price, the market clearing price concordance coefficient $\kappa(m)$ of that session should rise from one market period to the next. The $\kappa(m)$ for the market periods of all session are displayed in the tables IV.3 and IV.4. Visual inspection of the data indicates a tendency for κ(m) to rise with market repetition in most sessions. The convergence hypothesis is supported by the KENDALL rank correlation coefficients calculated for each session, relating the κ(m) of each market period to the period number. In all ten inexperienced and in six of eight experienced sessions the rank correlation coefficients are positive, i.e. κ(m) tends to rise from period to period in almost all sessions. Due to the small number (usually five) of observations in each session, however, not all rank correlation coefficients can be assumed to be reliable. But they are significantly different from zero at the 5% level (one-sided) in five of the ten inexperienced and four of the eight experienced groups.

In those cases in which the κ(m) already is at a high level, the rank correlation analysis seems to give no or little clue to a rise of κ(m) from period to period. The across session average κ(m) of these cases (.83) is larger than the across session average κ(m) of the significantly converging sessions (.76). Even though this result cannot be proven true statistically, it gives reason to believe that the closer contract prices of a session are to the market clearing price p*, the weaker the tendency of the prices to converge to p* over market periods. A quite similar observation is reported by SELTEN (1970) in the context of a different market experiment.

This result seems to be in line with the game theoretic analysis of the game with almost perfect information, which was presented in the previous chapter. The analysis had shown that in equilibrium most of the prices are one money unit away from the market clearing price. As mentioned above, in the game theoretic equilibrium κ(m) should take a value no greater than .96, since the prices occurring in the markets are very close, but not at p*. Thus, once the prices in a market are close to p*, κ(m) can be expected to remain high in the following market periods. But, κ(m) cannot be expected to converge to one. Hence, the experimental observation that the convergence of prices subsides in markets in which prices are already close to the market clearing price could be interpreted as a confirmation for the hypothesis that experience and static repetition of the game shift experimental outcomes in the direction of the game theoretic equilibrium outcome of the game with almost perfect information, rather than towards the Walrasian competitive equilibrium outcome. We must admit, however, that this conjecture is quite speculative, given the discrepancy between game theoretic equilibrium behavior and the observed behavior, which we later discuss in detail.

We conclude that the experimental outcomes of the alternating double auction sessions lead to efficient market clearing with a tendency of prices to converge towards the market clearing price over the market periods (static repetitions of the game). The price convergence, however, seems to slow down, when the market prices are close to market clearing price. In view of these results, one can conjecture that an experimental alternating double auction market generally exhibits properties comparable to those of common experimental continuous-time double auctions.

IV.C.4. Direction of Price Convergence

After having established the convergence of prices to the market clearing price, the question arises in which way the general level of prices changes from period to period, getting closer and closer to the market clearing price. Up to this point no evidence has been put forward, showing that the observed price convergence is a systematic process. A closer look at the average period prices of each session in tables IV.3 and IV.4, however, gives the impression, that these develop systematically in the direction of p^*.

Comparing the first and the last period average prices, four sessions with inexperienced and three sessions with experienced subjects seem to converge *from above* (M1840, M1C10, M2810, M2830, M1821, M1811, M2821), while all other sessions seem to converge *from below* the market clearing price p^* (M1810, M1820, M1830, M2820, M2840, M2A10, M1811, M1831, M1841, M2841,

M2A11). The signs of the KENDALL rank correlation coefficients of average period price to period number (see tables IV.3 and IV.4) confirm the conjectured direction of convergence in all but one case. (The rank correlation coefficient for the session M1831 is 0.) The listed rank correlation coefficients indicate more than 50% (either positive or negative) correlation in seven of ten sessions with inexperienced subjects (significant at the 5% level for five of ten) and in five of eight sessions with experienced subjects (significant at the 5% level for three of eight).

From this evidence, one can conclude that in most sessions there is a distinct direction for the convergence of period price levels to the market clearing price p*. Neither the Walrasian competitive equilibrium analysis nor the game theoretic analyses offer a prediction for the direction of price convergence to p*. Obviously, the zero-intelligence trader market simulations do not make a prediction on the convergence of prices over periods either, since an across periods effect is inherently contrary to the stochastic character of these simulations.

Both the game theoretic analysis of the game of almost perfect information and the zero-intelligence trader simulations, however, do provide benchmarks for the location of the average prices relative to the market clearing price p*. The game theoretic analysis predicts the location of average prices to be dependent on the opening market side of each market period. If the buyers open the market in a period, the game theoretic equilibrium expects most prices to occur below p*. If the sellers open the market in a period, the game theoretic equilibrium expects most prices to occur above p*. The information necessary to check this prediction is contained in the tables IV.1, IV.2, IV.3, and IV.4. To facilitate the inspection, the following table IV.5 displays the frequencies of periods in which buyers or sellers opened the market (lines of the table) and in which the average period price was below, equal, or above the market clearing price (columns of the table).

Obviously, the experimental data is contrary to the game theoretic prediction: The average price is below p* more frequently (above p* less frequently) when sellers open the market than when buyers open the market. In fact, the bias is so strongly in the opposite direction, that the application of a chi-squared test would lead to a rejection of the null hypothesis well below the 1% level. The test, however, cannot be correctly applied to this data, since the market periods of each subject group - both the periods of the group in the inexperienced and in the experienced session - are not independent observations. We, therefore, have added entries to the table IV.5. containing the distribution of only the first periods of the inexperienced sessions, i.e. ten independent observations. Applying Fisher's exact test to this data, the null hypothesis that the opening market side parameter has no effect on the average price of the first market period can be rejected at the 10% level (one-sided)[4]. This result, admittedly, is not strongly significant. But, taken together with the evidence on all market periods of all sessions, the conclusion that average period prices are to the disadvantage of the opening market side seems well established.

Table IV.5 - Distribution of market periods by market opening side and by the sign of average price minus p*.

Absolute frequency of periods (frequency relative to the sum of line)	average price < p*	average price ≐ p*	average price > p*	Sum of line
Buyers opened the market period	26 (.54)	2 (.04)	20 (.42)	48 (1.00)
Sellers opened the market period	36 (.88)	0 (.00)	5 (.12)	41 (1.00)
Sum of column	62 (.70)	2 (.02)	25 (.28)	89 (1.00)
... only the first period of the inexperienced sessions				
Buyers opened the market period	2 (.33)	0 (.00)	4 (.67)	6 (1.00)
Sellers opened the market period	4 (1.00)	0 (.00)	0 (.00)	4 (1.00)
Sum of column	6 (.60)	0 (.00)	4 (.40)	10 (1.00)

The zero-intelligence trader simulations provide a different benchmark for the location of average prices relative to the market clearing price p*. In the full price range simulations all average prices (and the median of the distribution of five period average prices) are above p*. (See the panels VII.C.1.1 - VII.C.1.18 for the

---

4) Given the reported data, the test rejects the null hypothesis at any significance level greater than .072 (one-sided).

simulation results. See the subsection IV.C.1 for the reason why prices are biased upwards.) The simulation results, evidently, contradict the experimental data. The results of the baseline simulations are different. (See the panels VII.C.0.1a - VII.C.0.2b for the simulation results.) The baseline zero-intelligence trader simulations suggest that the average prices in the market type M1 sessions are below p*, whereas the average prices in the market type M2 sessions are above p*. Table IV.6. displays the distribution of all periods (of only the first period of the inexperienced sessions) by the market side and the location of the average price relative to p*. No effect of the market type on the average period price is observed.

Table IV.6 - Distribution of market periods by market type and by the sign of average price minus p*.

Absolute frequency of periods (frequency relative to the sum of line)	average price < p*	average price = p*	average price > p*	Sum of line
Market type M1	28 (.64)	1 (.02)	15 (.34)	44 (1.00)
Market type M2	34 (.75)	1 (.02)	10 (.22)	45 (1.00)
Sum of column	62 (.70)	2 (.02)	25 (.28)	89 (1.00)
... only the first period of the inexperienced sessions				
Market type M1	3 (.60)	0 (.00)	2 (.40)	5 (1.00)
Market type M2	3 (.60)	0 (.00)	2 (.40)	5 (1.00)
Sum of column	6 (.60)	0 (.00)	4 (.40)	10 (1.00)

The comparison of the experimental data to the two benchmarks clearly reveals that the price formation in the alternating double auction market experiment can neither be explained with the game theoretic approach nor with the zero-intelligence traders approach. The analysis, however, did reveal a strong tendency - running contrary to the game theoretic prediction - for average prices to be to the disadvantage of the market opening side. Evidently, some mechanism is at work in the markets, which is compatible neither with full rationality, as required by the game theoretic approach, nor with the complete lack of rationality, as postulated by the zero-intelligence trader approach.

IV.C.5. Price Path Repetition

One possible influence for the overall direction of convergence in a session could be the chronological structure of contract prices within a market period, the price paths. The observed price paths in the examined alternating double auction markets are displayed in the figures VII.B.1 to VII.B.18 of the Appendix VII.B. A visual inspection of these figures discloses a striking structural regularity of the price paths of each session, the *price path repetition*. Price path repetition refers to the phenomenon that, the price paths of most market periods of an experimental session have an identical basic structure. It is an astonishing phenomenon, because - ad hoc - it would seem more plausible to believe that price paths would mostly have erratic random walk characteristics. Furthermore, even if a prominent structure should appear in one period, there is no obvious intuition, why this structure should be repeated in the next period. Finally, it is peculiar that the price path repetition can be observed, even though it often leads to situations in which an extreme current period first contract price follows a less extreme previous period last contract price. The immediate *learning* from the last contract price seems to stop as soon as a period ends. The market side that had made disadvantageous contracts in the first market period (buyers buying above, sellers selling below p*), shows a tendency to *make the same mistakes again and again* in the following periods. Obviously, these *mistakes* are reduced as the session proceeds, since the markets move closer to market clearing price from one period to the next. Nevertheless, the similarity of each period's price path to that of previous periods persists in many markets.

To establish the result the price path structures must first be classified. The visual inspection of the data exposes four typical types of price paths, *falling*, *rising*, *hanging*, and *standing* paths. A typical *falling* price path starts above the market clearing price p* and descends as the market period proceeds. A *rising* price path starts below p* and ascends. A *hanging* price path begins and ends at almost the same level, but distends downwards towards lower prices in the middle. Finally, a *standing* price path begins and ends at almost the same level, but distends upwards towards higher prices in the middle.

For the classification, the opening price p_{ot} and closing price p_{ct} of each market period t are compared to the average price of that period, $\hat{p}_t$. The structure of a price path is defined as:

> *falling*   if $p_{ot} > \hat{p}_t \geq p_{ct}$ or $p_{ot} \geq \hat{p}_t > p_{ct}$
>
> *rising*   if $p_{ot} < \hat{p}_t \leq p_{ct}$ or $p_{ot} \leq \hat{p}_t < p_{ct}$
>
> *hanging*   if $p_{ot} > \hat{p}_t$ and $\hat{p}_t < p_{ct}$
>
> *standing* if $p_{ot} < \hat{p}_t$ and $\hat{p}_t > p_{ct}$
>
> *unknown*   if $p_{ot} = \hat{p}_t = p_{ct}$

The *unknown* structure (the case that both p_{ot} equals $\hat{p}_t$ and p_{ct} equals $\hat{p}_t$) does not occur in the experimental data. It, however, does occur with a low probability in some of the simulations. Since this price path structure seems to be of little importance, it will not be explicitly considered any further in the following.

Tables IV.7 and IV.8 show the price path structure types of all periods of all sessions with inexperienced and experienced subjects, respectively. In the tables the letter f denotes a *falling*, r a *rising*, h a *hanging*, and s a *standing* price path.

Four of the ten sessions with inexperienced subjects show exactly the same price path structure in every period. Four of these ten sessions have a deviating price path structure in a single period only. Finally, the last two of these ten sessions, that display only three periods of one structure, have a single different structure in the remaining two periods. Thus, no inexperienced session exhibits more than two of the four possible price path structures.

In the eight sessions with experienced subjects, the results are less striking, since two sessions contain each of the four possible structures at least in one of the periods. Yet, price path repetition can be observed in the other six sessions. Two sessions contain a single structure in all periods and four in all but one period.

Table IV.7 - Price path structures in the sessions with inexperienced subjects.

Period	Session									
	M1810	M1820	M1830	M1840	M1C10	M2810	M2820	M2830	M2840	M2A10
1	r	h	r	s	h	f	r	f	r	r
2	r	h	r	h	f	f	r	h	r	r
3	r	h	s	h	f	f	r	h	r	r
4	r	h	s	h	f	s	r	f	r	r
5	r	h	r	h	-	f	r	h	r	h

Table IV.8 - Price path structures in the sessions with experienced subjects.

Period	Session							
	M1811	M1821	M1831	M1841	M2811	M2821	M2841	M2A11
1	r	h	s	r	s	r	s	r
2	f	h	f	r	f	s	f	r
3	s	h	s	r	f	r	h	r
4	h	r	s	r	f	r	r	r
5	h	h	s	r	f	r	s	r

Calculating the statistical significance of this evidence is not trivial. Since the markets are not symmetric, the probabilities with which the structures evolve randomly are not necessarily equal. A further complication is that the structures evolve as a result of multiple random draws of prices (or of bids and asks). Due to the rules of the market, the probability with which each price can be observed at a certain time, depends on the contracts made up to that time. It seems that the only practical method of obtaining the necessary probabilities is a Monte-Carlo simulation with zero-intelligence traders.

The results of the zero-intelligence trader market simulations are reported in the Appendix VII.C. Here we will only discuss the results of the baseline simulations (panels VII.C.0.0 - VII.C.0.2b), since the full price range simulations show extremely biased price path results, due to the reasons explained above, in the subsection VI.C.1. of this chapter.

Even in the baseline simulations the price path data are biased, depending on the market type. In market type M1 simulations (panels VII.C.0.1a - VII.C.0.1b) rising (40%) and hanging (26% or 29%) price paths occur more frequently than falling (22% or 19%) or standing (12% or 11%) price paths. This is due to the rent distribution that favors sellers in these markets. In market type M2 simulations the situation is (almost exactly) reversed. Panels VII.C.0.2a and VII.C.0.2b show that rising (25% or 23%) and hanging (12% or 13%) price paths occur less frequently than falling (40% or 39%) or standing (23% or 25%) price paths. This, again, is due to the rent distribution. In these markets buyers have the greater share of the market rent. The panel VII.C.0.0 displays the results for a completely symmetric market, market type M0, that was not used in the experiment. In this *ideal* market the frequencies of price path occurrence are symmetric, with rising and falling price paths each occurring in 32% of the cases and hanging and standing price paths each occurring in 18% of the cases. (Note that the difference in occurrence frequencies is due to the definitions of the price path types, which include equality cases in the rising and falling categories, but not so in the hanging and standing categories.)

Since the occurrence frequencies are diverse, a table containing information on the five period frequency distribution of the price path types is added at the end of each of the mentioned simulation result panels in the Appendix VII.C. In these tables, the relative (and cumulative) frequency distribution of simulation rounds in which n (or more) periods have price paths of the same type are displayed, where $n = 0...5$. This table is to be read in the following manner: The entry in the column *falling - f(x)* of the line $n = 4$ corresponds to the relative frequency with which simulation rounds had falling price paths in four of the five market periods. The entry in the column *falling - F(x)* of the same line corresponds to the cumulative frequency with which simulation rounds had falling price paths in four or more of the five periods. (Obviously, the line $n = 0$ must contain the value 1.0 in all *F(x)* columns, since the cumulative frequency with which simulation rounds had price paths of one type in zero or more of the five market periods must be one.)

To prove the significance of the extreme experimental results we use a simple and conservative scheme: First, note that the simulative frequency of observing four or more identical price path structures of any type is smaller or equal to .12 in all baseline simulations (panels VII.C.0.1a - VII.C.0.2b). Conservatively, we formulate the null hypothesis that the probability $P(n \geq 4)$ of observing four or more periods with an identical (but arbitrary) price path structure is .25 (i.e. we assume that $n \geq 4$ represents the bound of the lower quartile of the distribution). Next we run the binomial test with $P(n \geq 4) = .25$ and 8 of 10 hits, since in eight of the ten inexperienced sessions four or more periods have the same price path structure. The test rejects the null hypothesis at .1% level (one-sided). Thus, we can conclude that price path repetition is observed significantly more often than by chance.

The experimental results not only differ from the zero-intelligence simulation results in that price path repetition is observed significantly more often, but also in the fact that the market type (i.e. the distribution of rent) does not affect the frequency with which specific price paths are observed in the manner predicted by the simulations. All in all, more periods with rising price paths are observed than with any other structure. But, contrary to the results of the zero-intelligence simulation, relatively more periods with rising price path are observed in the market type M2 than in the market type M1. The difference, however, is not significant. It seems clear that the market type is not the parameter determining which type of price path structure evolves in a session.

The inspection of the connection between the direction of price convergence across the periods and the price path structure within each period of a session indicates a correlation between a convergence *from above* and *falling* price paths and a convergence *from below* and *rising* price paths. All of the sessions with a majority of falling price paths (M1C10, M2810, and M2811) converge from above to p*. All but one of the sessions with a majority of rising price paths (M1810, M1830, M2820, M2840, M2A10, M1841, M2821, and M2A11) converge from below to p*.

This is a strong result, but it has a quite straightforward explanation. Obviously, price paths starting well above p* will be more or less *forced* to drop to p* towards the end of the market period. This is simply due to the fact, that remaining trading possibilities narrow down more and more towards the marginal values, as the most profitable units are traded away. At the same time, because price paths start above p* and decline to p*, the average price in most periods with a falling price path will lie above p*. Since it was shown, that typically most market periods of a session have the same price path structure, it follows, that most average period prices in such a session will be greater than p*. This will lead to more possibilities for lowering than lifting the price level. Given the prices converge to p* from period to period, the convergence will be more likely to come from above p*. For analogous reasons, the opposite is true for sessions in which period price paths are rising. Both conjectures can be easily confirmed by comparing the average period prices of the sessions with inexperienced subjects in table IV.3 with the price path structures in table IV.7. The average prices of falling market periods are usually above, those of rising market periods below the market clearing price. Only three averages have a deviating sign.

Similarly one could argue, that hanging price paths - since the period's last prices normally get closer to p* - should correspond to average prices below p*, i.e. the price path *hangs* under the market clearing price. (This phenomenon is confirmed by the zero-intelligence simulations of the market type M1, where average prices are smaller than p* and hanging price paths are relatively frequent. See the panels VII.C.0.1a and VII.C.0.1b.) Again, for analogous reasons, standing price paths should correspond to average prices above p*, i.e. the price path *stands* over the market clearing price. (This phenomenon is confirmed by the zero-intelligence simulations of the market type M2, where average prices are greater than p* and standing price paths are relatively frequent. See the panels VII.C.0.2a and VII.C.0.2b.) Both correlations can be seen to be true in all but two of the market periods of the sessions with inexperienced subjects. But as the cases of the sessions

M1840, M2830, and M1831 show, the conclusion that *hanging* price paths necessarily coincide with convergence of prices *from below* and that *standing* price paths necessarily coincide with convergence *from above* is not always valid. The explanations given above fail here, on a number of different grounds.

The sessions M1840 and M2830 do not *hang* completely under p*, which means that the price paths start above p*, then move below p*, and finally *overshoot* p* on their way back up. This is possible, because the some of the trades involving buyers with high redemption values occur late in the market period. This leaves the buyers making the final contracts of a period enough *room* for contracting at prices well above p*, since their redemption values are greater than p*. Thus, price convergence in such sessions does not only involve an increase of the prices below p* (relative to the previous period's price), but also a decrease of the opening and closing prices, which are above p*. Hence, the price level closes in from both sides towards p* as the session moves from period to period. In such a case, it is quite possible that the greater part of the price level adjustment originates from the falling of prices above p*. In this way, a session with hanging price path structures may exhibit an across period price convergence from above.

The session M1831 has prices converging from below, even though the market periods exhibit standing price paths. The markets in this session are so close to p* that practically no further convergence to p* can be measured. The rank correlation coefficient comparing the periods' average prices to the corresponding period number is 0 and the one measuring the correlation of the periods' $\kappa(m)$ to the corresponding period number is .2. Since there is almost no convergence and since prices show no systematic decline, there is no reason to believe that a correlation between the price path structure and the direction of convergence exists. Generally, it must be noted, that those sessions with experienced subjects, that get very close to p*, tend to - more or less arbitrarily - fluctuate around the market clearing price, leaving little evidence for the price path structure and the convergence direction.

IV.C.6. Distribution of Contract Proposals

Since price paths are shown to be repeated persistently in most sessions and the average prices compared to the market clearing price (thus also the average payoffs compared to competitive equilibrium payoffs) are typically in favor of one of the market sides, the question arises, to which extent the disadvantaged market side is actively proposing the disadvantageous contracts. To be able to calculate the corresponding frequencies, contract proposal and acceptance in the alternating double auction markets must first be defined, since the market rules do not explicitly distinguish two different types of actions. The proposing actions can be separated from the accepting actions simply by checking, whether or not a submitted offer matches the other side's offer standing on the market. If it does not, submitting the offer represents a proposing, if it does, it represents an accepting action. Whenever a contract is made between two traders, one of them must have taken a proposing and one of them an accepting action. The term *contract proposer* will be used for those traders, whose proposing action led to a contract. The term *contract accepter* will be used for those traders, whose accepting action led to a contract. It must be noted that neither every proposing action is a contract proposal, nor every accepting action leads to accepting a contract. The former is true, because a proposal might not be the best offer of the corresponding market side and thus, not be placed on the market. If a proposal is placed on the market, it may be replaced in the next offer cycle by a better offer. Finally, even if it remains standing, it might not be the only bid or ask on the market when traders from the other side submit accepting offers. Thus, - if more offers are standing than accepting offers made - it might not be one of the randomly chosen ones that lead to contracts. Similarly, an accepting offer that is submitted simultaneously with a number of other accepting offers might not belong to those randomly chosen for contracts.

The contract proposal and acceptance frequencies are calculated and presented in the tables IV.9 and IV.10. The inspection of the relative frequencies with which buyers and with which sellers are contract proposers in each session reveals: In four of the seven sessions converging *from above* (M1840, M1C10, M2810, M2830, M1821, M2811, M2821) the buyers more frequently submit contract proposals, while in nine of the eleven sessions converging *from below* (M1810, M1820, M1830, M2820, M2840, M2A10, M1811, M1831, M1841, M2841, M2A11) the sellers more frequently submit contract proposals.

Table IV.9 - Relative contract proposal frequencies in the sessions with inexperienced subjects.

Relative proposal frequency	Session									
	M1810	M1820	M1830	M1840	M1C10	M2810	M2820	M2830	M2840	M2A10
Buyers	.08↓	.04↓	.55	.36↓	.69↑	.63↑	.03↓	.62↑	.30↓	.09↓
Sellers	.92↑	.96↑	.45	.64↑	.31↓	.37↓	.97↑	.38↓	.70↑	.91↑

↓ (↑) indicates that the value is smaller than the lower quartile bound (greater than the upper quartile bound) of the parameter's simulation distribution in the corresponding zero-intelligence trader simulation.

Table IV.10 - Relative contract proposal frequencies in the sessions with experienced subjects.

Relative proposal frequency	Session							
	M1811	M1821	M1831	M1841	M2811	M2821	M2841	M2A11
Buyers	.24↓	.24↓	.45	.24↓	.80↑	.24↓	.64↑	.13↓
Sellers	.76↑	.76↑	.55	.76↑	.20↓	.76↑	.36↓	.87↑

↓ (↑) indicates that the value is smaller than the lower quartile bound (greater than the upper quartile bound) of the parameter's simulation distribution in the corresponding zero-intelligence trader simulation.

All in all, eight of ten sessions with inexperienced subjects exhibit the property that the average price - compared to the market clearing price - is to the disadvantage of the market side making most contract proposals. The binomial test rejects the null hypothesis that the average price is equally likely to the advantage or disadvantage of the market side making most contract proposals at the 10% level (one-sided)[5]. Admittedly, the test does not establish the result very strongly. Additionally, the result cannot be established for the experienced groups, where only five of eight sessions exhibit the property. Nevertheless, it is a striking result that, in most sessions, the traders of the disadvantaged market side are proposing most of the disadvantageous contracts themselves. Evidently, the market side submitting most of the contract proposals, is not the *strong* market side, i.e the offering market side earns less of the competitive equilibrium rent than the accepting market side.

In the tables IV.9 and IV.10 the signs ↓ and ↑ have been added to those entries that are smaller than the lower quartile bound or greater than the upper quartile bound, correspondingly, of the parameter's simulation distribution in the zero-intelligence trader simulations. Note that the comparison is made to the baseline simulations of the corresponding market type with the corresponding number of traders. (See the panels VII.C.0.1a - VII.C.0.2b in the Appendix VII.C.) Since nine of ten inexperienced and seven of eight experienced sessions show extreme values for the frequencies of contract proposal, we conjecture: The experimentally observed distribution of contract proposals over market sides is more extreme than can be expected by chance. To test this hypothesis, we run the binomial test with the null hypothesis that observing proposal frequencies in the two extreme quartiles (the lower and the upper quartile) is equally likely to observing proposal frequencies in the two middle quartiles of the zero-intelligence trader simulation distributions. The test rejects the null hypothesis at the 2% level (one-sided) for the inexperienced and at the 5% level (one-sided) for the experienced sessions.

Thus, in the experimental market sessions, contract proposals are made by one market side only significantly more often than can be expected by chance. The activity on these markets is significantly biased, with one market side actively proposing

---

5) Given the reported data, the test rejects the null hypothesis at any significance level greater than .055 (one-sided).

most contracts, while the other market side mostly restricts its activity to contract acceptance. Moreover, there is evidence that the market side proposing most of the contracts in a session often proposes contracts that are to its own disadvantage.

IV.C.7. The Proposer's Curse Phenomenon

In the last subsection, we established that the sessions in the alternating double auction market experiment have a strong tendency to be biased concerning the frequency of contract proposals and average prices. A large majority of the contract proposals are made by only one market side, but the market side making most proposals is earning less than the accepting market side, when compared to competitive equilibrium payoffs. Two different explanations for this observation are conceivable: Either the contract proposing market side is disadvantaged, because the market is transformed into a de facto one-sided offer market. Or the contract proposing market side is disadvantaged, because contract proposers are generally disadvantaged, no matter on which market side they are. In the following, we first elaborate on the first explanation, then on the second. The analysis will leave little doubt that the second explanation fits the data better than the first.

If behavior in an experimental alternating double auction market has the tendency to convert the market into a market in which one market side (almost) exclusively submits offers, while the other market side only takes acceptance actions, then the results of the emerged market must be comparable to either posted-offer markets or one-sided offer markets. Posted-offer markets were studied by F. WILLIAMS (1973), PLOTT and SMITH (1978), and KETCHAM, SMITH, and A. WILLIAMS (1984). In such markets the traders of the offering market side each *post* a single price for all the units they are willing to trade. Then the traders of the other market side can choose to accept any number of the posted offers. This mechanism often results in tacit collusion of the traders who post prices, forcing the traders of other market side to accept contracts that are disadvantageous to them. Thus, the contract proposers in the posted-offer markets are generally better off than in competitive equilibrium.

In contrast, the disadvantaged market side is the contract proposing market side in one-sided-offer markets, as studied by SMITH (1964) and PLOTT and SMITH (1978). These markets are different from the posted-offer markets only in one aspect. Here traders of the offering market side do not post a single price for all their potential trades, but instead, can continuously submit offers. Specifically, this means, that they can change their offers at any time reacting to the offers of others on their own market side. PLOTT and SMITH (1978) argue that the posted-offer markets, on the one hand, tend to result in better prices for the offering side, because these markets resemble sealed-bid discriminatory auctions in their informational setting. The one-sided-offer markets, they argue, on the other hand, result in comparably more disadvantageous prices for the offering market side, since these markets are closer to competitive sealed-bid auctions. Although the authors do not use the phrase, in essence, the argument seems to be based on a notion of *commitment*. In the posted-offer markets, the procedure of posting prices causes the offering market side to credibly commit to a price, forcing the other market side into a *take-it-or-leave-it* position. In the one-sided-offer markets this commitment is lost, since the traders of the accepting market side know, that the offers of the offering side can be adjusted.

If there is a tendency for the alternating double auction markets to take on the character of a market in which only one market side is allowed to propose contracts, then the results should be expected to resemble those of one-sided-offer markets much more than those of posted-offer markets. This conjecture seems plausible, because - analogous to the one-sided-offer markets - the offering traders in the alternating double auction markets cannot credibly commit to their offers. In fact, the experimental observations hint in this direction: The contract proposers in both the one-sided-offer markets and in the alternating double auction markets seem to be competing exaggeratedly.

However, a closer look at the offering behavior of the subjects in the experiment uncovers that the alternating double auction markets do not transform into de facto one-sided-offer markets: Even though most contracts in a session are proposed by traders of only one market side, traders on both market sides are active in submitting offers. Tables IV.11 and IV.12 display the average number of submitted offers per contract for each market side. Comparing these tables to the tables IV.9 and IV.10, that contain the contract proposal frequencies, the reader can easily confirm that in many sessions, in which a majority of contracts are proposed by one market side, a greater number of offers is submitted by the other market side. This is true for seven of ten inexperienced sessions (M1810, M1820, M1830, M1840, M1C10, M2830, M2840) and three of eight experienced sessions (M1811, M1821, M1841). Thus, in many sessions one market side proposes more contracts than the other, but the other market side submits more offers than the first. Given this evidence, it seems unreasonable to conjecture that the behavior observed in the alternating double auction markets resembles the behavior observed in the one-sided-offer markets, in which only one of the market sides submits offers.

Table IV.11 - Average number of submitted offers per contract in the sessions with inexperienced subjects.

	Average number of submitted offers per contract ...								
by	Session								
	M1810	M1820	M1830	M1840	M1C10	M2810	M2820	M2830	M2840	M2A10
buyers	9.98	5.06	5.97	5.94	6.29	5.16	4.89	9.34	5.71	6.10
sellers	5.73	4.60	6.03	4.30	6.50	4.74	5.03	10.81	4.47	6.12

Table IV.12 - Average number of submitted offers per contract in the sessions with experienced subjects.

	Average number of submitted offers per contract ...							
by	Session							
	M1811	M1821	M1831	M1841	M2811	M2821	M2841	M2A11
buyers	4.34	4.09	4.01	4.23	5.74	3.99	6.10	5.00
sellers	3.69	3.82	4.31	4.11	4.14	5.82	5.06	5.67

The second approach - as mentioned above - for explaining the relative payoff disadvantage of the market side making most of the contract proposals in a session is that contract proposers are generally disadvantaged, no matter on which market side they are. We introduce the term *proposer's curse* to refer to a case in which a trader proposes a contract that is to his disadvantage when compared to the market clearing price p* (i.e. a buyer proposes a contract with a contract price greater than p* or a seller proposes a contract with a contract price smaller than p*).

In one-sided-offer markets proposer's curses - quite obviously - can only exist on one side of the market, the proposing side. Thus, observing proposer's curses in such markets does not represent feasible evidence for its general existence with respect to individual behavior. However, should proposer's curses be present and frequent on both market sides in a symmetric institution, as the alternating double auction market, we can conjecture that the phenomenon is a regularity of all individual traders' behavior, rather than the effect of the asymmetry observed or generated in certain trading institutions. The existence of a behavioral regularity would, evidently, explain the experimental results of the one-sided-offer markets as well as the experimental results of our alternating double auction markets. Furthermore, given the phenomenon is a behavioral regularity on individual level, then it should also be observable in other symmetrical double auction markets, in which offers can be revised at any time at a low cost. Even though the phenomenon has not been reported in any of the inspected papers concerning double auction market experiments, it is not completely clear, whether the underlying data has been analyzed with respect to this effect, since contradicting results have not been reported either.

To check for the existence of the proposer's curse phenomenon, we examine the frequency of proposer's curses relative to all contract proposals. Table IV.13 shows a clear picture for the sessions with inexperienced subjects: In nine of the ten sessions more than 50% of the contract proposals are proposers curses. This confirms that on average contract proposers in almost all sessions with inexperienced subjects are proposing more disadvantageous than advantageous contracts. Moreover, the proposer's curse frequencies separated by market side, that are also displayed in the table, show that traders on both market sides tend to propose disadvantageous contracts. (Note that the buyers of only one session, M2820, do not exhibit proposer's curses. In this session, however, the buyers only proposed 3% of all contracts. This extremely small sample size is most probably the reason for the lack of proposer's curses. Additionally, we will show later, that the contracts proposed by buyers in that session, nevertheless, were to the disadvantage of the buyers, when compared to the contracts proposed by sellers in the same session.)

Table IV.13 - Relative frequency of proposer's curses in the sessions with inexperienced subjects.

contracts proposed by	Frequency of proposer's curses relative to contract proposals ...									
	Session									
	M1810	M1820	M1830	M1840	M1C10	M2810	M2820	M2830	M2840	M2A10
buyers	.33	.50	.23	.44	.38	.83	.00	.41	.07	.83
sellers	.96	.79	.66	.71	.58	.45	.62	.85	.89	.75
all traders	.90	.78	.41	.58	.52	.67	.60	.55	.65	.76

For the most important result, however, the table IV.13 must be compared to the table IV.9., in which the contract proposal frequencies are displayed. The comparison discloses that in three sessions (M1830, M1C10, M2830) the traders on the market side making a majority of the contract proposals exhibit relatively less proposer's curses than the traders on the mainly accepting market side. Thus, it is clear that the proposer's curse is not a *proposing market side's curse*, i.e. is not a phenomenon that is solely connected to the asymmetry in contract proposal frequency. On the contrary, the contract proposing market side appears to be disadvantaged as a whole, simply because most contracts are proposed by traders of that market side and because traders generally tend to propose disadvantageous contracts.

The result is sharpened, when the deviation of average contract prices from the market clearing price p* for the buyer proposed and the seller proposed contracts are examined separately. (Note that, in a double auction market, this value is equal to the average profit of the proposers.) The observed values for the sessions with inexperienced subjects are reported in the table IV.15. In every session the average contract price of the contracts proposed by buyers is greater than the average price of the contracts proposed by sellers. The binomial test rejects the null hypothesis that the average contract price of the contracts proposed by buyers is equally likely to be greater or to be smaller than the average contract price of contracts proposed by sellers at the .1% level (one-sided). Thus, we can conclude that in the sessions with inexperienced subjects the average payoff of traders from contracts proposed by themselves is smaller than the average payoff from contracts proposed by traders from the other market side. It should once again be noted that this is true for both market sides in all these sessions, independent of the frequency with which either market side proposes contracts. Even in those sessions, in which the traders on the mainly accepting market side receive more than their competitive equilibrium payoffs from self proposed contracts, they - on average - earn less from these contracts than from the contracts proposed by the other market side.

Finally, more evidence for the general occurrence of proposer's curses on individual level in the sessions with inexperienced subjects is found in the panels VII.B.1 to VII.B.10 of the Appendix VII.B. From the 75 subjects, who made contract proposals in the ten sessions with inexperienced subjects, only 11 never exhibited a proposer's curse and only 21 exhibited proposer's curses in less than half of their contract proposals. Thus, a vast majority of inexperienced subjects (85%) exhibited at least one case of a proposer's curse and a large majority (72%) exhibited proposer's curses in more than half of their contract proposals. Checking the panels VII.B.11 to VII.B.18 discloses that the result is quite different for the experienced subjects. In the sessions with experienced subjects, only 41 (66%) of the 62 subjects, who proposed contracts, exhibited at least one proposer's curse and only 24 (39%) of them exhibited proposer's curses in more than half of their contract proposals. This indicates that experience in an earlier session reduces the tendency of subjects to exhibit proposer's curses. Below, we prove that this effect of experience on the occurrence of proposer's curses is significant. Before doing so, however, it should be pointed out that no correlation between the rank of a subject's player valuations and the frequency of proposer's curse occurrences can be seen in the panels of Appendix VII.B. In other words, the conjecture that the buyers with the greatest redemption values and the sellers with the smallest unit costs are more likely to exhibit a high frequency of proposer's curses is generally not valid; neither in the sessions with inexperienced, nor in those with experienced subjects.

From the evidence on individual level mentioned further up, it appears that the frequency of proposer's curses decreases with experience. The data in table IV.14 confirms this impression. There the proposer's curse frequencies in the sessions with experienced subjects are displayed. Only two of the eight experienced sessions exhibit proposer's curse frequencies of over 50% (in contrast to the nine of ten inexperienced sessions). The null hypothesis that the frequency of proposer's curses in the inexperienced session is equally likely to be smaller or to be greater than in the experienced session of the same subject group is rejected by the binomial test at the 5% level (one-side). Thus, we conclude that subjects' previous experience in an experimental market significantly reduces the frequency of proposer's curses.

Table IV.14 - Relative frequency of proposer's curses in the sessions with experienced subjects.

contracts proposed by	Frequency of proposer's curses relative to contract proposals ...							
	Session							
	M1811	M1821	M1831	M1841	M2811	M2821	M2841	M2A11
buyers	.26	.00	.65	.00	.47	.50	.12	.00
sellers	.24	.76	.03	.80	.72	.44	.58	.37
all traders	.38	.61	.30	.62	.49	.46	.28	.33

The result that experience in a previous session reduces the tendency of subjects to make disadvantageous contract proposals is also confirmed by the data presented in the table IV.16. The average contract price of contracts proposed by buyers is greater than the average contract price of contracts proposed by sellers in only two of the eight experienced sessions. Thus, in six of eight cases, traders of a market side are - on average - earning more from contracts proposed by themselves than from contracts proposed by traders of the other market side.

Table IV.15 - Deviation of average contract prices from p* in the sessions with inexperienced subjects.

contracts proposed by	Deviation of average contract prices from p* of ...									
	Session									
	M1810	M1820	M1830	M1840	M1C10	M2810	M2820	M2830	M2840	M2A10
buyers	-0.67	-0.50	-1.16	0.01	1.75	4.86	0.00	3.23	-6.60	3.17
sellers	-6.73	-4.38	-2.79	-1.17	-1.35	3.98	-4.41	-3.43	-7.24	-4.37

Table IV.16 - Deviation of average contract prices from p* in the sessions with experienced subjects.

contracts proposed by	Deviation of average contract prices from p* of ...							
	Session							
	M1811	M1821	M1831	M1841	M2811	M2821	M2841	M2A11
buyers	-0.20	-2.84	1.16	-3.67	3.27	1.20	-2.54	-0.75
sellers	-0.02	-2.52	2.31	-3.15	-2.44	-1.50	-1.20	-0.83

Note, however, that the results of the sessions with experienced subjects confirm the finding that the proposer's curse is a individual behavior phenomenon that is independent of the market side the trader is on. Comparing table IV.14 to table IV.10 reveals that in five of the eight sessions with experienced subjects (M1811, M1831, M2811, M2821, M2841) the traders on the market side making a majority of the contract proposals exhibit relatively less proposer's curses, than the subjects on the other market side. This result once again confirms that the reason why the market side proposing most of the contracts is disadvantaged in comparison to the market side mostly accepting contracts, is the prevalence of the proposer's curse phenomenon in individual behavior, rather than the mere existence of a bias in contract proposal frequency.

Not the mere occurrence, but observed high frequency of proposer's curses - especially in the sessions with inexperienced subjects - seems to be a puzzle on first sight. Given the informational setting of the double auction market institution, it seems quite natural to expect subjects to propose contracts that are to their own disadvantage once in a while. But, why do subjects so frequently tend to propose contracts that are to their own disadvantage? The main clue to the answer seems lie in an *adverse selection* property of the market, which bases on the feature that disadvantageous offer are more likely to be accepted by the other market side than advantageous offers. For this reason, the proposer's curse phenomenon seems very closely related to the *winner's curse* phenomenon observed in common value auctions. The winner's curse was first reported by CAPEN, CLAPP, and CAMPBELL (1971) and later confirmed in a number of experimental examinations. (For examples, see BAZERMAN and SAMUELSON 1983, KAGEL and LEVIN 1986, and SELTEN, ABBINK and COX 1997. For an overview see KAGEL 1995).

The winner's curse is found to be due to the subjects' tendency to condition their bids on the unconditional probability distribution of the signals for the value of the auctioned item, instead of on the probability distribution of the signals conditioned on the case (or probability) of winning. Hence, a winning bidder is *regularly surprised* by winning the auction more frequently when his signal had over-estimated the true value of the auctioned item than when his signal had under-estimated the true value. If all bidders employ the unconditional probability distribution of the signal for the formation of their bids, the bidder with the highest signal, i.e. the bidder whose signal over-estimates the true value most, turns out to be the winner of the auction. Thus, such behavior leads to an *adverse selection* of the winner of the auction, who usually finds that he has to pay an exaggerated price, because his own signal was the most optimistic signal drawn. Since the auction's winner only finds out that he had the highest signal, after he wins the auction, the *winner* seems to be *cursed*.

We believe that there is an analogy between the winner's and the proposer's curse. The market clearing price p* in a double auction market is the analog of the true value of the auction item in the common value auctions. In a double auction market, a trader may have some (and/or gather some) signals about the market clearing price p*. In the course of the market, each trader makes a guess about the value of p*. The offer submitted by a trader in a double auction market reveals his guess about p*. (In fact, the offer is equal to his estimation of p*. This is a slight difference to the case of common value auctions. There the offers typically also reveal the signals, but usually are not equal to the signals.)

Take, for example, a non-extra-marginal buyer in a double auction market. It is important for this buyer to guess the value of p* correctly, in order to submit the lowest possible bid that blocks the competition by the extra-marginal buyers. Given the buyers condition their bids on their estimates of p*, those buyers will trade most often, who have over-estimated p* most, because they submit the highest bids. This is the analogy to the winner's curse case: It is only after a buyer's submitted bid has been accepted, that he discovers that he had over-estimated p*. (Note that unlike the case in the common value auctions, the buyer does not necessarily discover his over-estimation immediately after the trade. But, as the market proceeds, the buyer receives more and more information hinting at the actual value of p*.) Thus, the proposer's curse phenomenon, similar to the winner's curse phenomenon, evolves, because of an *adverse selection* property of the institution: Those buyers, who over-estimate p* most, and those sellers, who under-estimate p* most, are those traders whose offers are most likely to be accepted by the other market side.

The occurrence of proposer's curses is incompatible with the results of the game theoretic analysis of the game with almost perfect information. In that setting, the players are informed on the player valuations of all other traders. Thus, there is no need for them to *guess* the market clearing price p*. Evidently, this means that there is no over- or under-estimation of p* in the theoretic setting. However, the fact that the frequency of proposer's curses declines as subjects gain experience, seems to underline that the subjects accumulate the necessary information with experience. In fact, since the market clearing prices of the inexperienced and the experienced sessions of each subject group were different and since the proposer's curse frequency decreased significantly in the experienced sessions, the subjects must have learned more than just the location of p* in their inexperienced session. They, obviously, also learned how to accumulate the information they needed in the new (experienced) market more quickly. Given the decline in proposer's curse frequencies, the experimental data is not in conflict with the hypothesis that outcomes will tend to come closer to the game theoretic prediction for the game with almost perfect information as subjects gain experience in the game.

In comparing the experimental observations to the benchmark of the zero-intelligence trader market simulations, a completely different question arises. Here, the question is how the observed proposer's curse frequencies compare to the proposer's curse frequencies generated by the zero-intelligence trading agents' random behavior. Note that we use the term *behavior* in referring to the zero-intelligence traders choices, because a zero-intelligence trader can be seen to represent a very simple model of naive behavior in such markets. If observed and random behavior do not differ much with respect to the proposer's curse, we can conjecture that subjects correct for the adverse selection property of the market just as little as the zero-intelligence traders do. The following analysis shows that outcomes of the zero-intelligence trader simulations are quite similar to those of the inexperienced subject groups, but very different from the outcomes of the experienced subject groups.

An inspection of the zero-intelligence trader market simulation outcomes in the Appendix VII.C. reveals that random behavior does lead to a high frequency of proposer's curses. This is both true in the baseline simulations (panels VII.C.0.0 to

VII.C.0.2a) and in the full price range simulations (panels VII.C.1.0 to VII.C.1.18). For the reasons given in the subsection IV.C.1 of this chapter, we will - once again - restrict the discussions to the baseline simulations only. Depending on the market type and the number of traders present in the market, the overall average frequency of proposer's curses ranges from .55 (market type M2 with eight traders, see panel VII.C.0.2a) to .57 (market type M1 with 8 traders, see panel VII.C.0.1a). Thus, the frequency of proposer's curses is above 50% in all zero-intelligence trader market simulations. Before we proceed to compare the experimental outcomes to the simulation outcomes, it seems helpful to give some insight into the reason for the frequent occurrence of proposer's curses in the simulations.

To see why a typical zero-intelligence trader market exhibits proposer's curses in more than half of the contract proposals, a comparative static examination of two situations helps. Consider a first situation in which the current market ask submitted by the zero-intelligence sellers is $\tilde{a}_1$ and a second situation in which the current market ask submitted by the zero-intelligence sellers is $\tilde{a}_2$. Assume, that $\tilde{a}_1$ is smaller than $\tilde{a}_2$. In either case, each zero-intelligence buyer submits a bid randomly and uniformly drawn from the range of feasible bids. Note that, keeping all other things equal in the two situations means that the lower bound of the range of feasible bids is the same in both cases. (It is either the lowest possible price or the current market bid plus one.) The upper bound of the range of feasible bids is at $\tilde{a}_1$, in the first case, and at $\tilde{a}_2$, in second case. Since $\tilde{a}_1 < \tilde{a}_2$, the range of feasible bids is smaller in the first case than in the second case. Hence, the probability that the smaller ask $\tilde{a}_1$ is accepted is greater than the probability that the greater ask $\tilde{a}_2$ is accepted by a buyer. An ask below p* is therefore more likely to be accepted than an ask above p*. Finally, for completely analogous reasons, bids above p* are more likely to be accepted than bids below p*.

The fact that disadvantageous offers are more likely to be accepted, however, is not sufficient to explain why proposer's curses occur in more than 50% of the cases, since the occurrence frequency also depends on the base rates of disadvantageous and advantageous offers. Thus, we will now show that these two base rates are almost equal in the cases we examine. First keep in mind that all contracts can only occur inside the market price range, i.e. with prices between the lowest unit cost and the highest redemption value. If p* is exactly at the mid-point of the market price range and if the technical price range is symmetric around p*, any offer that can possibly lead to a contract will be below p* with probability ½ and above p* with probability ½. But, since disadvantageous offers are more likely to be accepted than advantageous offers, the probability for proposer's curses must be greater than ½. If p* is above the mid-point of the market price range, then the probability for bids that can lead to disadvantageous contracts (i.e. prices above p*) is less than ½. Thus, the proposer's curse probability for buyers may also fall below ½. As the proposer's curse probability for buyers falls, however, the proposer's curse probability for sellers rises. This is true, since the probability of asks that can lead to disadvantageous contracts (i.e. prices below p*) becomes higher as p* is moved above the mid-point of the market price range. Hence, as long as the frequency with which sellers propose contracts does not decrease too much (relative to the frequency with which buyers propose contracts), the decline in buyers' proposer's curses is compensated with the increase in sellers' proposer's curses. For completely analogous reasons, a p* that is below the mid-point of the market price range, causes the probability for buyers' proposer's curses to rise and the probability for sellers' proposer's curses to fall.

The effect of the location of the market clearing price p* within the market price range can be verified in the simulation results. The frequency of buyer and seller proposer's curses is equal in the symmetric market M0. In the market M1, in which the sellers have a larger portion of the market rent than the buyers, i.e. in which p* is above the mid-point of the market price range, buyers exhibit proposer's curses less frequently than sellers. In the market M2, in which p* is below the mid-point of the market price range, buyers exhibit proposer's curses more frequently than sellers. Comparing these simulation results to the experimental outcomes, we find that the difference between the frequencies of buyers' and sellers' proposer's curses has the predicted sign in seven of ten inexperienced sessions, but only in three of eight experienced sessions. This evidence - together with the fact that nine of ten inexperienced sessions, but only two of eight experienced sessions, exhibit proposer's curse frequencies of over 50% - seems to hint that inexperienced subject's have a tendency to fall prey to the proposer's curse in much the same manner as the randomizing zero-intelligence traders do. However, subjects learn to avoid the curse with experience, while zero-intelligence traders - by definition - cannot.

To examine the conjecture that inexperienced subjects do no better than zero-intelligence traders with respect to the proposer's curse, but learn to avoid the curse with experience, we compare the average frequencies of observed proposer's curses to the medians of the zero-intelligence trader simulation distributions. The conjecture is supported. In seven of ten sessions with inexperienced subjects the observed frequency of proposer's curses is greater than the median of the corresponding simulation distribution. In contrast, only two of eight sessions with experienced subjects exhibit proposer's curse frequencies greater than the median of the corresponding simulation distribution. The comparison suggests that inexperienced subjects are equally (or even more) likely to propose disadvantageous contracts than zero-intelligence traders, whereas experienced subjects are less likely to do so. Although this result cannot be proven to be statistical true, the direction it hints at is clear and completely in line with the significant effect of experience on behavior presented before: Inexperienced subjects tend to ignore the problem of adverse selection in the market - just as the zero-intelligence trading agents do -, but experienced subjects learn to avoid the proposer's curse.

IV.C.8. Bounded Rationality Approaches

In the bounded rationality approaches to double auction market games, the traders are assumed to make rational decisions, but only in the bounds of their capabilities (see SIMON 1976 and SELTEN 1990). Four such approaches, which all can be considered to be behavioral theories, will be evaluated. Three of these theories (EASLEY and LEDYARD 1983 and 1993, FRIEDMAN 1991, and GJERSTAD and DICKHAUT 1997) were explicitly developed to explain data of double auction markets experiments. The fourth, the *learning direction theory*, was originally developed by SELTEN and STOECKER (1986) to explain the adjustment of end behavior in a finitely repeated prisoner's dilemma game. SELTEN and BUCHTA (1998) then applied the theory to a sealed-bid first price auction. Because of its general character, however, it seems sensible to also take (an adapted version of) this theory into consideration.

The theories of boundedly rational behavior differ in the bounds assumed on the rationality of traders and, hence, also in the character of their predictions. Two of the considered theories are optimization theories (FRIEDMAN 1991, and GJERSTAD and DICKHAUT 1997), in which individuals are assumed to optimize their choices, while the other two (EASLEY and LEDYARD 1993 and SELTEN and BUCHTA 1998) are non-optimization theories, in which subjects are assumed to follow simple rules of the thumb. Correspondingly, the predictions of the two former theories can be exactly quantified, while the predictions of the latter two are only qualitative. In the following, first the implications of the optimization theories are discussed, then those of the non-optimization theories.

The aspect of bounded rationality enters the two optimization theories, because the optimization by the players is non-strategic, i.e. each trader is assumed to be playing a *game against nature*. The game against nature can be characterized as follows: Each trader views the offers from the other traders as random draws from a given range of prices and chooses his own optimal offer conditioned on the probability distribution of those random draws. As the market period proceeds (or over repetitions of the game) the probability distribution of the random offers is updated in correspondence with the observed realizations.

The major difference between the two theories (next to a number of different details, that seem to be of minor importance for our discussion) is the updating process a trader employs for the updating of the probability distribution of the random offers of the other traders. FRIEDMAN (1991) assumes that subjects begin with non-dogmatic priors for the probability distribution and update these using the Bayes' rule. In contrast, GJERSTAD and DICKHAUT (1997) assume that subjects' prior for the probability distribution has a specific form (basically a cubic spline function) and is updated in a corresponding manner. Especially, the method of updating in the latter theory seems to be mainly motivated by calculability. This is understandable, since the Baysian updating needed in the model by FRIEDMAN (1991) requires very complicated calculations. (This, in fact, is noted by the author himself in a different paper: see CASON and FRIEDMAN 1994, p. 7.) It seems, however, that even the simpler calculations required in the model by GJERSTAD and DICKHAUT (1997) are far outside the scope of subjects' possibilities - especially during an experimental session. Furthermore, it seems little convincing that subjects in an experiment, on the one hand, should completely ignore the strategic dimension of the game, on the other hand, however, should put much effort into the calculations needed by either of the updating processes. (But, see FRIEDMAN 1991, p. 49 for an extensive justification of the duality in the approach.)

In any case, it does not seem necessary to examine the exactly quantified *point* predictions of the two theories, because of the following reason: In both approaches the optimization results in a trading sequence that is strictly correlated to the player valuations, i.e. the better a trader's player valuation is, the earlier this trader trades. In this aspect, the predictions of these two theories are identical to the prediction of WILSON's (1987) game theoretic analysis of the game of imperfect information. As mentioned in the subsection VI.C.1., it is well-known that the predicted order of trades is practically never observed in experiments. This also holds true for our results.

To examine the correlation between the sequence of trades and the rank of the player valuations involved in each trade, the average rank of the player valuation of a traded unit at trading time is calculated. (Remember that each trader in the experiment was assigned four player valuations, i.e. could trade up to four units. The four units, however, had to be traded one after the other. Thus, in any given offer cycle, each of the traders could only trade one unit, his *current unit*. The rank of the redemption value of a traded unit at trading time was one, if the redemption value was the greatest amongst the redemption values of the current units of all buyers in that offer cycle. Similarly, the rank of the unit cost of the traded unit at trading time was one, if the unit cost was the smallest amongst the unit costs of the current units of all sellers in that offer cycle.) The ranks of the redemption values and unit costs of the traded units at trading time are averaged across all trades and all periods. The values are reported in the following tables IV.17 and IV.18.

The tables clearly show that the correlation between the trading sequence and the rank of the player valuations of the traded units - at most - is only slightly positive. A strictly positive correlation would result in an average rank equal to one. A strictly negative correlation would result in an average rank equal to four (in markets with four buyers and four sellers), to five (in markets with five buyers and five sellers), or to six (in the market with six buyers and six sellers). No correlation would result in a value of 2, 2.5, or 3, correspondingly. The displayed average ranks - both for the sessions with inexperienced and with experienced subjects - are generally quite close to the no correlation values. Thus, we can conclude that (almost) no correlation between the sequence of trade and the rank of the player valuation of the traded units can be observed in the experimental sessions. As mentioned before, this result is contrary to both the game theoretic prediction for the game with imperfect information and to the predictions of the two boundedly rational optimization theories.

Table IV.17 - Average rank of the player valuations of the traded units at trading time in the sessions with inexperienced subjects.

	Average rank of the player valuations of the traded units at trading time									
	Session									
	M1810	M1820	M1830	M1840	M1C10	M2810	M2820	M2830	M2840	M2A10
redemption value	2.39	2.19	2.20	1.93	2.87	1.86	2.10	1.74	2.13	2.74
unit cost	1.77	1.96	1.83	1.93	2.54	2.21	1.88	1.78	1.70	1.80

Table IV.18 - Average rank of the player valuations of the traded units at trading time in the sessions with experienced subjects.

	Average rank of the player valuations of the traded units at trading time							
	Session							
	M1811	M1821	M1831	M1841	M2811	M2821	M2841	M2A11
redemption value	2.58	2.22	2.03	2.58	2.08	2.07	2.08	2.27
unit cost	2.59	2.02	2.26	1.99	2.14	1.92	2.04	2.22

Given the contradicting evidence presented above, we will not further elaborate on the quantitative predictions of two boundedly rational optimization approaches. One interesting aspect of these *game against nature* approaches, however, should be noted. FRIEDMAN points out, that if traders actually behave as though they were playing a game against nature, but do not consider the *a posteriori* information gain of having made a contract in their optimization, they will be susceptible to a "reminiscent of the 'winner's curse' problem" (FRIEDMAN 1991, p. 56, footnote 9). The effect described is the proposer's curse, which was indeed observed in our experiment and exhaustively examined in the previous subsection of this chapter. Thus, the game against nature approach does seem to be suited to explain the behavior in double auction markets to some extent. We tend to the view that the fundamental idea presented in the two described approaches is well suited to characterize behavior in double auction markets. It is the optimization assumption employed by both approaches that appears quite problematic in the light of experimental evidence.

In contrast to the optimization approaches, the theory presented by EASLEY and LEDYARD (1993) postulates that the traders follow very simple rules of the thumb, when deciding on the offers to submit. Every trader in the model is assumed to have an (unobservable) *reservation price*, i.e. a price up to which a buyer is willing to raise his bid and down to which a seller is willing to lower her asks. At any given moment, during the market period, the buyer with the highest reservation price amongst all active buyers holds the current market bid and the seller with the lowest reservation price amongst all active sellers holds the current market ask. Thus, trades can only occur involving the two traders with the best reservation prices. But, an important feature of the model is that the reservation prices are not necessarily correlated to the traders' player valuations. Hence, trading is not predicted to occur in a sequence that is strictly correlated to the player valuations of the traded units.

The rules of the thumb proposed by the authors basically describe the way the traders form and adjust their expectations and reservation prices. The traders expect that trades in the current market period will - generally - occur within the range of minimum and maximum contract prices of the previous period. Thus, their reservation prices - for the most part of the market period - are assumed to lie within this range. However, if the minimum price of the previous period was greater than a buyer's redemption value, that buyer's reservation price will be smaller than the previous period's minimum price. Analogously, if the maximum price of the previous period was smaller than a seller's unit cost, her reservation price is greater than the previous period's maximum price. (These two rules immediately follow from the *no-loss* rule.) Finally, as the market period proceeds, towards the end of the market period, those buyers who have not yet traded, are assumed to raise their reservation prices slowly, and possibly even above the previous period's maximum price. Similarly, sellers who have not traded, lower their reservation prices towards the end of the market period, possibly even below the previous period's minimum price. (This, in effect, means that traders get more and more anxious that they may not trade at all as the market period reaches its end.)

Although the theory presented by EASLEY and LEDYARD (1993) is more of a qualitative than a quantitative nature, it does have some testable implications. First, offers outside the range of minimum and maximum prices of the previous market period should rarely be submitted (and usually only towards the end of the current period). Second, the sequence of minimum prices across periods must be rising, if (and as long as) minimum prices are below the market clearing price. Analogously, the sequence of maximum prices across periods must be falling, if (and as long as) maximum prices are above the market clearing price. (Note that prices must not necessarily be converging to the market clearing price, because the prediction only concerns the minimum and maximum prices, but not all other prices. However, if the convergence of minimum and maximum prices to p* is fast enough, the total deviation of all prices from p* can be expected to also fall, i.e. prices generally can be expected to converge to p*.) Lastly, the theory also predicts that trading quantity should be (almost) equal to the number of trades in the competitive equilibrium of the *truncated* market, i.e. of the market, in which the demand and supply curves are truncated at the minimum and maximum prices of the previous period. For the regularly observed case, in which the previous period's minimum price is below p* and the maximum price is above p*, this simply means, that the predicted trading quantity is equal to the market clearing quantity q*.

Let us begin the comparison of the theory's implications to the experimental data by checking for the predicted increase of the minimum and decrease of the maximum price from period to period, given that the former are below and the latter are above the market clearing price p*. The panels VII.B.1 to VII.B.18 in the Appendix VII.B. display the minimum and maximum prices in every period of every session. Eight of the ten sessions with inexperienced subjects and five of the eight sessions with experienced subjects have a lower minimum price and a higher maximum price in the first period than in the last period. Next to this evidence, the overall impression is that the prediction is confirmed with the exception of those cases in which the minimum or the maximum prices are already *close* to p*. This effect seems to be related to the result concerning the convergence of prices to p* discussed earlier on. There we had found that prices generally converge to the market clearing price p*, unless they already are very close to p*.

Another prediction of the theory presented by EASLEY and LEDYARD (1993) is that traders will only *rarely* submit bids greater than the previous period's maximum price or asks smaller than the previous period's minimum price. The difficulty in examining the data with respect to this prediction stems from the term *rarely*: the theory is explicit about when these bounds will be crossed, but not all parameters needed for the examination are observable. For this reason, we will restrict our examination to a simple discussion of the data presented in the panels VII.B.1. to VII.B.18. Those panels display the absolute frequency of offers for all units in the periods 2 to 5 of each session. The relative frequency of those offers, amongst these, that were bids greater than the previous period's maximum price or asks smaller than the previous period's minimum price are also displayed. The result is that in all sessions 4% or less of the offers were outside this range. One half of the sessions with inexperienced and three quarters of the sessions with experienced subjects has only 1% or less offers outside this range. Even though one could tentatively conjecture that the prediction of the theory is confirmed, some doubts concerning the relevance of the bounds arise, when the corresponding numbers of the zero-intelligence trader market simulations are checked. The panels VII.C.0.1a to VII.C.0.2b display the relative frequency of offers that were bids greater than the previous period's maximum price or asks smaller than the previous period's minimum price in the baseline simulations. Only 1% to 3% of the offers submitted by the zero-intelligence traders were outside the range.

Finally, since the minimum price in a market period was usually below and the maximum usually above the market clearing price, the prediction of the theory presented by EASLEY and LEDYARD (1993) concerning the volume of trade is (almost) equal to the market clearing quantity q* for most market periods. This prediction seems to be confirmed, since - as demonstrated in a preceding subsection - the traded quantity in most markets was close to q*.

All in all, the theory presented by EASLEY and LEDYARD (1993), as far as testable, seems to be confirmed by the data. However, since the theory is quite broad, in the sense that many different types of behavior can be incorporated into its framework, the question whether subjects' behavior is actually well described by the suggested rules of the thumb can only be answered unsatisfactorily. It should be noted that the *Baysian game against nature* theory introduced by FRIEDMAN (1991) is compatible with the rules of the thumb of the theory presented by EASLEY and LEDYARD (1993) - in a way the former represents a quantification of the latter. But, as described further up, the predictions of the *Baysian game against nature* cannot be verified in the data. Nevertheless, since both theories have appealing features, it seems imaginable that a intermediate theory, which relaxes some of the strict assumptions of the former and sharpens some of the results of the latter, can evolve that is a better predictor of experimentally observed data than either of the two.

The second non-optimization approach, the *learning direction theory*, was originally introduced by SELTEN and STOECKER (1986). The basic idea of this approach is that subjects adjust their choice of a decision parameter through an ex-post examination of their preceding choice of that parameter. If they find that raising the parameter would have had led to a better outcome last time, they tend not to choose a lower value this time (or to do so less frequently than choosing an higher or equal value). Analogously, if they find that lowering the parameter would have had led to a better outcome last time, they tend not to choose a higher value this time (or to do so less frequently than choosing an lower or equal value). In this sense, the learning direction theory only predicts that the direction of adjustment of a choice parameter should correspond to the direction of greater success in the ex-post inspection of a preceding decision and outcome. The extent of the adjustments, however, are not predicted by the theory.

In their application of the learning direction theory to a sealed bid first price auction setting, SELTEN and BUCHTA (1998) find that subjects tend to decrease their bid, if their last bid was successful, i.e. if they had won the auction. In contrast, if their last bid was not successful, even though a higher successful bid would have been possible, subjects tended to increase their bid. This behavior is completely in line with the learning direction theory, since an adjustment of a bid in this manner corresponds to the direction of greater success in the ex-post inspection of the choice and outcome of the preceding bid.

Quite interestingly, EASLEY and LEDYARD (1993) present an intuition (that is not explicitly contained in the model) for the proposed adjustment of reservation prices in their theory, which also corresponds to the learning direction theory. A buyer in a double auction, they conjecture, should, on one hand, lower his reservation price (in effect: his bid), if he realizes that he was successful, but had "overpaid" in the last period. On the other hand, he should raise his reservation price, if he could have, but did not trade in the previous period. (EASLEY and LEDYARD 1993, p. 70.) Thus, it seems that an evaluation of the experimental data - specifically, the ob-served offering behavior - with regard to the learning direction theory can be informative.

The application of the learning direction theory to double auction markets, however, poses some problems. First, the choice task in a double auction market is at least two-dimensional, while the theory was originally devised for one-dimensional choice tasks. (The value and timing of the offer can be chosen by a trader. But, the task is even more complex, since the timing can be relative to different parameters, e.g. the number of past offer cycles or the number of observed contracts. The task is complicated even more by the fact that a trader can choose to refrain from submitting an offer.) Second, the ex-post examination of the success of the preceding choice and outcome is not as straightforward as in the case of single unit first price sealed bid auctions: If the examined preceding choice is the last submitted offer, not having traded does not imply failure, since the unit may be traded with some later offer in the same market period - possibly even with a larger profit. If the preceding choice is the offer that led to the last traded unit of the trader, there is no reason for the trader to believe that having traded means having paid too much (or received too little), since this trade was possibly made at the most profitable price.

To be able to apply the learning direction theory in the context of the experimental alternating double auction markets, we take the following path: We assume that each trader uses his own best previous market period's offer for the unit he is currently trying to trade as the default value of the parameter to be adjusted. (This means that we ignore the other dimensions of the choice task and concentrate on the value of a trader's best offer for each of his units.) To simplify the terminology, we introduce the two phrases *aggressive* and *moderate*. A buyer's bid for a certain unit is said to be **aggressive**, if it is greater than the buyer's greatest bid for the same unit in the previous period. It is said to be **moderate**, if it is smaller than the buyer's greatest bid for the same unit in the previous period. Analogously, a seller's ask for a certain unit is said to be **aggressive**, if it is smaller than the seller's smallest ask for the same unit in the previous period. It is said to be **moderate**, if it is greater than the seller's smallest ask for the same unit in the previous period.

If a unit was traded at a less profitable price than the previous period's average price, we assume that the trader *feels regret*: a buyer, who bought at a price above average in the previous period, *feels* that he has paid too much - a seller, who sold at a price below average in the previous period, *feels* that she has raised too little. To simplify terminology, we use the term **regret unit** for a unit that was traded at a price in the previous period that was less profitable than that period's average price. If our conjecture about the regret of the traders is true, then we should observe that traders, on average, less frequently submit aggressive offers for regret units than they submit aggressive offers for all units. This, in effect, means that buyers, who paid more than the average price in the previous period for a certain unit, will tend to bid lower for that unit in the current period, i.e. submit aggressive offers less frequently than this is done on average. Similarly, sellers, who received less than the average price in the previous period for a certain unit, will tend to ask higher prices for that unit in the current period, i.e. submit aggressive offers less frequently than this is done on average.

The following tables IV.19 and IV.20 display the relative frequencies of aggressive offers that were submitted both for all units and for the regret units. The results seem remarkably clear: The frequency of aggressive offers for regret units is smaller than the frequency of aggressive offers for all units in every session. No doubt that any statistical test would confirm the significance of this result.

Table IV.19 - Relative frequency of aggressive offers in the sessions with inexperienced subjects.

average value for	Relative frequency of aggressive offers								
	Session								
	M1810	M1820	M1830	M1840	M1C10	M2810	M2820	M2830	M2840 M2A10
all units	.20	.22	.16	.21	.20	.22	.21	.14	.26 .20
regret units	.08	.13	.07	.08	.15	.05	.11	.09	.13 .07

Table IV.20 - Relative frequency of aggressive offers in the sessions with experienced subjects.

average value for	Relative frequency of aggressive offers							
	Session							
	M1811	M1821	M1831	M1841	M2811	M2821	M2841	M2A11
all units	.16	.17	.20	.22	.18	.27	.20	.16
regret units	.06	.13	.02	.16	.09	.19	.03	.02

However, some serious caution in the interpretation of the evidence is necessary: First, remember that a regret unit is one which was traded in the previous period at a price less profitable than that period's average price. If we assume that the current period's average price will be close to the previous period's average price and that the prices are randomly distributed amongst the traded units, the probability that regret units have a more profitable price in this than in the previous period is greater than the probability that an arbitrary unit has a more profitable price in this than in the previous period. This is a feature of multiple draws of a stochastic variable from identical distributions sometimes referred to as the *regression towards the mean*. (For example, see TVERSKY and KAHNEMAN 1982).

To be able to suitably assess the significance of the obtained result, we have to correct for the *natural* extent of the regression towards the mean that lowers the frequency of aggressive offers for regret units. First, we calculate the *aggressive offer frequency ratio*, i.e. the ratio of the relative frequency of aggressive offers for regret units to the relative frequency of aggressive offers for all units. (The ratios for all sessions are reported in the panels VII.B.1 to VII.B.18 of the Appendix VII.B.) Next, we compare the sessions' aggressive offer frequency ratios to the median value of aggressive offer frequency ratios in the baseline zero-intelligence trader simulations. (The simulation results are reported in the panels VII.C.0.1a to VII.C.0.2b of the Appendix VII.C.) With this we can check our hypothesis:

Learning Direction Hypothesis H1

The experimentally observed aggressive offer frequency ratios are smaller than those generated by the zero-intelligence trader simulations; i.e. subjects less frequently exhibit aggressive offers for regret units, compared to aggressive offers for all units, than can be expected just by chance.

The comparison reveals that the experimentally observed aggressive offer frequency ratios are smaller than the median of the simulation distribution of the ratio in eight of the ten sessions with inexperienced subjects and in six of the eight sessions with experienced subjects. The binomial test rejects the null hypothesis that the observed ratio is equally likely to be greater or smaller than the simulation median of the ratio at the 10% level (one-sided)[6] for the sessions with inexperienced subjects. For the sessions with experienced subjects the null hypothesis cannot be rejected. Hence, we can only conclude that there is some tendency for subjects to reduce the frequency of their aggressive offers for units that were traded in the previous period at a price less profitable than that period's average price.

The essence of the second learning direction hypothesis is that, if a trader did not trade a unit in the previous period, he will reduce the frequency of moderate offers for that unit in the current period: a buyer will submit less bids below his previous period's greatest bid for that unit - a seller will submit less asks above her previous period's smallest ask. Thus, we expect to observe that, on average, traders less frequently submit moderate offers for units not traded in the previous period than they submit moderate offers for all units.

The following tables IV.21 and IV.22 display the relative frequencies of moderate offers that were submitted both for all units and for the units not traded in the previous period. The results again seem clear: The frequency of moderate offers for the units not traded in the previous period is smaller than the frequency of moderate offers for all units in all but one session.

Even though in this case there seems to be no reason - *a priori* - to assume a biasing effect, we will proceed to test the significance of the result in much the same manner as before: First, we calculate the **moderate offer frequency ratio**, i.e. the ratio of the relative frequency of moderate offers for the units not traded in the previous period to the relative frequency of moderate offers for all units. (The ratios for all sessions are reported in the panels VII.B.1 to VII.B.18 of the Appendix

---

6) Given the reported data, the test rejects the null hypothesis at any significance level greater than .055 (one-sided).

VII.B.) Next, we compare the sessions' moderate offer frequency ratios to the median value of moderate offer frequency ratios in the baseline zero-intelligence trader simulations. (The simulation results are reported in the panels VII.C.0.1a to VII.C.0.2b of the Appendix VII.C.) With this we can check our hypothesis:

Learning Direction Hypothesis H2

The experimentally observed moderate offer frequency ratios are smaller than those generated by the zero-intelligence trader simulations; i.e. subjects less frequently exhibit moderate offers for units not traded, compared to moderate offers for all units, than can be expected just by chance.

The comparison reveals that the experimentally observed moderate offer frequency ratios are smaller than the median of the simulation distribution of the ratio in nine of the ten sessions with inexperienced subjects and in all of the eight sessions with experienced subjects. The binomial test rejects the null hypothesis that the observed ratio is equally likely to be greater or smaller than the simulation median of the ratio at the 2% level (one-sided) for the sessions with inexperienced subjects and at the .5% level (one-sided) for the sessions with experienced subjects. Hence, we can conclude that subjects have a significantly lower frequency of moderate offers for units that were not traded in the previous period than for all units.

Table IV.21 - Relative frequency of moderate offers in the sessions with inexperienced subjects.

average value for	Relative frequency of moderate offers									
	Session									
	M1810	M1820	M1830	M1840	M1C10	M2810	M2820	M2830	M2840	M2A10
all units	.68	.66	.65	.64	.61	.69*	.66	.78	.64	.56
not traded units	.46	.57	.53	.49	.47	.69	.70	.75	.58	.34

* indicates that the values in the "all units" line and in the "not traded units" line are only equal, because of rounding errors. The actual values were .6875 and .685, correspondingly.

Table IV.22 - Relative frequency of moderate offers in the sessions with experienced subjects.

average value for	Relative frequency of moderate offers							
	Session							
	M1811	M1821	M1831	M1841	M2811	M2821	M2841	M2A11
all units	.50	.57	.44	.50	.60	.54	.50	.51
not traded units	.36	.43	.30	.47	.55	.48	.26	.09

Summarizing, we find that the learning direction theory is confirmed by the observed results of the experiment. It seems that the retrospective success control mechanism, postulated by the learning direction theory, actually governs the behavior of the subjects. The result is more strongly supported for the cases in which subjects regret having lost an opportunity to trade than for the cases in which they regret having traded at a price less profitable than the average price. The reason for this difference in reaction - perhaps - is that not trading a unit usually leads to a relatively greater lost profit than the profit lost by trading at a (marginally) less profitable price. A similar observation is reported by CASON and FRIEDMAN (1997, p. 21) in the context of single call markets.

IV.D. The Anchor Price Hypothesis
- A Concluding Summary and Evaluation

The experimental investigation of the alternating double auction market presented in this study has some anticipated and some very surprising results. It was anticipated that the experimental markets will exhibit the typical outcomes of double auction markets: high efficiency and convergence of the prices to the market clearing price p*. Indeed, both presumptions proved true.

The examined alternating double auction markets of the experiment were mostly very efficient, especially the markets in the sessions with experienced subjects. Some efficiency loss seemed to stem from untraded non-extra-marginal units, while some stemmed from *displacements* (i.e. trades with extra-marginal units). In a great majority of the markets, however, exactly the quantity predicted by the Walrasian competitive equilibrium was traded.

The anticipated convergence of contract prices to the market clearing price p* was also present in most markets: prices tended to be closer to p* in later market periods than in earlier market periods. The evidence for the convergence of prices to p*, however, was weaker in the markets in which prices were already very close to p*. The result that the convergence of prices to p* decelerate in sessions with prices close to p*, seemed to hint that the observed outcomes were closer to the outcome predicted by the game theoretic analysis of the game with almost perfect information (in which most prices were predicted to be one money unit away from p*) than to the Walrasian competitive equilibrium outcome. This was mainly the case in the sessions with experienced subjects, in which prices were significantly closer to p* than in the inexperienced sessions. Compared to the prices generated in the zero-intelligence trader simulations, however, the experimentally observed prices were significantly closer to p* both in the sessions with experienced and with inexperienced subjects.

The examination of the direction of convergence of the contract prices to the market clearing price p*, then, brought on the first surprising result. The sessions could be divided into two groups: those with a convergence of prices *from above* (decreasing average prices from period to period) and those with a convergence of prices *from below* (increasing average prices from period to period). The astonishing feature of the observation was that the constellation of sessions in the two groups, proved contrary not only to the game theoretic prediction, but also to earlier experimental evidence and the comparable zero-intelligence trader simulation results.

The game theoretic prediction (for the game with almost perfect information, as presented in the previous chapter) was that average prices should be slightly below p* in the markets opened by the buyers and slightly above p* in the markets opened by sellers. The experimental results were exactly opposite: the markets opened by buyers exhibited average prices above p* and those opened by sellers exhibited average prices below p* in the majority of the cases. This was shown to be significantly so for the data of the first market periods of the inexperienced sessions. (Note that the data suggested the result generally, but the statistical test was legitimate only for the first market period data, due to the lack of the statistical independence of the data of the later periods.)

The zero-intelligence trader simulations indicated that the direction of convergence would be related to the ratio of market rent distribution. In the markets of type M1, which allocates a larger share of the market rent to the sellers, the average simulation prices were below p*. In the markets of type M2, which allocates a larger share of the market rent to the buyers, the average simulation prices were above p*. Thus, the zero-intelligence trader simulations led to results similar to the experimental outcomes reported by SMITH (1976). In our data, however, no complying evidence could be found.

The second surprising phenomenon was discovered in the examination of the within market period price sequences, i.e. the price paths of the markets. Here, we found that, in most sessions, one type of price path (which we categorized into the four groups: *falling*, *rising*, *hanging*, and *standing*) was observed significantly more often than expected by chance. (Note that "more often than expected by chance" here means: "more often than can be explained by the random draws from the distribution generated with zero-intelligence trader simulations".) We introduced the term *price path repetition* for this phenomenon.

Moreover, a correlation between the direction of convergence and the prevalence of certain price path structures was observed. If the predominant price path structure in a session was *falling* or *standing* the convergence of the prices to p* tended to be *from above*. If the predominant price path structure in a session was *rising* or *hanging* the convergence of the prices to p* tended to be *from below*.

Next, we examined the frequency with which each market side proposed contracts, i.e. the frequency with which bids standing on the market were accepted by sellers or asks standing on the market were accepted by buyers. The inspection of the proposal frequencies disclosed two astonishing results. First, the proposal frequencies in most markets proved very extreme, i.e. either buyers proposed a vast majority (often even almost all) of the contracts or sellers did so. The experimentally observed proposal frequencies, in fact, were significantly more extreme that can be expected by a random draw from the distribution of zero-intelligence trader market simulation outcomes. Second, we found that, in the sessions with inexperienced subjects, the market side proposing most contracts was significantly more often disadvantaged by the average price. In other words, if buyers proposed most of the contracts, the average price tended to be above the market clearing price p^*, which means that the average payoff of the buyers tended to be smaller than in competitive equilibrium. In contrast, if sellers proposed most contracts, average price was typically below p^*, in which case the average payoff of the sellers was smaller than in competitive equilibrium.

Having observed this striking phenomenon on market side aggregate level, the next obvious inquiry was to look for the phenomenon on individual level. The result was clear: the *proposer's curse phenomenon*, i.e. the tendency of traders to propose contracts that are to their own disadvantage compared to the competitive equilibrium, was shown to be exhibited by subjects independent of the market side they were on. But, subjects' experience in a previous market session significantly reduced (but did not completely expunge) the occurrence of *proposer's curses*, i.e. of contracts that were to the disadvantage of the proposer when compared to the market clearing price p^*. The frequency of proposer's curses in the inexperienced sessions was close to (and mostly even greater than) the frequency of proposer's curses exhibited by zero-intelligence traders. In contrast, the occurrence frequency of proposer's curses in the experienced subject groups was generally well below the median of the distribution of that value in the zero-intelligence trader simulations.

This result seemed to confirm our conjecture that the proposer's curse phenomenon in our double auction markets has some analogy to the winner's curse phenomenon well-known in common value auctions. In both cases, neglecting the *adverse selection* property of the institution by traders, who must estimate an unknown value, leads to disadvantageous contract proposals. The unknown value that traders in double auction markets try to estimate is the market clearing price p*, i.e. the most profitable price that bars extra-marginal traders from trading. The adverse selection in double auction markets results from the fact that offers that are disadvantageous to the proposer are more likely to be accepted than offers that are disadvantageous to the accepter, even if accepters randomize their acceptance action (as in the case of zero-intelligence traders). With experience subjects learn to take the adverse selection property into account and the frequency of proposer's curses relative to all contract proposals falls below 50% and thus, below the frequency in the zero-intelligence trader simulations.

Finally, we concluded the experimental investigation with an examination of observed behavior with respect to a number of boundedly rational approaches suggested in the literature. Interestingly, the basic idea of the *game against nature* approach, introduced by FRIEDMAN (1991) and presented in a different version by GJERSTAD and DICKHAUT (1997), was found to give some insight into the observed proposer's curse phenomenon. The concept of the *game against nature* approach is that subjects in double auctions optimize their offers under the assumption that all other offers in the market are realizations of a stochastic process. Even though the predictions of the approach are generally not confirmed by the data, the occurrence of proposer's curses can be explained by naive expectation formation in a *game against nature*. This, in fact, was already hinted at by FRIEDMAN (1991) in a footnote of the paper (p. 56, footnote 9).

The qualitative theory presented by EASLEY and LEDYARD (1993) seems to be better in line with the experimental data. The predicted increase of minimum and decrease of maximum price from period to period is confirmed. Other implications of that theory, however, are somewhat vague and not always testable. Nevertheless, the theory seems appropriate as a starting point and baseline for the explanation of traders' behavior.

A much sharper result is obtained in the comparison of the *learning direction theory* (SELTEN and STOECKER 1986 and SELTEN and BUCHTA 1988) to the experimental outcomes. Two hypotheses implied by this theory are confirmed by the data. The first hypothesis is based on the idea that offers for units traded in the previous period at a price more disadvantageous than that period's average price will tend to be less aggressive than offers in general. The hypothesis is shown to be valid more often than can be expected by a random draw from the distribution of outcomes generated with zero-intelligence trader simulations (significantly so in the inexperienced sessions).

The second learning direction hypothesis is based on the idea that offers for units not traded in the previous period will tend to be less moderate than offers in general. This hypothesis is shown to be valid significantly more often than can be expected by a random draw from the distribution of outcomes generated with zero-intelligence trader simulations, in the sessions both with inexperienced and with inexperienced subjects.

Note that, if behavior is adjusted as assumed by the first learning direction hypothesis, subjects exhibiting a proposer's curse in the previous period will tend to avoid the curse in the current period. On the other hand, if behavior is adjusted as assumed by the second learning direction hypothesis, the number of non-extra-marginal traders who failed to trade in the previous market period will tend to decrease in the current period. Thus, the observed convergence of prices to the market clearing price seems to be the result of subjects' behavioral adjustment in the manner predicted by the learning direction theory.

None of the boundedly rational approaches to traders' behavior in double auction markets, however, can predict the direction of convergence of prices to the market clearing price and the corresponding predominance of a certain structure of price paths in a session. In other words, the question which market parameter is the key to the emergence of a session with convergence *from above* or *from below* has so far remained unanswered. The answer to this question, that we find most convincing, is the *anchor price hypothesis* introduced in the following.

The *anchor price* of a session is the very first contract price of the first market period. The anchor prices for all sessions are reported further up, in the tables IV.3 and IV.4. The anchor price hypothesis is that the anchor price of a session determines the direction of convergence of prices in that session: If the anchor price of a session is smaller than p*, the correlation of period's average prices to the period number will be positive, i.e. the session converges *from below*. If the anchor price of a session is greater than p*, the correlation of period's average prices to the period number will be negative, i.e. the session converges *from above*. The hypothesis is true in nine of the ten sessions with inexperienced subjects and in five of the eight sessions with experienced subjects.

But, why should the anchor price influence the price development of the entire session? Our conjecture is the following: Depending on the location of the anchor price, a certain type of price path structure emerges in the first session of the experiment. If the anchor price is smaller than p*, the price path that evolves tends to be *rising*. If the anchor price of a session is greater than p*, the price path that evolves tends to be *falling*. (As mentioned before, the evidence for *hanging* and *standing* price paths is mixed. Admitting the simplification, we restrict the arguments to the more frequently observed *rising* and *falling* price paths.) The emergence of the corresponding price path structures can easily be explained by a tendency of subjects - at least very early in the market period - to form their price aspiration level on the basis of the observed contract price. This together with the fact that the market *closes in* as the market period reaches its end, accounts for *rising* price paths, when the anchor price was below p*, and *falling* price paths, when the anchor price was above p*.

Then, in later market periods, subjects expect to observe prices that develop in a similar manner to the prices in the first market period. (In the post-experimental interviews, a number of subjects made comments of the following type: "I knew that I had to pay a little more in order to trade early, since the prices were bound to fall during the market period." Or: "The trick was to make low offers early on and then to go up with the price of later units, because buyers didn't accept high prices at the beginning of the market period.") The price path repetition that evolves in this way explains the general direction of the convergence of the prices. Sessions with *rising* price paths - quite obviously - have average prices smaller than p*, while sessions with *falling* price paths have average prices greater than p*.

The result on the distribution of contract proposals amongst buyers and sellers also neatly falls in line with this explanation: If price paths are mostly *rising*, buyers quickly accept the sellers' contract proposals to avoid later higher prices. If price paths are mostly *falling*, sellers quickly accept the buyers' contract proposals to avoid later lower prices. As the subjects gain experience - and by employing the simple rules implied by the learning direction theory - the slope of the price paths is reduced and the number disadvantageous proposals decreases.

The only open question left to answer is: How does the anchor price of a session emerge? Our conjecture here is that the anchor price emerges more or less randomly. The probability, however, for the anchor price to be smaller than p* seems to be greater if the sellers open the first period market than if the buyers open the first period market. All four sessions with inexperienced subjects that are opened by sellers exhibit anchor prices smaller or equal to the market clearing price p*, while in four of the remaining six sessions with inexperienced subjects the anchor price is greater than p*. This effect seems to be related to the proposer's curse: if traders on both market sides tend to exhibit proposer's curses, a trader on the market side proposing first, i.e. the opening market side, is more likely to be the first who exhibits a proposer's curse. Thus, the anchor price of a session is more likely to be the result of the proposer's curse of a trader on the opening market side than of a trader on the second moving market side.

In the following table IV.23, the two different scenarios predicted by the anchor price hypothesis are displayed. Note that we take the corresponding outcomes only to be more probable than other outcomes, given the results of the experimental investigation presented.

Table IV.23 - Scenarios as predicted by the anchor price hypothesis.

Outcomes as predicted by the anchor price hypothesis	Opening market side in the first period of the session	
	Buyers	Sellers
Anchor price tends to be	greater than p^*	smaller than p^*
Price paths tend to be	falling	rising
Average prices in the market periods tend to be	greater than p^*	smaller than p^*
Convergence of the prices to p^* tends to be	from above p^*	from below p^*
Most contracts tend to be proposed by	buyers	sellers

V. Concluding Remarks

We introduced a new market institution called *the alternating double auction market*. In contrast to the traditional (oral or computerized) continuous-time double auction market, this institution can be completely described as a game in extensive form. Thus, the advantage of the alternating double auction market, compared to the continuous-time double auction market, is that it allows a game theoretic analysis without the problem of ambiguity with respect to the *parameterization of time*.

A game theoretic analysis of the alternating double auction market game was presented. The analysis considered the game in a version with (almost) perfect information, because the game theoretic equilibria in that informational setting are of the greatest interest to the causes of experimental economics. The main result of the analysis was that in equilibrium the market will be 100% efficient and trades will occur at prices close to one of the bounds of the market clearing price range, either just within the range or just outside. (For a detailed summary of the results see the section III.F.)

This is a quite surprising result, since it holds for markets of any size. The general conjecture found in most of the theoretic literature up to date had been that markets will only perform as in the Walrasian competitive equilibrium, if the number of traders is *large enough*. (An important exception can be found in FRIEDMAN 1984.) Our analysis proved that three traders already constitute a market that is *large enough* to be competitive. The *failure of small markets*, that is politically often postulated, obviously does not occur, if traders are playing non-cooperatively. The analysis seems to underline that collusion amongst traders (either open or tacit), and not trader's strategic behavior, is the motor of *market failure*.

The result is also quite interesting for a second reason. The alternating double auction market game is a basically asymmetric game, because one market side has a genuine first mover advantage. Nevertheless, the asymmetry of power only suffices to allocate that part of the market rent which lies inside the market clearing price range to the advantage of the first moving market side. Thus, in markets in which the market clearing price range is small relative to the whole market price range (as in most experimentally examined markets) the first mover advantage only leads to a small advantage in the distribution of payoffs. In fact, since the Walrasian competitive equilibrium does not define a unique market clearing price, except for the degenerate case in which the bounds of the market clearing price range are equal, the distribution of rents as predicted by the game theoretic analysis generally does not contradict the Walrasian theory. (Strictly speaking, the game theoretically predicted distribution of rents might be slightly different, since the game equilibrium prices - in some cases - lie one money unit outside the market clearing price range.)

All in all, the presented game theoretic analysis of the alternating double auction market game revealed properties of the market institution which hinted that experimental outcomes could be close to the reported typical outcomes of continuous-time double auction markets: high efficiency and prices converging to the market clearing price. For the introduced market institution to be a scientifically useful alternative to the continuous-time double auction market, the confirmation of these outcomes in an experiment seems obligatory.

In the second part of this study, an experimental investigation of the alternating double auction market was presented. A number of anticipated features of the market institution were confirmed. The market outcomes were highly efficient and the convergence of prices to the market clearing price was established. (In fact, there even seemed to be some evidence that the game theoretic analysis predicted price convergence better than the Walrasian theory, since once prices had converged close to the market clearing price, the convergence seemed to decelerate or cease. This conjecture, however, was disconfirmed by the fact that the location of the average prices, relative to the market clearing price, often contradicted the game theoretic prediction.)

Next to the anticipated outcomes, a number of astonishing results were obtained. The two most amazing phenomena that were observed were the *price path repetition* and the *proposer's curse phenomenon*. The former refers to the observation that most sessions of the experiment exhibited price paths that were of a similar structure in almost every market period. The latter refers to the observation that most subjects, especially when inexperienced, proposed contracts that were to their own disadvantage, when compared to the market clearing price.

One goal of the experimental analysis was to find clues to the process of price formation in the market. To achieve this goal, a number of theories were compared to the collected data. The main result was that the convergence of prices to the market clearing price is driven by simple rules, as postulated by the *learning direction theory* (SELTEN and STOECKER 1986, SELTEN and BUCHTA 1998). The structure of price paths and the direction of convergence, we conjectured, can be best predicted by the *anchor price hypothesis*, introduced in section IV.D. (For a detailed summary of the experimental results see the section IV.D.)

Concluding, it seems that the alternating double auction market, which was introduced here, is an adequate alternative to the continuous-time double auction market. Both the game theoretic and the experimental results suggest that this new market institution combines the well-established characteristics of its continuous-time competitor with the property of being a completely specified game in extensive form. Thus, we believe that the obtained results can be viewed to have some general validity and hope that the presented investigation can be regarded as a small step towards the better understanding of complex competitive institutions. Perhaps this work can pave the trail for promising future investigations, both of theoretic nature (e.g. a game theoretic analysis of changes in the auction rules) and of experimental nature (e.g. *strategy method* experimentation as presented by SELTEN (1967) and SELTEN, MITZKEWITZ, and UHLICH (1997), in which computer simulations with strategies written by subjects are run).

VI. References

BAZERMAN, M. H. and W. F. SAMUELSON, 1983, *I Won the Auction But Don't Want the Prize* in: *Journal of Conflict Resolution*, Vol. 27, pp. 618-634.

BINMORE, Kenneth, 1987, *Modeling Rational Players: Part I* in: *Economics and Society*, Vol. 3(2), pp. 179-214.

CAPEN, E. C., R. V. CLAPP, and W. M. CAMPBELL, 1971, *Competitive Bidding in High-Risk Situations* in: *Journal of Petroleum Technology*, Vol. 23, pp. 641-653.

CASON, Timothy and Daniel FRIEDMAN, 1993, *An Empirical Analysis of Price Formation in Double Auction Markets* in: *The Double Auction Market*, Daniel FRIEDMAN and John RUST (eds.), Reading, pp. 253-283.

CASON, Timothy and Daniel FRIEDMAN, 1994, *Price Formation in Double Auction Markets*, *mimeo*, University of Southern California and University of California at Santa Cruz.

CASON, Timothy and Daniel FRIEDMAN, 1997, *Price Formation in Single Call Markets* in: *Econometrica*, forthcoming.

COX, James C., Jason SHACHAT, and Mark WALKER, 1997, *An Experiment to Evaluate Baysian Learning of Nash Equilibrium Play*, *mimeo*, University of Arizona.

EASLEY, David and John O. LEDYARD, 1983, *A Theory of Price Formation and Exchange in Oral Auctions* in: *Technical Report 249*, Cornell University.

EASLEY, David and John O. LEDYARD, 1993, *Theories of Price Formation and Exchange in Double Oral Auctions* in: *The Double Auction Market*, Daniel FRIEDMAN and John RUST (eds.), Reading, pp. 63-97.

FRIEDMAN, Daniel, 1984, *On the Efficiency of Experimental Double Auction Markets* in: *American Economic Review*, Vol. 74, pp. 60-72.

FRIEDMAN, Daniel, 1991, *A Simple Testable Model of Double Auction Markets* in: *Journal of Economic Behavior and Organization*, Vol. 15, pp. 47-70.

FRIEDMAN, Daniel, 1993, *The Double Auction Market Institution: A Survey* in: *The Double Auction Market*, Daniel FRIEDMAN and John RUST (eds.), Reading, pp. 3-25.

GJERSTAD, Steven and John DICKHAUT, 1997, *Price Formation in Double Auctions*, *mimeo*, Hewlett-Packard Laboratories and University of Minnesota.

GODE, Dhananjay K. and Shyam SUNDER, 1993, *Allocative Efficiency of Markets with Zero Intelligence Traders: Market as a Partial Subsitute for Individual Rationality* in: *Journal of Political Economy*, Vol. 101(1), pp. 119-137.

GODE, Dhananjay K. and Shyam SUNDER, 1997, *Double Auction Dynamics - Structural Consequenses of Non-Binding Price Controls*, *mimeo*, University of Rochester and Carnegie Mellon University.

JORDAN, James S., 1991, *Baysian Learning in Normal Form Games* in: *Games and Economic Behavior*, Vol. 3, pp. 60-81.

KAGEL, John H., 1995, *Auctions: A Survey of Experimental Research* in: *The Handbook of Experimental Economics*, John H. KAGEL and Alvin E. ROTH, Princeton University Press, pp. 501-585.

KAGEL, John H. and Dan LEVIN, 1986, *The Winner's Curse and Public Information in Common Value Auctions* in: *American Economic Review*, Vol. 76, pp. 894-920.

KAMECKE, Ulrich, 1998, *Competition, Cooperation, and Envy in a Simple English Auction*, *mimeo*, Humboldt-University at Berlin.

KETCHAM, Jon, Vernon L. SMITH, and Arlington W. WILLIAMS, 1984, *A Comparison of Posted-Offer and Double-Auction Pricing Institutions* in: *Review of Economic Studies*, Vol. 51, pp. 595-614.

PLOTT, Charles R., 1982, *Industrial Organization Theories and Experimental Economics* in: *Journal of Economic Literature*, Vol. 20(4), pp. 1484-1527.

PLOTT, Charles R. and Vernon L. SMITH, 1978, *An Experimental Examination of Two Exchange Institutions* in: *Review of Economic Studies*, Vol. 45(1), pp. 133-153.

SATTERTHWAITE, Mark and Steven WILLIAMS, 1993, *The Baysian Theory of the k-Double Auction* in: *The Double Auction Market*, Daniel FRIEDMAN and John RUST (eds.), Reading, pp. 99-123.

SELTEN, Reinhard, 1965, *Spieltheoretische Behandlung eines Oligopolmodells mit Nachfrageträgheit* in: *Zeitschrift für die Gesamte Staatswissenschaft*, Vol. 121, pp. 301-324 and pp. 667-689.

SELTEN, Reinhard, 1967, *Die Strategiemethode zur Erforschung des eingeschränkt rationalen Verhaltens im Rahmen eines Oligopolexperiments* in: *Beiträge zur experimentellen Wirtschaftsforschung*, Heinz SAUERMANN (ed.), Vol. I, Tübingen, pp. 136-168.

SELTEN, Reinhard, 1970, *Ein Marktexperiment* in: *Beiträge zur experimentellen Wirtschaftsforschung*, Heinz SAUERMANN (ed.), Vol. II, Tübingen, pp. 33-98.

SELTEN, Reinhard, 1975, *Reexamination of the Perfectness Concept for Equilibrium Points in Extensive Games* in: *International Journal of Game Theory*, Vol. 4, pp. 25-55.

SELTEN, Reinhard, 1990, *Bounded Rationality* in: *Journal of Institutional an Theoretical Economics*, Vol. 146, pp. 649-658.

SELTEN, Reinhard, 1995, *Multistage Game Models and Delay Supergames* in: *Les Prix Nobel - The Nobel Prizes*, Tore FRÄNGSMYR (ed.), Nobel Foundation, Stockholm.

SELTEN, Reinhard, Klaus ABBINK and Ricarda COX, 1997, *Winner's Curse and Learning Direction Theory* in: *SFB 303 Discussion Paper*, University of Bonn, in preparation.

SELTEN, Reinhard, and Joachim BUCHTA, 1998, *Experimental Sealed Bid First Price Auctions with Directly Observed Bid Functions* in: *Games and Human Behavior: Essays in the Honor of Amnon Rapoport*, David BUDESCU, Ido EREV, and Rami ZWICK (eds.), Lawrenz Erlbaum Associates, Mahwah NJ, forthcoming.

SELTEN, Reinhard, Michael MITZKEWITZ, and Gerald R. UHLICH, 1997, *Duopoly Strategies Programmed by Experienced Players* in: *Econometrica*, Vol. 65(3), pp. 517-555.

SELTEN, Reinhard, and Rolf STOECKER, 1986, *End Behavior in Sequences of Finite Prisiner's Dilemma Supergames, a Learning Theory Approach* in: *Journal of Economic Behavior*, pp. 47-70.

SIMON, Herbert, 1976, *From Substantive to Procedural Rationality* in: *Method and Appraisal in Economics*, S. LATSIS (ed.), Cambridge University Press, pp. 129-148.

SMITH, Vernon L., 1962, *An Experimental Study of Competitive Market Behaviour* in: *Journal of Political Economy*, Vol. 70(2), pp. 111-137.

SMITH, Vernon L., 1964, *Effect of Market Organization on Competitive Equilibrium* in: *The Quarterly Journal of Economics*, Vol. 78(2), pp. 387-393.

SMITH, Vernon L., 1976, *Bidding and Auctioning Institutions: Experimental Results* in: *Bidding and Auctioning for Procurement and Allocation*, Yakov AMIHUD (ed.), New York University Press, pp. 43-63.

SMITH, Vernon L., 1982, *Microeconomic Systems as an Experimental Science* in: *American Economic Review*, Vol. 72, pp. 923-955.

SMITH, Vernon L., Arlington W. WILLIAMS, Kenneth W. BRATTON, and Micheal G. VANNONI, 1982, *Competitive Market Institutions: Double Auction vs. Sealed Bid-Offer Auctions* in: *The American Economic Review*, Vol. 72(1), pp. 58-77.

TIROLE, Jean, 1988, *The Theory of Industrial Organization*, MIT Press.

TVERSKY, Amos, and Daniel KAHNEMAN, 1982, *Judgement Under Uncertainty: Heuristics and Biases* in: *Judgement Under Uncertainty: Heuristics and Biases*, Daniel KAHNEMAN, Paul SLOVIC, and Amos TVERSKY (eds.), Cambridge University Press.

WILLIAMS, Fredric, 1973, *Effect of Market Organization on Competitive Equilibrium: The Multiunit Case* in: *Review of Economic Studies*, Vol. 40, pp. 97-113.

WILSON, Robert, 1987, *On Equilibria of Bid-Ask Markets*, in: *Arrow and the Ascent of Modern Economic Theory: Essays in Honor of Kenneth J. Arrow*, George FEIWEL (ed.), Houndmills, pp. 375-414.

VII. Appendix

VII.A. Experimental Handouts and Screen Shots

The panels contained in this appendix display the information given to the subjects in the instructions which preceded every session of the experiment. Panel VII.A.1 contains the instruction sheet and panel VII.A.2 the screen description sheet. Both were handed out to the subjects on arrival. The subjects were asked to read the instruction sheet and study the screen description sheet. Then the entire instruction sheet was read aloud by the instructor. Finally, the screen shots displayed in the panels VII.A.3 and VII.A.4 were shown to the subjects using an overhead projector. With the help of these screen shots possible decision situations in the experiment were explained. The subjects were informed that the all screen shots were *merely fictitious examples containing only non-informative numbers*.

Subjects could ask questions during the course of the instructions. They, however, were requested to ask questions only concerning the rules of the game and the usage of the computer software, but not concerning the behavior in the game. After all questions were answered, the subjects were led to the cubicals of the computer laboratory. Communication during the sessions - other than through the submitted bids and asks - was prohibited.

Panel VII.A.1.a - Instruction Sheet (German Original)

MERKBLATT

Spielaufbau

- Sie befinden sich mit allen übrigen Teilnehmern in einem Markt.
- Jedem der Teilnehmer wird eine Rolle - Käufer oder Verkäufer - zufällig zugewiesen.
- Jeder Teilnehmer darf mit maximal vier Einheiten handeln.
- Nur die aktuelle Einheit ist zum An- oder Verkauf frei.
- Den Käufern werden Wiederverkaufspreise für jede ihrer Einheit zugewiesen.
- Den Verkäufern werden Stückkosten für jede ihrer Einheiten zugewiesen.
- Es werden zwei Runden gespielt.

Runde

- In jeder neuen Runde werden den Teilnehmern neue Wiederverkaufspreise bzw. Stückkosten zugewiesen.
- Jede Runde besteht aus fünf Perioden.

Periode

- Zu Beginn jeder Periode wird der Anfangszustand wiederhergestellt:

  es sind keine Kontrakte geschlossen;
  für jeden Teilnehmer ist die erste Einheit die aktuelle Einheit;
  eine Marktseite wird zufällig ausgewählt, als erste zu bieten;
  der erste Bietzyklus beginnt.

- Eine Periode endet, wenn in zwei aufeinanderfolgenden Bietzyklen keine Gebote abgegeben worden sind.

Bietzyklus

- Zu Beginn eines Bietzyklus wird eine Marktseite aufgefordert, Gebote abzugeben.
- Wenn alle Gebote der bietenden Marktseite abgegeben sind, wird das aktuelle Marktgebot - das höchste Kaufgebot bzw. das niedrigste Verkaufsgebot - festgestellt und allen Teilnehmern bekanntgegeben.
- Ist kein Gebot abgegeben worden, bleibt das bisherige aktuelle Marktgebot weiterhin bestehen.
- Ein Bietzyklus endet, wenn beide Marktseiten ihre Gebote abgegeben haben.

Gebote

- Jeder Teilnehmer hat die Möglichkeit, innerhalb einer festgelegten Entscheidungszeit entweder ein zulässiges Gebot abzugeben oder kein Gebot abzugeben.

- Kaufgebote sind zulässig, falls sie
  kleiner oder gleich dem Wiederverkaufspreis
  und   kleiner oder gleich dem aktuellen Verkaufsgebot
  und   größer als das aktuelle Kaufgebot sind.

- Verkaufsgebote sind zulässig, falls sie
  größer oder gleich den Stückkosten
  und   größer oder gleich dem aktuellen Kaufgebot
  und   kleiner als das aktuelle Verkaufsgebot sind.

- Ist die Entscheidungszeit verstrichen, ohne daß eine Wahl getroffen wurde, so wird davon ausgegangen, daß der Teilnehmer sich dafür entsschieden hat, kein Gebot abzugeben.

Kontrakte

- Kontrakte werden geschlossen, wenn Teilnehmer einer Marktseite Gebote in Höhe des aktuellen Marktgebotes der anderen Marktseite abgeben.

- Es werden jeweils so viele Kontrakte wie möglich geschlossen.

- Wenn mehr Kaufgebote als Verkaufsgebote gleicher Höhe existieren, wird zufällig ausgelost, welche der Käufer einen Kontrakt abschließen dürfen und umgekehrt.

- Alle übrigen Gebote verfallen, so daß danach weder ein aktuelles Kauf- noch Verkaufsgebot existiert.

- Jeder Kontraktpartner erhält nach Abschluß seinen Stückgewinn und eine Kommission.

- Der Stückgewinn berechnet sich für Käufer aus Wiederverkaufspreis abzüglich Kaufpries; für Verkäufer aus Verkaufspreis abzüglich Stückkosten.

- Die Kommission beträgt jeweils zwei Punkte.

Auszahlung

- Der Wechselkurs beträgt DM 0,15 pro Punkt.

- Jedem Teilnehmer wird ein Bonus für jede beendete Periode gezahlt.

Panel VII.A.1.b - Instruction Sheet (English Translation)

INSTRUCTIONSHEET

Structure of the Game

- You are in a single market together with all other participants.
- A trader role - buyer or seller - will be randomly assigned to each participant.
- Each participant may trade four units at maximum.
- Only the current unit can be traded at any instance.
- Each unit a buyer can trade is associated with a specific resale price.
- Each unit a seller can trade is associated with a specific unit cost.
- Two rounds will be played.

Round

- New resale prices and unit costs are assigned to the participants' units in each new round.
- Each round consists of five periods.

Period

- At the beginning of a period the initial state is restored:

    no contracts are made;
    the current unit of each trader is the first unit;
    one market side is randomly chosen to be first to submit offers;
    the first bidding cycle starts.

- A period ends when no offers are made in two subsequent offer cycles.

Offer Cycles

- At the openning of an offer cycle, the participants on one of the market sides - the active market side - can submit their offers.
- After all offers of the active market side are submitted, the current market offer - the highest bid or lowest ask - is announced to be standing on te market.
- If no offer was submitted, the currently standing market offer remains standing.
- An offer cycle ends, when both market sides have submitted their offers.

Offers

- The participants are given a fixed amount of time, in which they may decide to submit a feasible offer or decide not to submit an offer.
- An offer to buy (bid) is feasible, if it is
             smaller than or equal to the resale price
  and   smaller than or equal to the current market ask
  and   greater than the current market bid.
- An offer to sell (ask) is feasible, if it is
             greater than or equal to the unit cost
  and   greater than or equal to the current market bid
  and   smaller than the current market ask.
- If the decision time expires before a decision is made, it will be assumed that the participant has decided not to submit an offer.

Contracts

- When participants on one market side submit offers equal to the current market offer of the other market side, contracts are made.
- When contracts are made, as many contracts as possible are made.
- If there are more matching bids than asks, those buyers who will trade are randomly selected. If there are more matching asks than bids, those sellers who will trade are randomly selected.
- After all possible contracts are made, the other current offers expire, so that neither a current market bid nor a current market ask is left standing on the market.
- After having made a contract, each contracting trader recieves his unit payoff and a commission.
- The unit payoff of a buyer is calculated by subtracting the price paid from the resale price of the traded unit. The unit payoff of a seller is calculated by subtracting the unit cost of the traded unit from the price received.
- Each trading partner receives a commission of two points.

Payment

- The exchange rate is DM 0,15 per point.
- Each participant will receive a bonus for every period that is completed.

Panel VII.A.2 - Screen Description Sheet

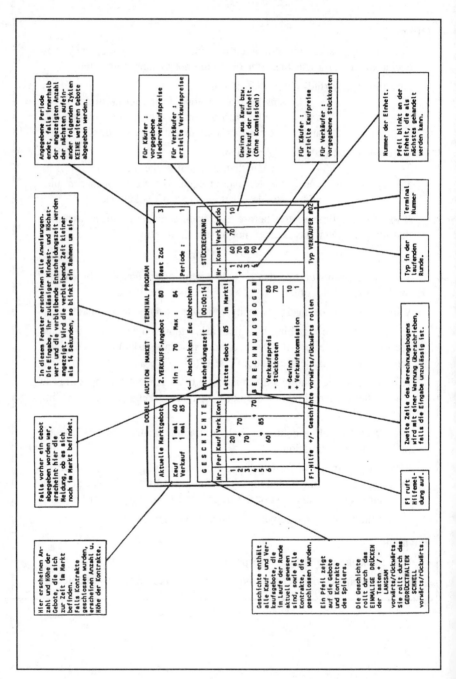

Panel VII.A.3 - Screen Shot Overhead Slide 1

```
1. KAUF -Angebot :   0        Aktuelle Marktgebote      1.VERKAUFS-Angebot machen ?

Min :   1   Max :  80         Kauf                      Min :  60    Max : 9999
                              Verkauf
Während Sie warten, können                              Bitte J oder N drücken!

Sie Ihr Angebot vorbereiten!                            Entscheidungszeit   00:00:30
```

```
1. KAUF -Angebot :   0        Aktuelle Marktgebote      1.VERKAUFS-Angebot :    0

Min :   1   Max :  80         Kauf                      Min :  60    Max : 9999
                              Verkauf
Während Sie warten, können                              Sie wollen KEIN Angebot machen!
                                                        Sind Sie sicher? J oder N drücken!
Sie Ihr Angebot vorbereiten!                            Entscheidungszeit   00:00:29
```

```
1. KAUF -Angebot machen ?     Aktuelle Marktgebote      1.VERKAUFS-Angebot :    0

Min :   1   Max :  80         Kauf                      Min :  60    Max : 9999
                              Verkauf
Bitte J oder N drücken!                                 Während Sie warten, können

Entscheidungszeit   00:00:30                            Sie Ihr Angebot vorbereiten!
```

```
1. KAUF -Angebot :  20        Aktuelle Marktgebote      1.VERKAUFS-Angebot :    0

Min :   1   Max :  80         Kauf                      Min :  60    Max : 9999
                              Verkauf
<┘ Abschicken  Esc-Abbrechen                            Während Sie warten, können

Entscheidungszeit   00:00:29                            Sie Ihr Angebot vorbereiten!
```

```
1. KAUF -Angebot :  20        Aktuelle Marktgebote      1.VERKAUFS-Angebot :    0

Min :   1   Max :  80         Kauf                      Min :  60    Max : 9999
                              Verkauf
Ihr Gebot :  KAUFEN zu  20                              Während Sie warten, können
Sind Sie sicher? J oder N drücken
Entscheidungszeit   00:00:28                            Sie Ihr Angebot vorbereiten!
```

```
1. KAUF -Angebot :  20        Aktuelle Marktgebote      1.VERKAUFS-Angebot machen ?
                              Kauf    1 mal   20         Min :  60    Max : 9999
Angebot wird bearbeitet!      Verkauf
                                                        Bitte J oder N drücken!

Bitte   Warten!                                         Entscheidungszeit   00:00:30

Letztes Gebot   20 im Markt!
```

Panel VII.A.4 - Screen Shot Overhead Slide 2

```
┌─────────────────────────┐  ┌───────────────────────┐  ┌───────────────────────────┐
│ 1.  KAUF  -Angebot :  20│  │ Aktuelle Marktgebote  │  │ 1.VERKAUFS-Angebot :   70 │
│                         │  │ Kauf      1 mal   20  │  │ Min :   60      Max : 9999│
│ Angebot wird bearbeitet!│  │ Verkauf               │  │                           │
│                         │  │                       │  │ ←┘ Abschicken Esc-Abbrechen│
│      Bitte  Warten!     │  │                       │  │ Entscheidungszeit  00:00:29│
└─────────────────────────┘  └───────────────────────┘  └───────────────────────────┘
┌─────────────────────────┐
│ Letztes Gebot   20 im Markt!│
└─────────────────────────┘


┌─────────────────────────┐  ┌───────────────────────┐  ┌───────────────────────────┐
│ 2.  KAUF  -Angebot machen ?│ Aktuelle Marktgebote  │  │ 1.VERKAUFS-Angebot :   70 │
│ Min :   21.   Max :   70│  │ Kauf      1 mal   20  │  │                           │
│ Bitte J oder N drücken! │  │ Verkauf   1 mal   70  │  │ Angebot wird bearbeitet!  │
│ Entscheidungszeit 00:00:30│ │                       │  │      Bitte   Warten!      │
└─────────────────────────┘  └───────────────────────┘  └───────────────────────────┘
┌─────────────────────────┐                             ┌───────────────────────────┐
│ Letztes Gebot   20 im Markt!│                         │ Letztes Gebot   70 im Markt!│
└─────────────────────────┘                             └───────────────────────────┘


┌─────────────────────────┐  ┌───────────────────────┐  ┌───────────────────────────┐
│ 2.  KAUF  -Angebot :  70│  │ Aktuelle Marktgebote  │  │ 1.VERKAUFS-Angebot :   70 │
│ Min :   21    Max :   70│  │ Kauf      1 mal   20  │  │                           │
│ ←┘ Abschicken  Esc-Abbrechen│ Verkauf  1 mal   70  │  │ Angebot wird bearbeitet!  │
│ Entscheidungszeit 00:00:29│ │                       │  │      Bitte   Warten!      │
└─────────────────────────┘  └───────────────────────┘  └───────────────────────────┘
┌─────────────────────────┐                             ┌───────────────────────────┐
│ Letztes Gebot   20 im Markt!│                         │ Letztes Gebot   70 im Markt!│
└─────────────────────────┘                             └───────────────────────────┘


┌─────────────────────────┐  ┌───────────────────────┐  ┌───────────────────────────┐
│ 1.  KAUF  -Angebot :   0│  │ K O N T R A K T E     │  │ 1.VERKAUFS-Angebot :    0 │
│ Min :   1     Max :   70│  │    1 mal    70        │  │ Min :   70      Max : 9999│
│ Während Sie warten, können│ │ Sie haben gekauft!    │  │ Bitte J oder N drücken!   │
│ Sie Ihr Angebot vorbereiten!│ Sie haben verkauft!   │  │ Entscheidungszeit 00:00:30│
└─────────────────────────┘  └───────────────────────┘  └───────────────────────────┘
┌─────────────────────────┐                             ┌───────────────────────────┐
│ Letztes Gebot   70 :Kontrakt!│                        │ Letztes Gebot   70 :Kontrakt!│
└─────────────────────────┘                             └───────────────────────────┘
```

VII.B. Experimental Data Panels and Figures

The figures contained in this appendix display the price paths, i.e. the sequence and prices of the contracts, of every market period of every session of the experiment. To enable a quick reference, the figures are sorted in the following manner: First, the figures for all inexperienced sessions are displayed, where these are ordered (as in all preceding tables) from the market type M1 to the market type M2 sessions and from sessions with less players to those with more players. Then the figures for the experienced sessions follow. These are ordered in a similar manner. Note that the sessions with names only differing in the last character (inexperienced = 0 and experienced = 1) were run with an identical group of subjects.

Following the figures, panels are contained in this appendix that display the summery results of the experimental sessions. The panels are sorted in the same manner as the figures.

Most of the parameters' labels in the panels are self-explanatory.

The following abbreviations are used in the panels:

&	and
avg.	average
p-avg	previous period's average price
p-max	previous period's maximum price
p-min	previous period's minimum price
p-own	trader's best offer for the unit in the previous period
p*	market clearing price p*
q*	market clearing quantity q*
red.value	redemption value
rel. freq.	relative frequency

Figure VII.B.1

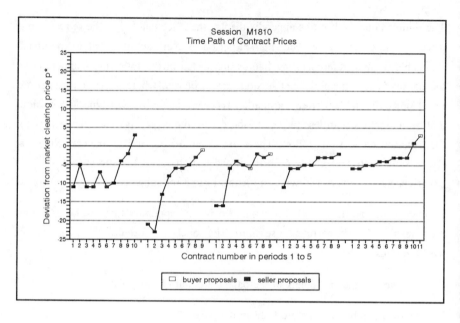

Figure VII.B.2

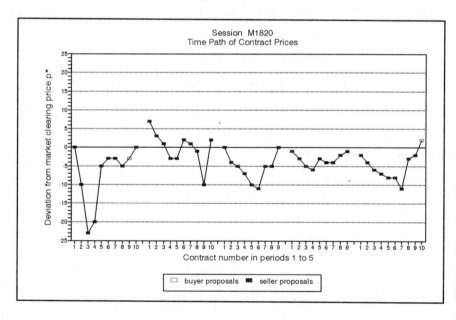

Figure VII.B.3

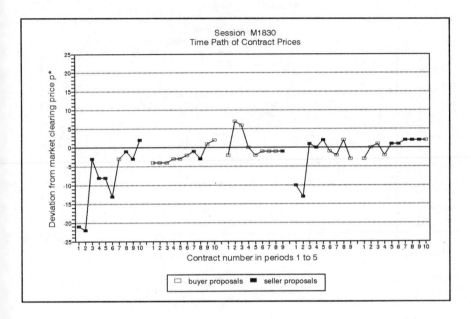

Figure VII.B.4

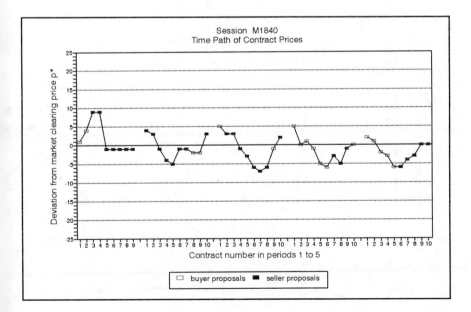

Figure VII.B.5

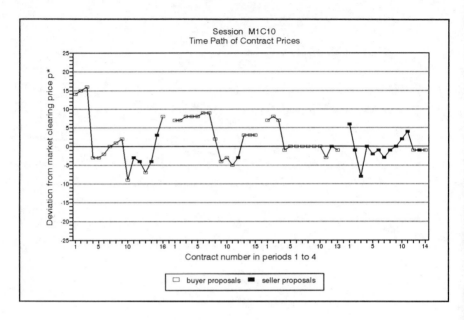

Figure VII.B.6

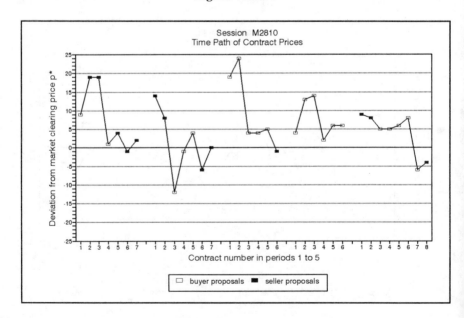

Figure VII.B.7

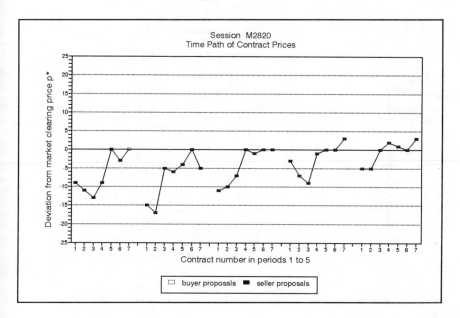

Figure VII.B.8

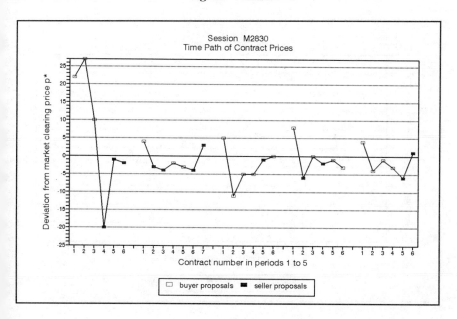

Figure VII.B.9

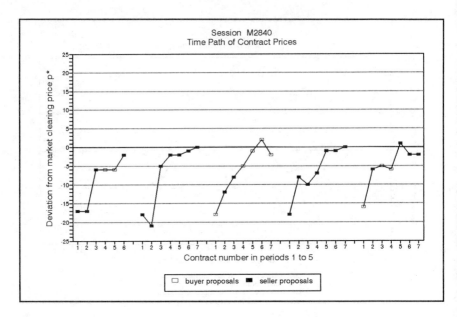

Figure VII.B.10

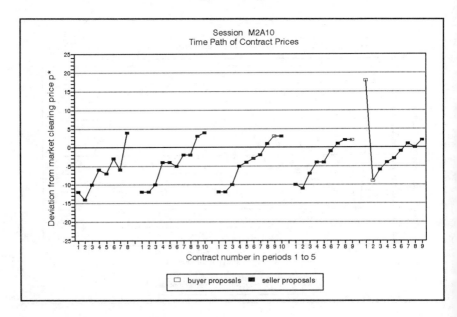

Figure VII.B.11

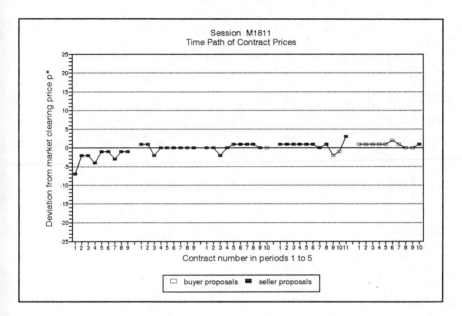

Figure VII.B.12

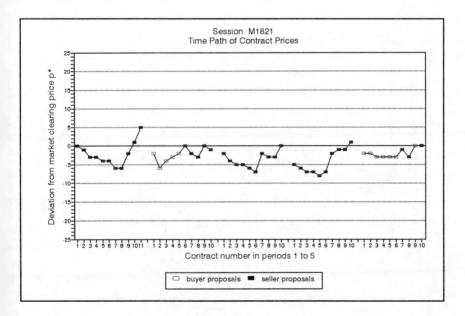

Figure VII.B.13

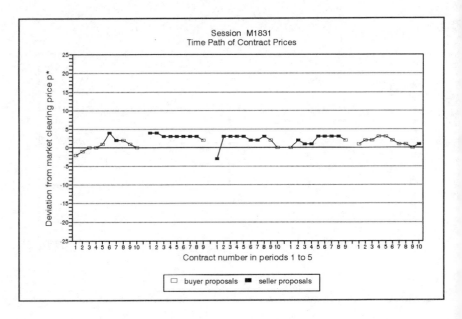

Figure VII.B.14

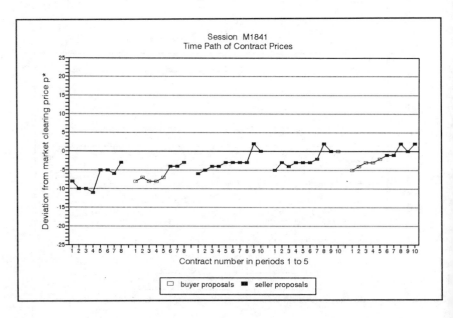

Figure VII.B.15

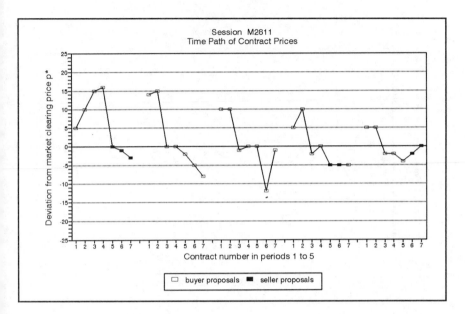

Figure VII.B.16

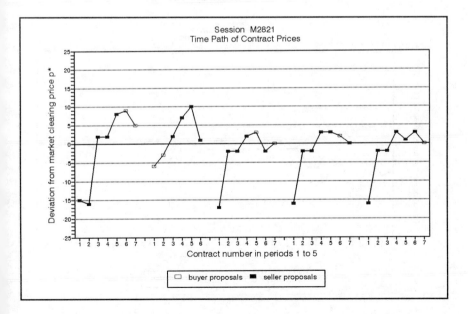

Figure VII.B.17

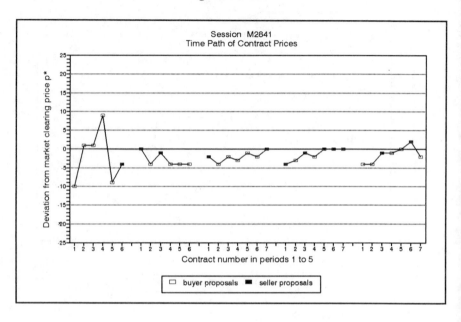

Figure VII.B.18

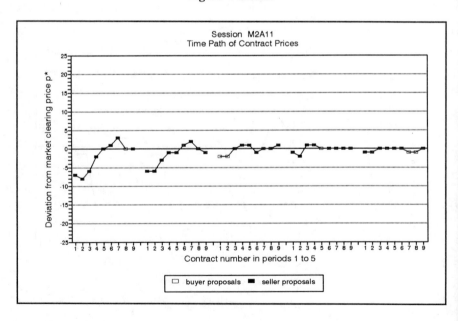

Panel VII.B.1

Double Auction Market Session Results   Session: M1810   Number of traders: 8   File: DAMM1810.SES

Market clearing quantity q^* :   10
Market clearing price p^* :   261

	Average of Periods	Period 1	Period 2	Period 3	Period 4	Period 5
Deviation of traded quantity from q^*	-0.40	0	-1	-1	-1	1
Efficiency	0.95	0.97	0.92	0.92	0.98	0.99
Average contract price (deviation from p^*)	-6.24	-6.90	-9.56	-6.67	-4.89	-3.18
Minimum contract price (deviation from p^*)	-13.40	-11.00	-23.00	-16.00	-11.00	-6.00
Maximum contract price (deviation from p^*)	0.20	3.00	-1.00	-2.00	-2.00	3.00
Market clearing price concordance coefficient	0.64	0.58	0.48	0.64	0.73	0.77
Rel. freq. of falling price paths	0.00	0	0	0	0	0
Rel. freq. of rising price paths	1.00	1	1	1	1	1
Rel. freq. of hanging price paths	0.00	0	0	0	0	0
Rel. freq. of standing price paths	0.00	0	0	0	0	0
Rel. freq. of unknown price paths	0.00	0	0	0	0	0
Rel. freq. of contract proposals by buyers	0.08	0.00	0.11	0.22	0.00	0.09
... rel. freq. of these with price < p^*	0.67	-	1.00	1.00	-	0.00
... rel. freq. of these with price = p^*	0.00	-	0.00	0.00	-	0.00
... rel. freq. of these with price > p^*	0.33	-	0.00	0.00	-	1.00
... avg. buyer proposed contract price - p^*	-0.67	-	-1.00	-4.00	-	3.00
Rel. freq. of contract proposals by sellers	0.92	1.00	0.89	0.78	1.00	0.91
... rel. freq. of these with price < p^*	0.96	0.90	1.00	1.00	1.00	0.90
... rel. freq. of these with price = p^*	0.00	0.00	0.00	0.00	0.00	0.00
... rel. freq. of these with price > p^*	0.04	0.10	0.00	0.00	0.00	0.10
... avg. seller proposed contract price - p^*	-6.73	-6.90	-10.63	-7.43	-4.89	-3.80
Rel. freq. of proposer's curses - all traders	0.90	0.90	0.89	0.78	1.00	0.91
Rel. freq. of proposer's curses - all buyers	0.33	-	0.00	0.00	-	1.00
Rel. freq. of proposer's curses - all sellers	0.96	0.90	1.00	1.00	1.00	0.90
Rel. freq. of proposer's curses - buyer 1	-	-	-	-	-	-
Rel. freq. of proposer's curses - buyer 2	0.00	-	0.00	0.00	-	-
Rel. freq. of proposer's curses - buyer 3	-	-	-	-	-	-
Rel. freq. of proposer's curses - buyer 4	1.00	-	-	-	-	1.00
Rel. freq. of proposer's curses - seller 5	1.00	1.00	1.00	1.00	1.00	1.00
Rel. freq. of proposer's curses - seller 6	0.87	0.67	1.00	1.00	1.00	0.67
Rel. freq. of proposer's curses - seller 7	1.00	1.00	1.00	1.00	1.00	1.00
Rel. freq. of proposer's curses - seller 8	1.00	1.00	1.00	1.00	1.00	1.00
Number of market ask and market bid matches	9.20	10.00	9.00	8.00	8.00	11.00
Number of contracts per market ask-bid match	1.04	1.00	1.00	1.13	1.13	1.00
Avg. rank of traded red.value at trading time	2.39	2.50	2.56	2.63	2.38	1.91
Avg. rank of traded unit cost at trading time	1.77	1.20	1.78	2.25	2.00	1.64
Number of submitted bids per contract	9.98	8.10	12.67	11.56	8.56	9.00
... rel. freq. of bids < p^* - 200	0.01	0.01	0.00	0.01	0.01	0.00
... rel. freq. of bids < p^*	0.98	0.94	1.00	1.00	1.00	0.94
... rel. freq. of bids = p^*	0.01	0.02	0.00	0.00	0.00	0.02
... rel. freq. of bids > p^*	0.02	0.04	0.00	0.00	0.00	0.04
Number of submitted asks per contract	5.73	6.70	6.11	5.22	4.44	6.18
... rel. freq. of asks < p^*	0.52	0.42	0.55	0.57	0.63	0.43
... rel. freq. of asks = p^*	0.11	0.00	0.04	0.17	0.18	0.19
... rel. freq. of asks > p^*	0.37	0.58	0.42	0.26	0.20	0.38
... rel. freq. of asks > p^* + 200	0.03	0.10	0.02	0.02	0.00	0.00
Number of offers for all units in periods > 1	151	-	169	151	117	167
(a) rel. freq. of bids > p-max & asks < p-min	0.02	-	0.02	0.00	0.00	0.04
(b) rel. freq. of bids > p-avg & asks < p-avg	0.15	-	0.08	0.16	0.21	0.15
(c) rel. freq. of bids > p-own & asks < p-own	0.20	-	0.30	0.10	0.16	0.24
(d) rel. freq. of bids < p-own & asks > p-own	0.68	-	0.67	0.79	0.69	0.55
Number of offers for units bought above p-avg or sold below p-avg in the previous period	51	-	65	55	36	46
(C) rel. freq. of bids > p-own & asks < p-own	0.08	-	0.06	0.07	0.08	0.09
Number of offers for units that were not traded in the previous period	54	-	45	56	45	70
(D) rel. freq. of bids < p-own & asks > p-own	0.46	-	0.31	0.68	0.53	0.33
Aggressive offer frequency ratio (c)/(C)	0.45	-	0.21	0.73	0.51	0.36
Moderate offer frequency ratio (d)/(D)	0.67	-	0.46	0.86	0.77	0.60

Panel VII.B.2

Double Auction Market Session Results   Session: M1820   Number of traders: 8   File: DAMM1820.SES

	Average of Periods	Period 1	Period 2	Period 3	Period 4	Period 5
Market clearing quantity q* : 10						
Market clearing price p* : 390						
Deviation of traded quantity from q*	-0.40	0	0	-1	-1	0
Efficiency	0.99	1.00	0.96	0.99	0.99	1.00
Average contract price (deviation from p*)	-4.13	-7.20	-0.10	-5.22	-3.22	-4.90
Minimum contract price (deviation from p*)	-12.20	-23.00	-10.00	-11.00	-6.00	-11.00
Maximum contract price (deviation from p*)	1.60	0.00	7.00	0.00	-1.00	2.00
Market clearing price concordance coefficient	0.73	0.59	0.81	0.72	0.82	0.70
Rel. freq. of falling price paths	0.00	0	0	0	0	0
Rel. freq. of rising price paths	0.00	0	0	0	0	0
Rel. freq. of hanging price paths	1.00	1	1	1	1	1
Rel. freq. of standing price paths	0.00	0	0	0	0	0
Rel. freq. of unknown price paths	0.00	0	0	0	0	0
Rel. freq. of contract proposals by buyers	0.04	0.10	0.00	0.00	0.00	0.10
... rel. freq. of these with price < p*	0.50	1.00	-	-	-	0.00
... rel. freq. of these with price = p*	0.00	0.00	-	-	-	0.00
... rel. freq. of these with price > p*	0.50	0.00	-	-	-	1.00
... avg. buyer proposed contract price - p*	-0.50	-3.00	-	-	-	2.00
Rel. freq. of contract proposals by sellers	0.96	0.90	1.00	1.00	1.00	0.90
... rel. freq. of these with price < p*	0.79	0.78	0.40	0.78	1.00	1.00
... rel. freq. of these with price = p*	0.09	0.22	0.00	0.22	0.00	0.00
... rel. freq. of these with price > p*	0.12	0.00	0.60	0.00	0.00	0.00
... avg. seller proposed contract price - p*	-4.38	-7.67	-0.10	-5.22	-3.22	-5.67
Rel. freq. of proposer's curses - all traders	0.78	0.70	0.40	0.78	1.00	1.00
Rel. freq. of proposer's curses - all buyers	0.50	0.00	-	-	-	1.00
Rel. freq. of proposer's curses - all sellers	0.79	0.78	0.40	0.78	1.00	1.00
Rel. freq. of proposer's curses - buyer 1	-	-	-	-	-	-
Rel. freq. of proposer's curses - buyer 2	0.00	0.00	-	-	-	-
Rel. freq. of proposer's curses - buyer 3	-	-	-	-	-	-
Rel. freq. of proposer's curses - buyer 4	1.00	-	-	-	-	1.00
Rel. freq. of proposer's curses - seller 5	0.87	1.00	0.67	0.67	1.00	1.00
Rel. freq. of proposer's curses - seller 6	0.80	0.67	0.33	1.00	1.00	1.00
Rel. freq. of proposer's curses - seller 7	0.80	1.00	0.50	0.50	1.00	1.00
Rel. freq. of proposer's curses - seller 8	0.70	0.50	0.00	1.00	1.00	1.00
Number of market ask and market bid matches	8.80	9.00	9.00	9.00	8.00	9.00
Number of contracts per market ask-bid match	1.09	1.11	1.11	1.00	1.13	1.11
Avg. rank of traded red.value at trading time	2.19	1.89	2.33	2.00	2.50	2.22
Avg. rank of traded unit cost at trading time	1.96	1.78	2.00	1.89	2.13	2.00
Number of submitted bids per contract	5.06	5.20	5.60	5.89	4.22	4.40
... rel. freq. of bids < p* - 200	0.00	0.00	0.00	0.00	0.00	0.02
... rel. freq. of bids < p*	0.92	0.96	0.77	0.92	0.97	0.98
... rel. freq. of bids = p*	0.04	0.04	0.04	0.08	0.03	0.00
... rel. freq. of bids > p*	0.04	0.00	0.20	0.00	0.00	0.02
Number of submitted asks per contract	4.60	4.70	4.60	5.22	3.78	4.70
... rel. freq. of asks < p*	0.28	0.28	0.13	0.32	0.35	0.34
... rel. freq. of asks = p*	0.10	0.06	0.00	0.11	0.18	0.17
... rel. freq. of asks > p*	0.61	0.66	0.87	0.57	0.47	0.49
... rel. freq. of asks > p* + 200	0.09	0.09	0.11	0.11	0.12	0.02
Number of offers for all units in periods > 1	91	-	102	100	72	91
(a) rel. freq. of bids > p-max & asks < p-min	0.04	-	0.11	0.01	0.00	0.05
(b) rel. freq. of bids > p-avg & asks < p-avg	0.20	-	0.26	0.15	0.22	0.16
(c) rel. freq. of bids > p-own & asks < p-own	0.22	-	0.29	0.22	0.17	0.20
(d) rel. freq. of bids < p-own & asks > p-avg	0.66	-	0.61	0.69	0.69	0.66
Number of offers for units bought above p-avg						
or sold below p-avg in the previous period	30	-	45	35	18	22
(C) rel. freq. of bids > p-own & asks < p-own	0.13	-	0.27	0.14	0.00	0.09
Number of offers for units						
that were not traded in the previous period	28	-	29	22	26	35
(D) rel. freq. of bids < p-own & asks > p-own	0.57	-	0.52	0.77	0.42	0.57
Aggressive offer frequency ratio (c)/(C)	0.50	-	0.91	0.65	0.00	0.46
Moderate offer frequency ratio (d)/(D)	0.86	-	0.85	1.12	0.61	0.87

Panel VII.B.3

Double Auction Market Session Results   Session: M1830   Number of traders: 8   File: DAMM1830.SES

	Average of Periods	Period 1	Period 2	Period 3	Period 4	Period 5
Market clearing quantity q* : 10						
Market clearing price p* : 893						
Deviation of traded quantity from q*	-0.40	0	0	-1	-1	0
Efficiency	0.98	0.99	0.99	0.97	0.95	0.99
Average contract price (deviation from p*)	-2.32	-8.00	-2.10	0.56	-2.67	0.60
Minimum contract price (deviation from p*)	-8.80	-22.00	-4.00	-2.00	-13.00	-3.00
Maximum contract price (deviation from p*)	3.00	2.00	2.00	7.00	2.00	2.00
Market clearing price concordance coefficient	0.79	0.53	0.85	0.87	0.79	0.91
Rel. freq. of falling price paths	0.00	0	0	0	0	0
Rel. freq. of rising price paths	0.60	1	1	0	0	1
Rel. freq. of hanging price paths	0.00	0	0	0	0	0
Rel. freq. of standing price paths	0.40	0	0	1	1	0
Rel. freq. of unknown price paths	0.00	0	0	0	0	0
Rel. freq. of contract proposals by buyers	0.55	0.10	0.80	0.89	0.44	0.50
... rel. freq. of these with price < p*	0.71	1.00	0.75	0.63	0.75	0.40
... rel. freq. of these with price = p*	0.07	0.00	0.00	0.13	0.00	0.20
... rel. freq. of these with price > p*	0.23	0.00	0.25	0.25	0.25	0.40
... avg. buyer proposed contract price - p*	-1.16	-3.00	-2.13	0.75	-1.00	-0.40
Rel. freq. of contract proposals by sellers	0.45	0.90	0.20	0.11	0.56	0.50
... rel. freq. of these with price < p*	0.66	0.89	1.00	1.00	0.40	0.00
... rel. freq. of these with price = p*	0.04	0.00	0.00	0.00	0.20	0.00
... rel. freq. of these with price > p*	0.30	0.11	0.00	0.00	0.40	1.00
... avg. seller proposed contract price - p*	-2.79	-8.56	-2.00	-1.00	-4.00	1.60
Rel. freq. of proposer's curses - all traders	0.41	0.80	0.40	0.33	0.33	0.20
Rel. freq. of proposer's curses - all buyers	0.23	0.00	0.25	0.25	0.25	0.40
Rel. freq. of proposer's curses - all sellers	0.66	0.89	1.00	1.00	0.40	0.00
Rel. freq. of proposer's curses - buyer 1	0.17	-	0.50	0.00	0.00	-
Rel. freq. of proposer's curses - buyer 2	0.20	0.00	0.50	0.50	0.00	0.00
Rel. freq. of proposer's curses - buyer 3	0.17	-	0.00	0.33	0.00	0.33
Rel. freq. of proposer's curses - buyer 4	0.67	-	0.00	-	1.00	1.00
Rel. freq. of proposer's curses - seller 5	0.63	1.00	1.00	-	0.50	0.00
Rel. freq. of proposer's curses - seller 6	0.56	0.67	1.00	-	-	0.00
Rel. freq. of proposer's curses - seller 7	0.75	1.00	-	-	0.50	-
Rel. freq. of proposer's curses - seller 8	0.50	1.00	-	1.00	0.00	0.00
Number of market ask and market bid matches	9.40	9.00	10.00	9.00	9.00	10.00
Number of contracts per market ask-bid match	1.02	1.11	1.00	1.00	1.00	1.00
Avg. rank of traded red.value at trading time	2.20	2.11	2.00	2.33	2.44	2.10
Avg. rank of traded unit cost at trading time	1.83	1.67	1.70	1.89	2.11	1.80
Number of submitted bids per contract	5.97	7.50	6.00	5.11	5.33	5.90
... rel. freq. of bids < p* - 200	0.04	0.12	0.03	0.00	0.00	0.03
... rel. freq. of bids < p*	0.88	0.96	0.92	0.93	0.83	0.76
... rel. freq. of bids = p*	0.04	0.01	0.03	0.02	0.04	0.08
... rel. freq. of bids > p*	0.08	0.03	0.05	0.04	0.13	0.15
Number of submitted asks per contract	6.03	6.20	6.10	6.44	6.00	5.40
... rel. freq. of asks < p*	0.18	0.35	0.26	0.10	0.13	0.04
... rel. freq. of asks = p*	0.04	0.02	0.02	0.07	0.04	0.06
... rel. freq. of asks > p*	0.78	0.63	0.72	0.83	0.83	0.91
... rel. freq. of asks > p* + 200	0.01	0.02	0.00	0.00	0.00	0.02
Number of offers for all units in periods > 1	110	-	121	104	102	113
(a) rel. freq. of bids > p-max & asks < p-min	0.01	-	0.00	0.02	0.03	0.00
(b) rel. freq. of bids > p-avg & asks < p-avg	0.18	-	0.26	0.11	0.12	0.25
(c) rel. freq. of bids > p-own & asks < p-own	0.16	-	0.13	0.16	0.17	0.19
(d) rel. freq. of bids < p-own & asks > p-own	0.65	-	0.68	0.68	0.65	0.58
Number of offers for units bought above p-avg or sold below p-avg in the previous period	39	-	39	40	44	32
(C) rel. freq. of bids > p-own & asks < p-own	0.07	-	0.10	0.00	0.09	0.09
Number of offers for units that were not traded in the previous period	35	-	39	25	32	45
(D) rel. freq. of bids < p-own & asks > p-own	0.53	-	0.51	0.68	0.53	0.40
Aggressive offer frequency ratio (c)/(C)	0.45	-	0.78	0.00	0.55	0.48
Moderate offer frequency ratio (d)/(D)	0.81	-	0.76	1.00	0.82	0.68

Panel VII.B.4

Double Auction Market Session Results	Session: M1840	Number of traders: 8	File: DAMM1840.SES		
Market clearing quantity q* : 10	Average	Period	Period	Period	Period	Period
Market clearing price p* : 211	of Periods	1	2	3	4	5

	Average of Periods	Period 1	Period 2	Period 3	Period 4	Period 5
Deviation of traded quantity from q*	-0.20	-1	0	0	0	0
Efficiency	1.00	0.99	0.98	1.00	1.00	1.00
Average contract price (deviation from p*)	-0.66	2.00	-0.60	-1.10	-1.50	-2.10
Minimum contract price (deviation from p*)	-5.00	-1.00	-5.00	-7.00	-6.00	-6.00
Maximum contract price (deviation from p*)	5.00	9.00	4.00	5.00	5.00	2.00
Market clearing price concordance coefficient	0.83	0.83	0.85	0.79	0.85	0.85
Rel. freq. of falling price paths	0.00	0	0	0	0	0
Rel. freq. of rising price paths	0.00	0	0	0	0	0
Rel. freq. of hanging price paths	0.80	0	1	1	1	1
Rel. freq. of standing price paths	0.20	1	0	0	0	0
Rel. freq. of unknown price paths	0.00	0	0	0	0	0
Rel. freq. of contract proposals by buyers	0.36	0.22	0.20	0.20	0.70	0.50
... rel. freq. of these with price < p*	0.51	0.00	1.00	0.50	0.43	0.60
... rel. freq. of these with price = p*	0.06	0.00	0.00	0.00	0.29	0.00
... rel. freq. of these with price > p*	0.44	1.00	0.00	0.50	0.29	0.40
... avg. buyer proposed contract price - p*	0.01	2.50	-2.00	2.00	-0.86	-1.60
Rel. freq. of contract proposals by sellers	0.64	0.78	0.80	0.80	0.30	0.50
... rel. freq. of these with price < p*	0.71	0.71	0.63	0.63	1.00	0.60
... rel. freq. of these with price = p*	0.08	0.00	0.00	0.00	0.00	0.40
... rel. freq. of these with price > p*	0.21	0.29	0.38	0.38	0.00	0.00
... avg. seller proposed contract price - p*	-1.17	1.86	-0.25	-1.88	-3.00	-2.60
Rel. freq. of proposer's curses - all traders	0.58	0.78	0.50	0.60	0.50	0.50
Rel. freq. of proposer's curses - all buyers	0.44	1.00	0.00	0.50	0.29	0.40
Rel. freq. of proposer's curses - all sellers	0.71	0.71	0.63	0.63	1.00	0.60
Rel. freq. of proposer's curses - buyer 1	0.50	1.00	.	0.00	0.50	0.50
Rel. freq. of proposer's curses - buyer 2	0.57	1.00	0.00	1.00	0.33	0.50
Rel. freq. of proposer's curses - buyer 3	0.00	.	0.00	.	.	.
Rel. freq. of proposer's curses - buyer 4	0.00	.	.	.	0.00	0.00
Rel. freq. of proposer's curses - seller 5	0.73	0.67	1.00	0.50	1.00	0.50
Rel. freq. of proposer's curses - seller 6	0.71	1.00	0.67	0.67	.	0.50
Rel. freq. of proposer's curses - seller 7	0.33	0.00	0.50	0.50	.	.
Rel. freq. of proposer's curses - seller 8	0.90	1.00	0.50	1.00	1.00	1.00
Number of market ask and market bid matches	9.20	8.00	9.00	9.00	10.00	10.00
Number of contracts per market ask-bid match	1.07	1.13	1.11	1.11	1.00	1.00
Avg. rank of traded red.value at trading time	1.93	1.50	2.44	2.11	1.70	1.90
Avg. rank of traded unit cost at trading time	1.93	2.00	2.22	2.11	1.60	1.70
Number of submitted bids per contract	5.94	4.89	5.60	5.80	7.30	6.10
... rel. freq. of bids < p* - 200	0.00	0.00	0.00	0.00	0.00	0.00
... rel. freq. of bids < p*	0.87	0.75	0.86	0.84	0.95	0.93
... rel. freq. of bids = p*	0.03	0.02	0.04	0.03	0.03	0.03
... rel. freq. of bids > p*	0.10	0.23	0.11	0.12	0.03	0.03
Number of submitted asks per contract	4.30	4.00	4.00	3.90	5.30	4.30
... rel. freq. of asks < p*	0.26	0.14	0.33	0.26	0.26	0.33
... rel. freq. of asks = p*	0.03	0.03	0.00	0.00	0.08	0.05
... rel. freq. of asks > p*	0.71	0.83	0.68	0.74	0.66	0.63
... rel. freq. of asks > p* + 200	0.00	0.00	0.00	0.00	0.00	0.00
Number of offers for all units in periods > 1	106	-	96	97	126	104
(a) rel. freq. of bids > p-max & asks < p-min	0.02	-	0.04	0.04	0.00	0.00
(b) rel. freq. of bids > p-avg & asks < p-avg	0.14	-	0.19	0.14	0.10	0.13
(c) rel. freq. of bids > p-own & asks < p-own	0.21	-	0.24	0.21	0.26	0.14
(d) rel. freq. of bids < p-own & asks > p-own	0.64	-	0.63	0.64	0.57	0.71
Number of offers for units bought above p-avg or sold below p-avg in the previous period	25	-	31	18	24	27
(C) rel. freq. of bids > p-own & asks < p-own	0.08	-	0.13	0.17	0.00	0.04
Number of offers for units that were not traded in the previous period	26	-	31	27	26	20
(D) rel. freq. of bids < p-own & asks > p-own	0.49	-	0.58	0.52	0.46	0.40
Aggressive offer frequency ratio (c)/(C)	0.40	-	0.54	0.81	0.00	0.26
Moderate offer frequency ratio (d)/(D)	0.78	-	0.93	0.81	0.81	0.56

Panel VII.B.5

Double Auction Market Session Results   Session: M1C10   Number of traders: 12   File: DAMM1C10.SES

	Average of Periods	Period 1	Period 2	Period 3	Period 4	Period 5
Market clearing quantity q* : 14						
Market clearing price p* : 763						
Deviation of traded quantity from q*	0.50	2	1	-1	0	
Efficiency	0.98	0.97	0.98	0.99	0.99	
Average contract price (deviation from p*)	1.44	1.50	3.47	1.31	-0.50	
Minimum contract price (deviation from p*)	-6.25	-9.00	-5.00	-3.00	-8.00	
Maximum contract price (deviation from p*)	9.75	16.00	9.00	8.00	6.00	
Market clearing price concordance coefficient	0.84	0.75	0.78	0.92	0.91	
Rel. freq. of falling price paths	0.75	0	1	1	1	
Rel. freq. of rising price paths	0.00	0	0	0	0	
Rel. freq. of hanging price paths	0.25	1	0	0	0	
Rel. freq. of standing price paths	0.00	0	0	0	0	
Rel. freq. of unknown price paths	0.00	0	0	0	0	
Rel. freq. of contract proposals by buyers	0.69	0.75	0.93	0.92	0.14	
... rel. freq. of these with price < p*	0.47	0.42	0.21	0.25	1.00	
... rel. freq. of these with price = p*	0.15	0.08	0.00	0.50	0.00	
... rel. freq. of these with price > p*	0.38	0.50	0.79	0.25	0.00	
... avg. buyer proposed contract price - p*	1.75	2.67	3.93	1.42	-1.00	
Rel. freq. of contract proposals by sellers	0.31	0.25	0.07	0.08	0.86	
... rel. freq. of these with price < p*	0.58	0.75	1.00	0.00	0.58	
... rel. freq. of these with price = p*	0.29	0.00	0.00	1.00	0.17	
... rel. freq. of these with price > p*	0.13	0.25	0.00	0.00	0.25	
... avg. seller proposed contract price - p*	-1.35	-2.00	-3.00	0.00	-0.42	
Rel. freq. of proposer's curses - all traders	0.52	0.56	0.80	0.23	0.50	
Rel. freq. of proposer's curses - all buyers	0.38	0.50	0.79	0.25	0.00	
Rel. freq. of proposer's curses - all sellers	0.58	0.75	1.00	0.00	0.58	
Rel. freq. of proposer's curses - buyer 1	0.89	1.00	0.67	1.00	-	
Rel. freq. of proposer's curses - buyer 2	0.21	0.33	0.50	0.00	0.00	
Rel. freq. of proposer's curses - buyer 3	0.50	0.50	1.00	0.00	-	
Rel. freq. of proposer's curses - buyer 4	0.72	0.67	1.00	0.50	-	
Rel. freq. of proposer's curses - buyer 5	0.50	0.50	1.00	0.00	-	
Rel. freq. of proposer's curses - buyer 6	0.22	0.00	0.67	0.00	-	
Rel. freq. of proposer's curses - seller 7	0.50	1.00	-	-	0.00	
Rel. freq. of proposer's curses - seller 8	1.00	-	-	-	1.00	
Rel. freq. of proposer's curses - seller 9	0.67	1.00	-	0.00	1.00	
Rel. freq. of proposer's curses - seller 10	0.89	1.00	1.00	-	0.67	
Rel. freq. of proposer's curses - seller 11	0.33	-	-	-	0.33	
Rel. freq. of proposer's curses - seller 12	0.25	0.00	-	-	0.50	
Number of market ask and market bid matches	14.00	15.00	14.00	13.00	14.00	
Number of contracts per market ask-bid match	1.04	1.07	1.07	1.00	1.00	
Avg. rank of traded red.value at trading time	2.87	2.80	3.21	2.38	3.07	
Avg. rank of traded unit cost at trading time	2.54	2.27	2.93	2.23	2.71	
Number of submitted bids per contract	6.29	7.56	6.00	5.31	6.29	
... rel. freq. of bids < p* - 200	0.03	0.12	0.01	0.00	0.00	
... rel. freq. of bids < p*	0.85	0.91	0.73	0.83	0.92	
... rel. freq. of bids = p*	0.05	0.01	0.02	0.12	0.03	
... rel. freq. of bids > p*	0.11	0.08	0.24	0.06	0.05	
Number of submitted asks per contract	6.50	7.13	6.53	5.69	6.64	
... rel. freq. of asks < p*	0.12	0.18	0.06	0.09	0.13	
... rel. freq. of asks = p*	0.08	0.04	0.00	0.26	0.04	
... rel. freq. of asks > p*	0.80	0.78	0.94	0.65	0.83	
... rel. freq. of asks > p* + 200	0.01	0.00	0.01	0.00	0.01	
Number of offers for all units in periods > 1	171	-	188	143	181	
(a) rel. freq. of bids > p-max & asks < p-min	0.00	-	0.00	0.00	0.01	
(b) rel. freq. of bids > p-avg & asks < p-avg	0.16	-	0.11	0.26	0.11	
(c) rel. freq. of bids > p-own & asks < p-own	0.20	-	0.22	0.22	0.18	
(d) rel. freq. of bids < p-own & asks > p-own	0.61	-	0.69	0.57	0.57	
Number of offers for units bought above p-avg						
or sold below p-avg in the previous period	44	-	56	36	39	
(C) rel. freq. of bids > p-own & asks < p-own	0.15	-	0.09	0.17	0.18	
Number of offers for units						
that were not traded in the previous period	44	-	40	18	73	
(D) rel. freq. of bids < p-own & asks > p-own	0.47	-	0.38	0.56	0.49	
Aggressive offer frequency ratio (c)/(C)	0.73	-	0.41	0.77	1.02	
Moderate offer frequency ratio (d)/(D)	0.79	-	0.55	0.97	0.86	

Panel VII.B.6

Double Auction Market Session Results	Session: M2810	Number of traders: 8		File: DAMM2810.SES		
Market clearing quantity q* : 7	Average of Periods	Period 1	Period 2	Period 3	Period 4	Period 5
Market clearing price p* : 351						
Deviation of traded quantity from q*	-0.20	0	0	-1	-1	1
Efficiency	0.99	1.00	1.00	0.99	0.99	0.96
Average contract price (deviation from p*)	5.82	7.57	1.00	9.17	7.50	3.88
Minimum contract price (deviation from p*)	-3.60	-1.00	-12.00	-1.00	2.00	-6.00
Maximum contract price (deviation from p*)	16.00	19.00	14.00	24.00	14.00	9.00
Market clearing price concordance coefficient	0.69	0.67	0.73	0.63	0.70	0.72
Rel. freq. of falling price paths	0.80	1	1	1	0	1
Rel. freq. of rising price paths	0.00	0	0	0	0	0
Rel. freq. of hanging price paths	0.00	0	0	0	0	0
Rel. freq. of standing price paths	0.20	0	0	0	1	0
Rel. freq. of unknown price paths	0.00	0	0	0	0	0
Rel. freq. of contract proposals by buyers	0.63	0.29	0.43	0.83	1.00	0.63
... rel. freq. of these with price < p*	0.17	0.00	0.67	0.00	0.00	0.20
... rel. freq. of these with price = p*	0.00	0.00	0.00	0.00	0.00	0.00
... rel. freq. of these with price > p*	0.83	1.00	0.33	1.00	1.00	0.80
... avg. buyer proposed contract price - p*	4.86	5.00	-3.00	11.20	7.50	3.60
Rel. freq. of contract proposals by sellers	0.37	0.71	0.57	0.17	0.00	0.38
... rel. freq. of these with price < p*	0.45	0.20	0.25	1.00	-	0.33
... rel. freq. of these with price = p*	0.06	0.00	0.25	0.00	-	0.00
... rel. freq. of these with price > p*	0.49	0.80	0.50	0.00	-	0.67
... avg. seller proposed contract price - p*	3.98	8.60	4.00	-1.00	-	4.33
Rel. freq. of proposer's curses - all traders	0.67	0.43	0.29	1.00	1.00	0.63
Rel. freq. of proposer's curses - all buyers	0.83	1.00	0.33	1.00	1.00	0.80
Rel. freq. of proposer's curses - all sellers	0.45	0.20	0.25	1.00	-	0.33
Rel. freq. of proposer's curses - buyer 1	0.80	1.00	0.00	1.00	1.00	1.00
Rel. freq. of proposer's curses - buyer 2	1.00	-	1.00	1.00	1.00	1.00
Rel. freq. of proposer's curses - buyer 3	1.00	-	-	1.00	1.00	1.00
Rel. freq. of proposer's curses - buyer 4	0.67	-	-	1.00	1.00	0.00
Rel. freq. of proposer's curses - seller 5	0.00	0.00	0.00	-	-	-
Rel. freq. of proposer's curses - seller 6	0.50	0.50	0.00	-	-	1.00
Rel. freq. of proposer's curses - seller 7	0.25	0.00	0.50	-	-	-
Rel. freq. of proposer's curses - seller 8	0.50	-	-	1.00	-	0.00
Number of market ask and market bid matches	6.00	6.00	7.00	5.00	5.00	7.00
Number of contracts per market ask-bid match	1.13	1.17	1.00	1.20	1.20	1.14
Avg. rank of traded red.value at trading time	1.86	1.83	1.86	2.00	1.60	2.00
Avg. rank of traded unit cost at trading time	2.21	2.17	1.71	2.20	2.40	2.57
Number of submitted bids per contract	5.16	4.86	5.71	5.33	5.50	4.38
... rel. freq. of bids < p* - 200	0.04	0.09	0.05	0.03	0.00	0.03
... rel. freq. of bids < p*	0.73	0.65	0.78	0.78	0.67	0.77
... rel. freq. of bids = p*	0.01	0.03	0.03	0.00	0.00	0.00
... rel. freq. of bids > p*	0.26	0.32	0.20	0.22	0.33	0.23
Number of submitted asks per contract	4.74	5.57	6.14	4.17	4.33	3.50
... rel. freq. of asks < p*	0.05	0.03	0.09	0.04	0.00	0.07
... rel. freq. of asks = p*	0.01	0.00	0.02	0.04	0.00	0.00
... rel. freq. of asks > p*	0.94	0.97	0.88	0.92	1.00	0.93
... rel. freq. of asks > p* + 200	0.01	0.05	0.00	0.00	0.00	0.00
Number of offers for all units in periods > 1	65	-	83	57	59	63
(a) rel. freq. of bids > p-max & asks < p-min	0.03	-	0.04	0.04	0.00	0.03
(b) rel. freq. of bids > p-avg & asks < p-avg	0.19	-	0.22	0.16	0.22	0.16
(c) rel. freq. of bids > p-own & asks < p-own	0.22	-	0.19	0.19	0.25	0.25
(d) rel. freq. of bids < p-own & asks > p-own	0.69	-	0.75	0.77	0.64	0.59
Number of offers for units bought above p-avg or sold below p-avg in the previous period	17	-	21	21	12	14
(C) rel. freq. of bids > p-own & asks < p-own	0.05	-	0.14	0.05	0.00	0.00
Number of offers for units that were not traded in the previous period	27	-	38	13	25	31
(D) rel. freq. of bids < p-own & asks > p-own	0.69	-	0.76	0.85	0.68	0.45
Aggressive offer frequency ratio (c)/(C)	0.25	-	0.74	0.25	0.00	0.00
Moderate offer frequency ratio (d)/(D)	0.99	-	1.02	1.10	1.06	0.77

Panel VII.B.7

Double Auction Market Session Results	Session: M2820	Number of traders: 8		File: DAMM2820.SES	
Market clearing quantity q* : 7	Average	Period	Period	Period	Period	Period
Market clearing price p* : 739	of Periods	1	2	3	4	5
Deviation of traded quantity from q*	0.00	0	0	0	0	0
Efficiency	1.00	1.00	1.00	1.00	1.00	1.00
Average contract price (deviation from p*)	-4.20	-6.43	-7.43	-4.14	-2.43	-0.57
Minimum contract price (deviation from p*)	-11.00	-13.00	-17.00	-11.00	-9.00	-5.00
Maximum contract price (deviation from p*)	1.20	0.00	0.00	0.00	3.00	3.00
Market clearing price concordance coefficient	0.80	0.73	0.69	0.83	0.86	0.90
Rel. freq. of falling price paths	0.00	0	0	0	0	0
Rel. freq. of rising price paths	1.00	1	1	1	1	1
Rel. freq. of hanging price paths	0.00	0	0	0	0	0
Rel. freq. of standing price paths	0.00	0	0	0	0	0
Rel. freq. of unknown price paths	0.00	0	0	0	0	0
Rel. freq. of contract proposals by buyers	0.03	0.14	0.00	0.00	0.00	0.00
... rel. freq. of these with price < p*	0.00	0.00	-	-	-	-
... rel. freq. of these with price = p*	1.00	1.00	-	-	-	-
... rel. freq. of these with price > p*	0.00	0.00	-	-	-	-
... avg. buyer proposed contract price - p*	0.00	0.00	-	-	-	-
Rel. freq. of contract proposals by sellers	0.97	0.86	1.00	1.00	1.00	1.00
... rel. freq. of these with price < p*	0.62	0.83	0.86	0.57	0.57	0.29
... rel. freq. of these with price = p*	0.26	0.17	0.14	0.43	0.29	0.29
... rel. freq. of these with price > p*	0.11	0.00	0.00	0.00	0.14	0.43
... avg. seller proposed contract price - p*	-4.41	-7.50	-7.43	-4.14	-2.43	-0.57
Rel. freq. of proposer's curses - all traders	0.60	0.71	0.86	0.57	0.57	0.29
Rel. freq. of proposer's curses - all buyers	0.00	0.00	-	-	-	-
Rel. freq. of proposer's curses - all sellers	0.62	0.83	0.86	0.57	0.57	0.29
Rel. freq. of proposer's curses - buyer 1	-	-	-	-	-	-
Rel. freq. of proposer's curses - buyer 2	-	-	-	-	-	-
Rel. freq. of proposer's curses - buyer 3	0.00	0.00	-	-	-	-
Rel. freq. of proposer's curses - buyer 4	-	-	-	-	-	-
Rel. freq. of proposer's curses - seller 5	0.40	0.50	1.00	0.00	0.50	0.00
Rel. freq. of proposer's curses - seller 6	0.70	1.00	1.00	1.00	0.50	0.00
Rel. freq. of proposer's curses - seller 7	0.60	1.00	0.50	0.50	0.50	0.50
Rel. freq. of proposer's curses - seller 8	1.00	1.00	1.00	1.00	1.00	1.00
Number of market ask and market bid matches	6.60	7.00	7.00	7.00	6.00	6.00
Number of contracts per market ask-bid match	1.06	1.00	1.00	1.00	1.17	1.17
Avg. rank of traded red.value at trading time	2.10	1.71	2.14	2.00	2.33	2.33
Avg. rank of traded unit cost at trading time	1.88	1.14	1.71	1.71	2.50	2.33
Number of submitted bids per contract	4.89	5.14	6.14	5.29	3.86	4.00
... rel. freq. of bids < p* - 200	0.00	0.00	0.00	0.00	0.00	0.00
... rel. freq. of bids < p*	0.90	0.94	0.98	0.92	0.89	0.75
... rel. freq. of bids = p*	0.07	0.06	0.02	0.08	0.07	0.11
... rel. freq. of bids > p*	0.04	0.00	0.00	0.00	0.04	0.14
Number of submitted asks per contract	5.03	5.57	5.00	5.29	4.71	4.57
... rel. freq. of asks < p*	0.20	0.18	0.37	0.19	0.21	0.06
... rel. freq. of asks = p*	0.15	0.08	0.17	0.30	0.12	0.06
... rel. freq. of asks > p*	0.65	0.74	0.46	0.51	0.67	0.88
... rel. freq. of asks > p* + 200	0.01	0.05	0.00	0.00	0.00	0.00
Number of offers for all units in periods > 1	68	-	78	74	60	60
(a) rel. freq. of bids > p-max & asks < p-min	0.01	-	0.03	0.00	0.02	0.00
(b) rel. freq. of bids > p-avg & asks < p-avg	0.14	-	0.14	0.09	0.17	0.15
(c) rel. freq. of bids > p-own & asks < p-own	0.21	-	0.28	0.19	0.18	0.17
(d) rel. freq. of bids < p-own & asks > p-own	0.66	-	0.58	0.66	0.65	0.73
Number of offers for units bought above p-avg						
or sold below p-avg in the previous period	16	-	21	15	13	14
(C) rel. freq. of bids > p-own & asks < p-own	0.11	-	0.10	0.13	0.00	0.21
Number of offers for units						
that were not traded in the previous period	28	-	29	34	26	24
(D) rel. freq. of bids < p-own & asks > p-own	0.70	-	0.52	0.76	0.77	0.75
Aggressive offer frequency ratio (c)/(C)	0.58	-	0.34	0.70	0.00	1.29
Moderate offer frequency ratio (d)/(D)	1.06	-	0.90	1.15	1.18	1.02

Panel VII.B.8

Double Auction Market Session Results	Session: M2830	Number of traders: 8		File: DAMM2830.SES	
Market clearing quantity q* : 7	Average	Period	Period	Period	Period	Period
Market clearing price p* : 281	of Periods	1	2	3	4	5
Deviation of traded quantity from q*	-0.80	-1	0	-1	-1	-1
Efficiency	0.99	0.99	1.00	0.99	0.99	0.97
Average contract price (deviation from p*)	-0.06	6.00	-1.29	-2.83	-0.67	-1.50
Minimum contract price (deviation from p*)	-9.40	-20.00	-4.00	-11.00	-6.00	-6.00
Maximum contract price (deviation from p*)	9.60	27.00	4.00	5.00	8.00	4.00
Market clearing price concordance coefficient	0.78	0.46	0.86	0.82	0.87	0.88
Rel. freq. of falling price paths	0.40	1	0	0	1	0
Rel. freq. of rising price paths	0.00	0	0	0	0	0
Rel. freq. of hanging price paths	0.60	0	1	1	0	1
Rel. freq. of standing price paths	0.00	0	0	0	0	0
Rel. freq. of unknown price paths	0.00	0	0	0	0	0
Rel. freq. of contract proposals by buyers	0.62	0.50	0.43	0.83	0.67	0.67
... rel. freq. of these with price < p*	0.50	0.00	0.67	0.60	0.50	0.75
... rel. freq. of these with price = p*	0.09	0.00	0.00	0.20	0.25	0.00
... rel. freq. of these with price > p*	0.41	1.00	0.33	0.20	0.25	0.25
... avg. buyer proposed contract price - p*	3.23	19.67	-0.33	-3.20	1.00	-1.00
Rel. freq. of contract proposals by sellers	0.38	0.50	0.57	0.17	0.33	0.33
... rel. freq. of these with price < p*	0.85	1.00	0.75	1.00	1.00	0.50
... rel. freq. of these with price = p*	0.00	0.00	0.00	0.00	0.00	0.00
... rel. freq. of these with price > p*	0.15	0.00	0.25	0.00	0.00	0.50
... avg. seller proposed contract price - p*	-3.43	-7.67	-2.00	-1.00	-4.00	-2.50
Rel. freq. of proposer's curses - all traders	0.55	1.00	0.57	0.33	0.50	0.33
Rel. freq. of proposer's curses - all buyers	0.41	1.00	0.33	0.20	0.25	0.25
Rel. freq. of proposer's curses - all sellers	0.85	1.00	0.75	1.00	1.00	0.50
Rel. freq. of proposer's curses - buyer 1	0.70	1.00	1.00	0.50	0.50	0.50
Rel. freq. of proposer's curses - buyer 2	0.20	1.00	0.00	0.00	0.00	0.00
Rel. freq. of proposer's curses - buyer 3	-	-	-	-	-	-
Rel. freq. of proposer's curses - buyer 4	0.00	-	-	0.00	0.00	-
Rel. freq. of proposer's curses - seller 5	-	-	-	-	-	-
Rel. freq. of proposer's curses - seller 6	1.00	1.00	1.00	1.00	1.00	-
Rel. freq. of proposer's curses - seller 7	0.00	-	0.00	-	-	0.00
Rel. freq. of proposer's curses - seller 8	1.00	1.00	1.00	-	1.00	1.00
Number of market ask and market bid matches	6.20	6.00	7.00	6.00	6.00	6.00
Number of contracts per market ask-bid match	1.00	1.00	1.00	1.00	1.00	1.00
Avg. rank of traded red.value at trading time	1.74	1.67	1.71	1.67	1.67	2.00
Avg. rank of traded unit cost at trading time	1.78	1.67	1.57	1.83	2.00	1.83
Number of submitted bids per contract	9.34	11.33	12.71	8.83	7.00	6.83
... rel. freq. of bids < p* - 200	0.01	0.03	0.00	0.00	0.00	0.00
... rel. freq. of bids < p*	0.94	0.90	0.94	0.96	0.95	0.95
... rel. freq. of bids = p*	0.01	0.00	0.01	0.02	0.02	0.00
... rel. freq. of bids > p*	0.05	0.10	0.04	0.02	0.02	0.05
Number of submitted asks per contract	10.81	9.50	11.71	14.67	9.50	8.67
... rel. freq. of asks < p*	0.09	0.05	0.17	0.06	0.07	0.10
... rel. freq. of asks = p*	0.03	0.02	0.02	0.05	0.05	0.00
... rel. freq. of asks > p*	0.88	0.93	0.80	0.90	0.88	0.90
... rel. freq. of asks > p* + 200	0.03	0.07	0.05	0.02	0.00	0.02
Number of offers for all units in periods > 1	126	-	171	141	99	93
(a) rel. freq. of bids > p-max & asks < p-min	0.01	-	0.00	0.03	0.01	0.00
(b) rel. freq. of bids > p-avg & asks < p-avg	0.09	-	0.19	0.04	0.06	0.05
(c) rel. freq. of bids > p-own & asks < p-own	0.14	-	0.15	0.13	0.18	0.08
(d) rel. freq. of bids < p-own & asks > p-own	0.78	-	0.82	0.81	0.71	0.78
Number of offers for units bought above p-avg or sold below p-avg in the previous period	32	-	45	44	19	20
(C) rel. freq. of bids > p-own & asks < p-own	0.09	-	0.11	0.09	0.11	0.05
Number of offers for units that were not traded in the previous period	46	-	75	32	48	30
(D) rel. freq. of bids < p-own & asks > p-own	0.75	-	0.79	0.81	0.67	0.73
Aggressive offer frequency ratio (c)/(C)	0.66	-	0.73	0.67	0.58	0.66
Moderate offer frequency ratio (d)/(D)	0.96	-	0.95	1.00	0.94	0.93

Panel VII.B.9

Double Auction Market Session Results	Session: M2840	Number of traders: 8		File: DAMM2840.SES	
Market clearing quantity q* : 7	Average	Period	Period	Period	Period	Period
Market clearing price p* : 836	of Periods	1	2	3	4	5

	Average of Periods	Period 1	Period 2	Period 3	Period 4	Period 5
Deviation of traded quantity from q*	-0.20	-1	0	0	0	0
Efficiency	0.99	0.94	1.00	1.00	1.00	1.00
Average contract price (deviation from p*)	-6.77	-9.00	-7.00	-6.29	-6.43	-5.14
Minimum contract price (deviation from p*)	-18.00	-17.00	-21.00	-18.00	-18.00	-16.00
Maximum contract price (deviation from p*)	0.20	-2.00	0.00	2.00	0.00	1.00
Market clearing price concordance coefficient	0.71	0.64	0.71	0.71	0.73	0.77
Rel. freq. of falling price paths	0.00	0	0	0	0	0
Rel. freq. of rising price paths	1.00	1	1	1	1	1
Rel. freq. of hanging price paths	0.00	0	0	0	0	0
Rel. freq. of standing price paths	0.00	0	0	0	0	0
Rel. freq. of unknown price paths	0.00	0	0	0	0	0
Rel. freq. of contract proposals by buyers	0.30	0.33	0.00	0.71	0.00	0.43
... rel. freq. of these with price < p*	0.93	1.00	-	0.80	-	1.00
... rel. freq. of these with price = p*	0.00	0.00	-	0.00	-	0.00
... rel. freq. of these with price > p*	0.07	0.00	-	0.20	-	0.00
... avg. buyer proposed contract price - p*	-6.60	-6.00	-	-4.80	-	-9.00
Rel. freq. of contract proposals by sellers	0.70	0.67	1.00	0.29	1.00	0.57
... rel. freq. of these with price < p*	0.89	1.00	0.86	1.00	0.86	0.75
... rel. freq. of these with price = p*	0.06	0.00	0.14	0.00	0.14	0.00
... rel. freq. of these with price > p*	0.05	0.00	0.00	0.00	0.00	0.25
... avg. seller proposed contract price - p*	-7.24	-10.50	-7.00	-10.00	-6.43	-2.25
Rel. freq. of proposer's curses - all traders	0.65	0.67	0.86	0.43	0.86	0.43
Rel. freq. of proposer's curses - all buyers	0.07	0.00	-	0.20	-	0.00
Rel. freq. of proposer's curses - all sellers	0.89	1.00	0.86	1.00	0.86	0.75
Rel. freq. of proposer's curses - buyer 1	0.00	-	-	0.00	-	0.00
Rel. freq. of proposer's curses - buyer 2	0.50	-	-	0.50	-	-
Rel. freq. of proposer's curses - buyer 3	0.00	0.00	-	0.00	-	0.00
Rel. freq. of proposer's curses - buyer 4	0.00	0.00	-	0.00	-	-
Rel. freq. of proposer's curses - seller 5	1.00	1.00	1.00	1.00	1.00	1.00
Rel. freq. of proposer's curses - seller 6	1.00	1.00	1.00	1.00	1.00	1.00
Rel. freq. of proposer's curses - seller 7	0.50	1.00	0.50	-	0.50	0.00
Rel. freq. of proposer's curses - seller 8	1.00	-	1.00	1.00	1.00	1.00
Number of market ask and market bid matches	6.40	5.00	6.00	7.00	7.00	7.00
Number of contracts per market ask-bid match	1.06	1.20	1.17	1.00	1.00	1.00
Avg. rank of traded red.value at trading time	2.13	2.60	2.33	2.00	1.86	1.86
Avg. rank of traded unit cost at trading time	1.70	1.60	2.17	1.57	1.29	1.86
Number of submitted bids per contract	5.71	5.00	5.14	5.86	6.00	6.57
... rel. freq. of bids < p* - 200	0.01	0.03	0.00	0.00	0.00	0.00
... rel. freq. of bids < p*	0.98	1.00	0.97	0.98	0.98	0.96
... rel. freq. of bids = p*	0.01	0.00	0.03	0.00	0.02	0.00
... rel. freq. of bids > p*	0.01	0.00	0.00	0.02	0.00	0.04
Number of submitted asks per contract	4.47	5.50	4.14	3.71	4.29	4.71
... rel. freq. of asks < p*	0.37	0.36	0.38	0.42	0.37	0.33
... rel. freq. of asks = p*	0.06	0.00	0.03	0.00	0.17	0.09
... rel. freq. of asks > p*	0.57	0.64	0.59	0.58	0.47	0.58
... rel. freq. of asks > p* + 200	0.01	0.03	0.00	0.00	0.00	0.00
Number of offers for all units in periods > 1	71	-	65	67	72	79
(a) rel. freq. of bids > p-max & asks < p-min	0.03	-	0.06	0.01	0.00	0.03
(b) rel. freq. of bids > p-avg & asks < p-avg	0.19	-	0.22	0.19	0.19	0.14
(c) rel. freq. of bids > p-own & asks < p-own	0.26	-	0.37	0.25	0.21	0.20
(d) rel. freq. of bids < p-own & asks > p-own	0.64	-	0.60	0.63	0.65	0.68
Number of offers for units bought above p-avg or sold below p-avg in the previous period	19	-	12	21	24	19
(C) rel. freq. of bids > p-own & asks < p-own	0.13	-	0.33	0.10	0.08	0.00
Number of offers for units that were not traded in the previous period	25	-	34	22	23	23
(D) rel. freq. of bids < p-own & asks > p-own	0.58	-	0.44	0.55	0.61	0.74
Aggressive offer frequency ratio (c)/(C)	0.42	-	0.90	0.38	0.40	0.00
Moderate offer frequency ratio (d)/(D)	0.90	-	0.74	0.87	0.93	1.08

Panel VII.B.10

```
Double Auction Market Session Results   Session: M2A10   Number of traders: 10   File: DAMM2A10.SES
```

	Average of Periods	Period 1	Period 2	Period 3	Period 4	Period 5
Market clearing quantity q* : 9						
Market clearing price p* : 342						
Deviation of traded quantity from q*	0.20	-1	1	1	0	0
Efficiency	0.97	0.92	0.97	0.98	0.98	0.99
Average contract price (deviation from p*)	-3.81	-6.75	-4.40	-4.10	-3.56	-0.22
Minimum contract price (deviation from p*)	-11.60	-14.00	-12.00	-12.00	-11.00	-9.00
Maximum contract price (deviation from p*)	6.20	4.00	4.00	3.00	2.00	18.00
Market clearing price concordance coefficient	0.64	0.53	0.62	0.64	0.70	0.69
Rel. freq. of falling price paths	0.00	0	0	0	0	0
Rel. freq. of rising price paths	0.80	1	1	1	1	0
Rel. freq. of hanging price paths	0.20	0	0	0	0	1
Rel. freq. of standing price paths	0.00	0	0	0	0	0
Rel. freq. of unknown price paths	0.00	0	0	0	0	0
Rel. freq. of contract proposals by buyers	0.09	0.00	0.00	0.10	0.11	0.22
... rel. freq. of these with price < p*	0.17	-	-	0.00	0.00	0.50
... rel. freq. of these with price = p*	0.00	-	-	0.00	0.00	0.00
... rel. freq. of these with price > p*	0.83	-	-	1.00	1.00	0.50
... avg. buyer proposed contract price - p*	3.17	-	-	3.00	2.00	4.50
Rel. freq. of contract proposals by sellers	0.91	1.00	1.00	0.90	0.89	0.78
... rel. freq. of these with price < p*	0.75	0.88	0.80	0.78	0.75	0.57
... rel. freq. of these with price = p*	0.03	0.00	0.00	0.00	0.00	0.14
... rel. freq. of these with price > p*	0.22	0.13	0.20	0.22	0.25	0.29
... avg. seller proposed contract price - p*	-4.37	-6.75	-4.40	-4.89	-4.25	-1.57
Rel. freq. of proposer's curses - all traders	0.76	0.88	0.80	0.80	0.78	0.56
Rel. freq. of proposer's curses - all buyers	0.83	-	-	1.00	1.00	0.50
Rel. freq. of proposer's curses - all sellers	0.75	0.88	0.80	0.78	0.75	0.57
Rel. freq. of proposer's curses - buyer 1	1.00	-	-	1.00	-	-
Rel. freq. of proposer's curses - buyer 2	0.50	-	-	-	1.00	0.00
Rel. freq. of proposer's curses - buyer 3	1.00	-	-	-	-	1.00
Rel. freq. of proposer's curses - buyer 4	-	-	-	-	-	-
Rel. freq. of proposer's curses - buyer 5	-	-	-	-	-	-
Rel. freq. of proposer's curses - seller 6	0.73	1.00	0.67	1.00	0.50	0.50
Rel. freq. of proposer's curses - seller 7	1.00	1.00	1.00	1.00	1.00	1.00
Rel. freq. of proposer's curses - seller 8	0.60	1.00	1.00	0.50	0.50	0.00
Rel. freq. of proposer's curses - seller 9	0.80	1.00	0.50	0.50	1.00	1.00
Rel. freq. of proposer's curses - seller 10	0.80	0.50	1.00	1.00	1.00	0.50
Number of market ask and market bid matches	8.40	8.00	7.00	10.00	8.00	9.00
Number of contracts per market ask-bid match	1.10	1.00	1.43	1.00	1.13	1.00
Avg. rank of traded red.value at trading time	2.74	2.75	3.57	2.70	3.00	1.67
Avg. rank of traded unit cost at trading time	1.80	1.63	2.57	1.40	1.50	1.89
Number of submitted bids per contract	6.10	7.50	4.90	6.00	5.22	6.89
... rel. freq. of bids < p* - 200	0.01	0.02	0.02	0.00	0.00	0.00
... rel. freq. of bids < p*	0.90	0.95	0.90	0.90	0.85	0.92
... rel. freq. of bids = p*	0.02	0.02	0.00	0.02	0.04	0.02
... rel. freq. of bids > p*	0.08	0.03	0.10	0.08	0.11	0.06
Number of submitted asks per contract	6.12	7.50	4.70	6.40	5.56	6.44
... rel. freq. of asks < p*	0.32	0.33	0.36	0.36	0.34	0.21
... rel. freq. of asks = p*	0.03	0.02	0.00	0.02	0.08	0.03
... rel. freq. of asks > p*	0.65	0.65	0.64	0.63	0.58	0.76
... rel. freq. of asks > p* + 200	0.02	0.05	0.00	0.02	0.02	0.00
Number of offers for all units in periods > 1	109	-	96	124	97	120
(a) rel. freq. of bids > p-max & asks < p-min	0.00	-	0.00	0.00	0.00	0.01
(b) rel. freq. of bids > p-avg & asks < p-avg	0.19	-	0.22	0.15	0.16	0.22
(c) rel. freq. of bids > p-own & asks < p-own	0.20	-	0.26	0.12	0.20	0.22
(d) rel. freq. of bids < p-own & asks > p-own	0.56	-	0.60	0.53	0.55	0.56
Number of offers for units bought above p-avg						
or sold below p-avg in the previous period	21	-	16	19	29	20
(C) rel. freq. of bids > p-own & asks < p-own	0.07	-	0.13	0.00	0.14	0.00
Number of offers for units						
that were not traded in the previous period	37	-	50	39	25	33
(D) rel. freq. of bids < p-own & asks > p-own	0.34	-	0.52	0.33	0.12	0.39
Aggressive offer frequency ratio (c)/(C)	0.30	-	0.48	0.00	0.70	0.00
Moderate offer frequency ratio (d)/(D)	0.60	-	0.86	0.63	0.22	0.71

Panel VII.B.11

Double Auction Market Session Results   Session: M1811   Number of traders: 8   File: DAMM1811.SES

	Average of Periods	Period 1	Period 2	Period 3	Period 4	Period 5
Market clearing quantity q* : 10						
Market clearing price p* : 829						
Deviation of traded quantity from q*	-0.20	-1	-1	0	1	0
Efficiency	0.99	0.99	0.95	1.00	0.99	1.00
Average contract price (deviation from p*)	-0.14	-2.44	0.00	0.20	0.64	0.90
Minimum contract price (deviation from p*)	-2.60	-7.00	-2.00	-2.00	-2.00	0.00
Maximum contract price (deviation from p*)	1.20	-1.00	1.00	1.00	3.00	2.00
Market clearing price concordance coefficient	0.94	0.87	0.98	0.97	0.93	0.95
Rel. freq. of falling price paths	0.20	0	1	0	0	0
Rel. freq. of rising price paths	0.20	1	0	0	0	0
Rel. freq. of hanging price paths	0.40	0	0	0	1	1
Rel. freq. of standing price paths	0.20	0	0	1	0	0
Rel. freq. of unknown price paths	0.00	0	0	0	0	0
Rel. freq. of contract proposals by buyers	0.24	0.00	0.00	0.10	0.18	0.90
... rel. freq. of these with price < p*	0.33	-	-	0.00	1.00	0.00
... rel. freq. of these with price = p*	0.41	-	-	1.00	0.00	0.22
... rel. freq. of these with price > p*	0.26	-	-	0.00	0.00	0.78
... avg. buyer proposed contract price - p*	-0.20	-	-	0.00	-1.50	0.89
Rel. freq. of contract proposals by sellers	0.76	1.00	1.00	0.90	0.82	0.10
... rel. freq. of these with price < p*	0.24	1.00	0.11	0.11	0.00	0.00
... rel. freq. of these with price = p*	0.24	0.00	0.67	0.44	0.11	0.00
... rel. freq. of these with price > p*	0.51	0.00	0.22	0.44	0.89	1.00
... avg. seller proposed contract price - p*	-0.02	-2.44	0.00	0.22	1.11	1.00
Rel. freq. of proposer's curses - all traders	0.38	1.00	0.11	0.10	0.00	0.70
Rel. freq. of proposer's curses - all buyers	0.26	-	-	0.00	0.00	0.78
Rel. freq. of proposer's curses - all sellers	0.24	1.00	0.11	0.11	0.00	0.00
Rel. freq. of proposer's curses - buyer 1	0.00	-	-	-	0.00	0.00
Rel. freq. of proposer's curses - buyer 2	0.50	-	-	-	0.00	1.00
Rel. freq. of proposer's curses - buyer 3	0.33	-	-	0.00	-	0.67
Rel. freq. of proposer's curses - buyer 4	1.00	-	-	-	-	1.00
Rel. freq. of proposer's curses - seller 5	0.38	1.00	0.50	0.00	0.00	-
Rel. freq. of proposer's curses - seller 6	0.38	1.00	0.00	0.50	0.00	-
Rel. freq. of proposer's curses - seller 7	0.25	1.00	0.00	0.00	0.00	-
Rel. freq. of proposer's curses - seller 8	0.20	1.00	0.00	0.00	0.00	0.00
Number of market ask and market bid matches	7.60	7.00	6.00	9.00	8.00	8.00
Number of contracts per market ask-bid match	1.29	1.29	1.50	1.11	1.38	1.25
Avg. rank of traded red.value at trading time	2.58	2.57	2.83	2.00	2.88	2.63
Avg. rank of traded unit cost at trading time	2.59	2.71	3.00	2.22	2.38	2.63
Number of submitted bids per contract	4.34	6.33	3.78	3.50	4.27	3.80
... rel. freq. of bids < p* - 200	0.04	0.11	0.03	0.09	0.00	0.00
... rel. freq. of bids < p*	0.74	0.98	0.74	0.60	0.68	0.68
... rel. freq. of bids = p*	0.12	0.02	0.21	0.26	0.06	0.08
... rel. freq. of bids > p*	0.14	0.00	0.06	0.14	0.26	0.24
Number of submitted asks per contract	3.69	4.56	2.89	3.90	3.73	3.40
... rel. freq. of asks < p*	0.10	0.34	0.08	0.03	0.05	0.00
... rel. freq. of asks = p*	0.16	0.17	0.35	0.15	0.07	0.06
... rel. freq. of asks > p*	0.74	0.49	0.58	0.82	0.88	0.94
... rel. freq. of asks > p* + 200	0.00	0.00	0.00	0.00	0.00	0.00
Number of offers for all units in periods > 1	73	-	60	74	88	72
(a) rel. freq. of bids > p-max & asks < p-min	0.04	-	0.15	0.00	0.02	0.00
(b) rel. freq. of bids > p-avg & asks < p-avg	0.11	-	0.17	0.08	0.16	0.04
(c) rel. freq. of bids > p-own & asks < p-own	0.16	-	0.20	0.15	0.24	0.06
(d) rel. freq. of bids < p-own & asks > p-own	0.50	-	0.60	0.51	0.45	0.43
Number of offers for units bought above p-avg or sold below p-avg in the previous period	11	-	16	9	12	8
(C) rel. freq. of bids > p-own & asks < p-own	0.06	-	0.25	0.00	0.00	0.00
Number of offers for units that were not traded in the previous period	22	-	14	26	31	16
(D) rel. freq. of bids < p-own & asks > p-own	0.36	-	0.36	0.42	0.23	0.44
Aggressive offer frequency ratio (c)/(C)	0.31	-	1.25	0.00	0.00	0.00
Moderate offer frequency ratio (d)/(D)	0.73	-	0.60	0.82	0.50	1.02

Panel VII.B.12

```
Double Auction Market Session Results    Session: M1821    Number of traders: 8    File: DAMM1821.SES
```

	Average of Periods	Period 1	Period 2	Period 3	Period 4	Period 5
Market clearing quantity q* : 10						
Market clearing price p* : 740						
Deviation of traded quantity from q*	0.20	1	0	0	0	0
Efficiency	0.99	0.96	1.00	0.99	1.00	1.00
Average contract price (deviation from p*)	-2.88	-2.09	-2.30	-3.70	-4.30	-2.00
Minimum contract price (deviation from p*)	-6.00	-6.00	-6.00	-7.00	-8.00	-3.00
Maximum contract price (deviation from p*)	1.20	5.00	0.00	0.00	1.00	0.00
Market clearing price concordance coefficient	0.82	0.81	0.87	0.79	0.75	0.89
Rel. freq. of falling price paths	0.00	0	0	0	0	0
Rel. freq. of rising price paths	0.40	0	0	0	1	1
Rel. freq. of hanging price paths	0.60	1	1	1	0	0
Rel. freq. of standing price paths	0.00	0	0	0	0	0
Rel. freq. of unknown price paths	0.00	0	0	0	0	0
Rel. freq. of contract proposals by buyers	0.24	0.00	0.50	0.00	0.00	0.70
... rel. freq. of these with price < p*	0.93	-	1.00	-	-	0.86
... rel. freq. of these with price = p*	0.07	-	0.00	-	-	0.14
... rel. freq. of these with price > p*	0.00	-	0.00	-	-	0.00
... avg. buyer proposed contract price - p*	-2.84	-	-3.40	-	-	-2.29
Rel. freq. of contract proposals by sellers	0.76	1.00	0.50	1.00	1.00	0.30
... rel. freq. of these with price < p*	0.76	0.73	0.60	0.90	0.90	0.67
... rel. freq. of these with price = p*	0.18	0.09	0.40	0.10	0.00	0.33
... rel. freq. of these with price > p*	0.06	0.18	0.00	0.00	0.10	0.00
... avg. seller proposed contract price - p*	-2.52	-2.09	-1.20	-3.70	-4.30	-1.33
Rel. freq. of proposer's curses - all traders	0.61	0.73	0.30	0.90	0.90	0.20
Rel. freq. of proposer's curses - all buyers	0.00	-	0.00	-	-	0.00
Rel. freq. of proposer's curses - all sellers	0.76	0.73	0.60	0.90	0.90	0.67
Rel. freq. of proposer's curses - buyer 1	0.00	-	-	-	-	0.00
Rel. freq. of proposer's curses - buyer 2	0.00	-	0.00	-	-	0.00
Rel. freq. of proposer's curses - buyer 3	0.00	-	0.00	-	-	0.00
Rel. freq. of proposer's curses - buyer 4	0.00	-	0.00	-	-	0.00
Rel. freq. of proposer's curses - seller 5	0.90	1.00	0.50	1.00	1.00	1.00
Rel. freq. of proposer's curses - seller 6	0.93	0.67	1.00	1.00	1.00	1.00
Rel. freq. of proposer's curses - seller 7	0.40	0.67	0.00	0.67	0.67	0.00
Rel. freq. of proposer's curses - seller 8	0.89	0.67	-	1.00	1.00	-
Number of market ask and market bid matches	9.40	9.00	10.00	10.00	9.00	9.00
Number of contracts per market ask-bid match	1.09	1.22	1.00	1.00	1.11	1.11
Avg. rank of traded red.value at trading time	2.22	2.78	1.80	2.30	2.44	1.78
Avg. rank of traded unit cost at trading time	2.02	1.89	2.10	1.80	2.22	2.11
Number of submitted bids per contract	4.09	4.27	4.10	4.00	3.70	4.40
... rel. freq. of bids < p* - 200	0.01	0.02	0.00	0.03	0.00	0.00
... rel. freq. of bids < p*	0.93	0.85	0.93	0.95	0.95	0.95
... rel. freq. of bids = p*	0.06	0.09	0.07	0.05	0.03	0.05
... rel. freq. of bids > p*	0.02	0.06	0.00	0.00	0.03	0.00
Number of submitted asks per contract	3.82	4.00	3.20	4.70	3.30	3.90
... rel. freq. of asks < p*	0.44	0.32	0.47	0.45	0.58	0.41
... rel. freq. of asks = p*	0.11	0.09	0.13	0.09	0.06	0.18
... rel. freq. of asks > p*	0.45	0.59	0.41	0.47	0.36	0.41
... rel. freq. of asks > p* + 200	0.01	0.02	0.00	0.00	0.00	0.03
Number of offers for all units in periods > 1	78	-	73	87	70	83
(a) rel. freq. of bids > p-max & asks < p-min	0.01	-	0.00	0.01	0.03	0.00
(b) rel. freq. of bids > p-avg & asks < p-avg	0.19	-	0.18	0.14	0.20	0.23
(c) rel. freq. of bids > p-own & asks < p-own	0.17	-	0.15	0.18	0.19	0.16
(d) rel. freq. of bids > p-own & asks > p-own	0.57	-	0.58	0.55	0.54	0.60
Number of offers for units bought above p-avg or sold below p-avg in the previous period	26	-	29	23	18	34
(C) rel. freq. of bids > p-own & asks < p-own	0.13	-	0.07	0.22	0.22	0.03
Number of offers for units that were not traded in the previous period	20	-	10	27	23	22
(D) rel. freq. of bids < p-own & asks > p-own	0.43	-	0.50	0.30	0.57	0.36
Aggressive offer frequency ratio (c)/(C)	0.76	-	0.46	1.18	1.20	0.19
Moderate offer frequency ratio (d)/(D)	0.76	-	0.87	0.54	1.04	0.60

Panel VII.B.13

Double Auction Market Session Results	Session: M1831	Number of traders: 8		File: DAMM1831.SES	
Market clearing quantity q* : 10	Average	Period	Period	Period	Period	Period
Market clearing price p* : 436	of Periods	1	2	3	4	5
Deviation of traded quantity from q*	-0.40	0	-1	0	-1	0
Efficiency	0.98	1.00	0.96	1.00	0.96	1.00
Average contract price (deviation from p*)	1.84	0.70	3.11	1.80	2.00	1.60
Minimum contract price (deviation from p*)	-0.60	-2.00	2.00	-3.00	0.00	0.00
Maximum contract price (deviation from p*)	3.40	4.00	4.00	3.00	3.00	3.00
Market clearing price concordance coefficient	0.88	0.93	0.83	0.86	0.89	0.91
Rel. freq. of falling price paths	0.20	0	1	0	0	0
Rel. freq. of rising price paths	0.20	0	0	0	1	0
Rel. freq. of hanging price paths	0.00	0	0	0	0	0
Rel. freq. of standing price paths	0.60	1	0	1	0	1
Rel. freq. of unknown price paths	0.00	0	0	0	0	0
Rel. freq. of contract proposals by buyers	0.45	0.80	0.11	0.20	0.22	0.90
... rel. freq. of these with price < p*	0.05	0.25	0.00	0.00	0.00	0.00
... rel. freq. of these with price = p*	0.30	0.38	0.00	0.50	0.50	0.11
... rel. freq. of these with price > p*	0.65	0.38	1.00	0.50	0.50	0.89
... avg. buyer proposed contract price - p*	1.16	0.13	2.00	1.00	1.00	1.67
Rel. freq. of contract proposals by sellers	0.55	0.20	0.89	0.80	0.78	0.10
... rel. freq. of these with price < p*	0.03	0.00	0.00	0.13	0.00	0.00
... rel. freq. of these with price = p*	0.00	0.00	0.00	0.00	0.00	0.00
... rel. freq. of these with price > p*	0.98	1.00	1.00	0.88	1.00	1.00
... avg. seller proposed contract price - p*	2.31	3.00	3.25	2.00	2.29	1.00
Rel. freq. of proposer's curses - all traders	0.30	0.30	0.11	0.20	0.11	0.80
Rel. freq. of proposer's curses - all buyers	0.65	0.38	1.00	0.50	0.50	0.89
Rel. freq. of proposer's curses - all sellers	0.03	0.00	0.00	0.13	0.00	0.00
Rel. freq. of proposer's curses - buyer 1	1.00	1.00	-	-	-	1.00
Rel. freq. of proposer's curses - buyer 2	0.75	0.50	-	-	-	1.00
Rel. freq. of proposer's curses - buyer 3	0.17	0.00	-	0.00	0.00	0.67
Rel. freq. of proposer's curses - buyer 4	0.80	0.00	1.00	1.00	1.00	1.00
Rel. freq. of proposer's curses - seller 5	0.00	0.00	0.00	0.00	0.00	-
Rel. freq. of proposer's curses - seller 6	0.00	-	-	0.00	-	-
Rel. freq. of proposer's curses - seller 7	0.13	-	0.00	0.50	0.00	0.00
Rel. freq. of proposer's curses - seller 8	0.00	0.00	0.00	0.00	0.00	-
Number of market ask and market bid matches	8.60	9.00	8.00	8.00	9.00	9.00
Number of contracts per market ask-bid match	1.12	1.11	1.13	1.25	1.00	1.11
Avg. rank of traded red.value at trading time	2.03	1.89	1.88	2.38	2.00	2.00
Avg. rank of traded unit cost at trading time	2.26	2.11	2.38	2.50	2.44	1.89
Number of submitted bids per contract	4.01	4.10	4.22	3.80	4.11	3.80
... rel. freq. of bids < p* - 200	0.00	0.00	0.00	0.00	0.00	0.00
... rel. freq. of bids < p*	0.42	0.63	0.24	0.53	0.27	0.45
... rel. freq. of bids = p*	0.17	0.17	0.16	0.08	0.27	0.18
... rel. freq. of bids > p*	0.40	0.20	0.61	0.39	0.46	0.37
Number of submitted asks per contract	4.31	4.50	4.00	4.00	4.67	4.40
... rel. freq. of asks < p*	0.02	0.07	0.00	0.03	0.00	0.00
... rel. freq. of asks = p*	0.03	0.07	0.00	0.03	0.02	0.02
... rel. freq. of asks > p*	0.95	0.87	1.00	0.95	0.98	0.98
... rel. freq. of asks > p* + 200	0.01	0.00	0.00	0.03	0.00	0.00
Number of offers for all units in periods > 1	78	-	74	78	79	82
(a) rel. freq. of bids > p-max & asks < p-min	0.01	-	0.00	0.04	0.00	0.00
(b) rel. freq. of bids > p-avg & asks < p-avg	0.16	-	0.31	0.10	0.10	0.11
(c) rel. freq. of bids > p-own & asks < p-own	0.20	-	0.26	0.12	0.25	0.18
(d) rel. freq. of bids < p-own & asks > p-own	0.44	-	0.45	0.46	0.35	0.49
Number of offers for units bought above p-avg						
or sold below p-avg in the previous period	19	-	23	10	16	26
(C) rel. freq. of bids > p-own & asks < p-own	0.02	-	0.09	0.00	0.00	0.00
Number of offers for units						
that were not traded in the previous period	24	-	15	32	19	31
(D) rel. freq. of bids < p-own & asks > p-own	0.30	-	0.40	0.44	0.05	0.29
Aggressive offer frequency ratio (c)/(C)	0.08	-	0.34	0.00	0.00	0.00
Moderate offer frequency ratio (d)/(D)	0.65	-	0.90	0.95	0.15	0.60

Panel VII.B.14

```
Double Auction Market Session Results    Session: M1841    Number of traders: 8    File: DAMM1841.SES
```

	Average of Periods	Period 1	Period 2	Period 3	Period 4	Period 5
Market clearing quantity q* : 10						
Market clearing price p* : 628						
Deviation of traded quantity from q*	-0.80	-2	-2	0	0	0
Efficiency	0.94	0.83	0.91	0.99	0.99	0.99
Average contract price (deviation from p*)	-3.97	-7.25	-6.13	-2.90	-2.10	-1.50
Minimum contract price (deviation from p*)	-7.00	-11.00	-8.00	-6.00	-5.00	-5.00
Maximum contract price (deviation from p*)	0.00	-3.00	-3.00	2.00	2.00	2.00
Market clearing price concordance coefficient	0.77	0.62	0.68	0.81	0.86	0.87
Rel. freq. of falling price paths	0.00	0	0	0	0	0
Rel. freq. of rising price paths	1.00	1	1	1	1	1
Rel. freq. of hanging price paths	0.00	0	0	0	0	0
Rel. freq. of standing price paths	0.00	0	0	0	0	0
Rel. freq. of unknown price paths	0.00	0	0	0	0	0
Rel. freq. of contract proposals by buyers	0.25	0.00	0.63	0.00	0.10	0.50
... rel. freq. of these with price < p*	0.67	.	1.00	.	0.00	1.00
... rel. freq. of these with price = p*	0.33	.	0.00	.	1.00	0.00
... rel. freq. of these with price > p*	0.00	.	0.00	.	0.00	0.00
... avg. buyer proposed contract price - p*	-3.67	.	-7.60	.	0.00	-3.40
Rel. freq. of contract proposals by sellers	0.76	1.00	0.38	1.00	0.90	0.50
... rel. freq. of these with price < p*	0.80	1.00	1.00	0.80	0.78	0.40
... rel. freq. of these with price = p*	0.08	0.00	0.00	0.10	0.11	0.20
... rel. freq. of these with price > p*	0.12	0.00	0.00	0.10	0.11	0.40
... avg. seller proposed contract price - p*	-3.15	-7.25	-3.67	-2.90	-2.33	0.40
Rel. freq. of proposer's curses - all traders	0.62	1.00	0.38	0.80	0.70	0.20
Rel. freq. of proposer's curses - all buyers	0.00	.	0.00	.	0.00	0.00
Rel. freq. of proposer's curses - all sellers	0.80	1.00	1.00	0.80	0.78	0.40
Rel. freq. of proposer's curses - buyer 1	0.00	.	0.00	.	0.00	.
Rel. freq. of proposer's curses - buyer 2	0.00	.	0.00	.	.	0.00
Rel. freq. of proposer's curses - buyer 3	.	.	.	.	.	.
Rel. freq. of proposer's curses - buyer 4	0.00	.	0.00	.	.	0.00
Rel. freq. of proposer's curses - seller 5	0.50	1.00	.	0.50	0.50	0.00
Rel. freq. of proposer's curses - seller 6	0.90	1.00	1.00	1.00	1.00	0.50
Rel. freq. of proposer's curses - seller 7	0.58	1.00	.	0.67	0.67	0.00
Rel. freq. of proposer's curses - seller 8	1.00	1.00	1.00	1.00	1.00	1.00
Number of market ask and market bid matches	8.20	8.00	7.00	8.00	8.00	10.00
Number of contracts per market ask-bid match	1.12	1.00	1.14	1.25	1.25	1.00
Avg. rank of traded red.value at trading time	2.58	2.75	2.86	2.63	2.75	1.90
Avg. rank of traded unit cost at trading time	1.99	1.38	2.14	2.38	2.25	1.80
Number of submitted bids per contract	4.23	4.50	4.63	4.00	3.70	4.30
... rel. freq. of bids < p* - 200	0.00	0.00	0.00	0.00	0.00	0.00
... rel. freq. of bids < p*	0.94	1.00	1.00	0.90	0.89	0.91
... rel. freq. of bids = p*	0.04	0.00	0.00	0.08	0.05	0.05
... rel. freq. of bids > p*	0.03	0.00	0.00	0.03	0.05	0.05
Number of submitted asks per contract	4.11	4.50	4.63	3.60	3.50	4.30
... rel. freq. of asks < p*	0.49	0.56	0.59	0.50	0.46	0.35
... rel. freq. of asks = p*	0.04	0.00	0.00	0.03	0.06	0.09
... rel. freq. of asks > p*	0.47	0.44	0.41	0.47	0.49	0.56
... rel. freq. of asks > p* + 200	0.00	0.00	0.00	0.00	0.00	0.00
Number of offers for all units in periods > 1	77	.	74	76	72	86
(a) rel. freq. of bids > p-max & asks < p-min	0.04	.	0.04	0.11	0.00	0.00
(b) rel. freq. of bids > p-avg & asks < p-avg	0.23	.	0.24	0.34	0.15	0.20
(c) rel. freq. of bids > p-own & asks < p-own	0.22	.	0.34	0.34	0.11	0.10
(d) rel. freq. of bids < p-own & asks > p-own	0.50	.	0.50	0.46	0.54	0.51
Number of offers for units bought above p-avg or sold below p-avg in the previous period	21	.	25	23	12	24
(C) rel. freq. of bids > p-own & asks < p-own	0.16	.	0.20	0.39	0.00	0.04
Number of offers for units that were not traded in the previous period	28	.	32	34	23	24
(D) rel. freq. of bids < p-own & asks > p-own	0.47	.	0.44	0.50	0.61	0.33
Aggressive offer frequency ratio (c)/(C)	0.53	.	0.59	1.14	0.00	0.40
Moderate offer frequency ratio (d)/(D)	0.93	.	0.88	1.09	1.12	0.65

Panel VII.B.15

Double Auction Market Session Results	Session: M2811	Number of traders: 8		File: DAMM2811.SES		
Market clearing quantity q* : 7	Average	Period	Period	Period	Period	Period
Market clearing price p* : 765	of Periods	1	2	3	4	5
Deviation of traded quantity from q*	0.00	0	0	0	0	0
Efficiency	1.00	1.00	1.00	1.00	1.00	1.00
Average contract price (deviation from p*)	1.71	6.00	2.00	0.86	-0.29	0.00
Minimum contract price (deviation from p*)	-6.40	-3.00	-8.00	-12.00	-5.00	-4.00
Maximum contract price (deviation from p*)	11.20	16.00	15.00	10.00	10.00	5.00
Market clearing price concordance coefficient	0.79	0.70	0.74	0.80	0.81	0.88
Rel. freq. of falling price paths	0.80	0	1	1	1	1
Rel. freq. of rising price paths	0.00	0	0	0	0	0
Rel. freq. of hanging price paths	0.00	0	0	0	0	0
Rel. freq. of standing price paths	0.20	1	0	0	0	0
Rel. freq. of unknown price paths	0.00	0	0	0	0	0
Rel. freq. of contract proposals by buyers	0.80	0.57	1.00	1.00	0.71	0.71
... rel. freq. of these with price < p*	0.37	0.00	0.43	0.43	0.40	0.60
... rel. freq. of these with price = p*	0.15	0.00	0.29	0.29	0.20	0.00
... rel. freq. of these with price > p*	0.47	1.00	0.29	0.29	0.40	0.40
... avg. buyer proposed contract price - p*	3.27	11.50	2.00	0.86	1.60	0.40
Rel. freq. of contract proposals by sellers	0.20	0.43	0.00	0.00	0.29	0.29
... rel. freq. of these with price < p*	0.72	0.67	.	.	1.00	0.50
... rel. freq. of these with price = p*	0.28	0.33	.	.	0.00	0.50
... rel. freq. of these with price > p*	0.00	0.00	.	.	0.00	0.00
... avg. seller proposed contract price - p*	-2.44	-1.33	.	.	-5.00	-1.00
Rel. freq. of proposer's curses - all traders	0.49	0.86	0.29	0.29	0.57	0.43
Rel. freq. of proposer's curses - all buyers	0.47	1.00	0.29	0.29	0.40	0.40
Rel. freq. of proposer's curses - all sellers	0.72	0.67	.	.	1.00	0.50
Rel. freq. of proposer's curses - buyer 1	0.25	1.00	0.00	0.00	.	0.00
Rel. freq. of proposer's curses - buyer 2	0.60	1.00	0.50	0.50	0.50	0.50
Rel. freq. of proposer's curses - buyer 3	0.60	1.00	0.50	0.50	0.50	0.50
Rel. freq. of proposer's curses - buyer 4	0.25	1.00	0.00	0.00	0.00	.
Rel. freq. of proposer's curses - seller 5	.	.	.	.	.	.
Rel. freq. of proposer's curses - seller 6	0.00	.	.	.	.	0.00
Rel. freq. of proposer's curses - seller 7	0.50	0.00	.	.	.	1.00
Rel. freq. of proposer's curses - seller 8	1.00	1.00	.	.	1.00	.
Number of market ask and market bid matches	6.80	7.00	7.00	7.00	7.00	6.00
Number of contracts per market ask-bid match	1.03	1.00	1.00	1.00	1.00	1.17
Avg. rank of traded red.value at trading time	2.08	1.29	2.00	2.29	2.14	2.67
Avg. rank of traded unit cost at trading time	2.14	2.14	2.29	2.00	2.29	2.00
Number of submitted bids per contract	5.74	5.86	5.00	6.14	6.29	5.43
... rel. freq. of bids < p* - 200	0.01	0.02	0.00	0.00	0.00	0.03
... rel. freq. of bids < p*	0.86	0.78	0.83	0.86	0.93	0.92
... rel. freq. of bids = p*	0.05	0.05	0.09	0.07	0.02	0.03
... rel. freq. of bids > p*	0.08	0.08	0.09	0.07	0.05	0.05
Number of submitted asks per contract	4.14	5.14	3.43	4.57	3.71	3.86
... rel. freq. of asks < p*	0.16	0.08	0.17	0.16	0.19	0.22
... rel. freq. of asks = p*	0.11	0.03	0.13	0.16	0.12	0.11
... rel. freq. of asks > p*	0.73	0.89	0.71	0.69	0.69	0.67
... rel. freq. of asks > p* + 200	0.01	0.03	0.00	0.03	0.00	0.00
Number of offers for all units in periods > 1	67	.	59	75	70	65
(a) rel. freq. of bids > p-max & asks < p-min	0.01	.	0.03	0.01	0.00	0.00
(b) rel. freq. of bids > p-avg & asks < p-avg	0.16	.	0.19	0.17	0.14	0.12
(c) rel. freq. of bids > p-own & asks < p-own	0.18	.	0.17	0.23	0.16	0.18
(d) rel. freq. of bids < p-own & asks > p-own	0.60	.	0.68	0.61	0.59	0.51
Number of offers for units bought above p-avg						
or sold below p-avg in the previous period	14	.	20	16	12	9
(C) rel. freq. of bids > p-own & asks < p-own	0.09	.	0.15	0.13	0.08	0.00
Number of offers for units						
that were not traded in the previous period	26	.	21	26	30	26
(D) rel. freq. of bids < p-own & asks > p-own	0.55	.	0.57	0.65	0.40	0.58
Aggressive offer frequency ratio (c)/(C)	0.49	.	0.89	0.55	0.53	0.00
Moderate offer frequency ratio (d)/(D)	0.93	.	0.84	1.07	0.68	1.14

Panel VII.B.16

Double Auction Market Session Results	Session: M2821	Number of traders: 8	File: DAMM2821.SES			
Market clearing quantity q* : 7	Average	Period	Period	Period	Period	Period
Market clearing price p* : 416	of Periods	1	2	3	4	5
Deviation of traded quantity from q*	-0.20	0	-1	0	0	0
Efficiency	1.00	1.00	0.99	1.00	1.00	1.00
Average contract price (deviation from p*)	-1.00	-0.71	1.83	-2.57	-1.71	-1.86
Minimum contract price (deviation from p*)	-14.20	-16.00	-6.00	-17.00	-16.00	-16.00
Maximum contract price (deviation from p*)	5.60	9.00	10.00	3.00	3.00	3.00
Market clearing price concordance coefficient	0.80	0.66	0.81	0.83	0.83	0.84
Rel. freq. of falling price paths	0.00	0	0	0	0	0
Rel. freq. of rising price paths	0.80	1	0	1	1	1
Rel. freq. of hanging price paths	0.00	0	0	0	0	0
Rel. freq. of standing price paths	0.20	0	1	0	0	0
Rel. freq. of unknown price paths	0.00	0	0	0	0	0
Rel. freq. of contract proposals by buyers	0.24	0.29	0.33	0.29	0.14	0.14
... rel. freq. of these with price < p*	0.20	0.00	1.00	0.00	0.00	0.00
... rel. freq. of these with price = p*	0.30	0.00	0.00	0.50	0.00	1.00
... rel. freq. of these with price > p*	0.50	1.00	0.00	0.50	1.00	0.00
... avg. buyer proposed contract price - p*	1.20	7.00	-4.50	1.50	2.00	0.00
Rel. freq. of contract proposals by sellers	0.76	0.71	0.67	0.71	0.86	0.86
... rel. freq. of these with price < p*	0.44	0.40	0.00	0.80	0.50	0.50
... rel. freq. of these with price = p*	0.03	0.00	0.00	0.00	0.17	0.00
... rel. freq. of these with price > p*	0.53	0.60	1.00	0.20	0.33	0.50
... avg. seller proposed contract price - p*	-1.50	-3.80	5.00	-4.20	-2.33	-2.17
Rel. freq. of proposer's curses - all traders	0.46	0.57	0.00	0.71	0.57	0.43
Rel. freq. of proposer's curses - all buyers	0.50	1.00	0.00	0.50	1.00	0.00
Rel. freq. of proposer's curses - all sellers	0.44	0.40	0.00	0.80	0.50	0.50
Rel. freq. of proposer's curses - buyer 1	1.00	1.00	.	.	.	.
Rel. freq. of proposer's curses - buyer 2	0.00	.	.	0.00	.	0.00
Rel. freq. of proposer's curses - buyer 3	0.75	1.00	0.00	1.00	1.00	.
Rel. freq. of proposer's curses - buyer 4	.	.	.	.	.	.
Rel. freq. of proposer's curses - seller 5	0.25	.	0.00	1.00	0.00	0.00
Rel. freq. of proposer's curses - seller 6	0.00	0.00	0.00	0.00	0.00	0.00
Rel. freq. of proposer's curses - seller 7	0.70	0.50	0.00	1.00	1.00	1.00
Rel. freq. of proposer's curses - seller 8	0.70	0.50	0.00	1.00	1.00	1.00
Number of market ask and market bid matches	6.40	6.00	6.00	7.00	7.00	6.00
Number of contracts per market ask-bid match	1.06	1.17	1.00	1.00	1.00	1.17
Avg. rank of traded red.value at trading time	2.07	2.33	1.83	1.86	2.14	2.17
Avg. rank of traded unit cost at trading time	1.92	2.00	2.00	1.86	1.57	2.17
Number of submitted bids per contract	3.99	4.43	4.67	4.43	3.86	2.57
... rel. freq. of bids < p* - 200	0.01	0.03	0.00	0.00	0.00	0.00
... rel. freq. of bids < p*	0.72	0.61	0.68	0.84	0.74	0.72
... rel. freq. of bids = p*	0.03	0.00	0.04	0.03	0.04	0.06
... rel. freq. of bids > p*	0.25	0.39	0.29	0.13	0.22	0.22
Number of submitted asks per contract	5.82	5.71	6.83	6.00	6.43	4.14
... rel. freq. of asks < p*	0.09	0.08	0.07	0.12	0.09	0.10
... rel. freq. of asks = p*	0.02	0.00	0.02	0.02	0.02	0.03
... rel. freq. of asks > p*	0.89	0.93	0.90	0.86	0.89	0.86
... rel. freq. of asks > p* + 200	0.00	0.00	0.00	0.00	0.00	0.00
Number of offers for all units in periods > 1	65	-	69	73	72	47
(a) rel. freq. of bids > p-max & asks < p-min	0.01	-	0.01	0.01	0.00	0.00
(b) rel. freq. of bids > p-avg & asks < p-avg	0.16	-	0.17	0.11	0.24	0.13
(c) rel. freq. of bids > p-own & asks < p-own	0.27	-	0.39	0.30	0.18	0.21
(d) rel. freq. of bids < p-own & asks > p-own	0.54	-	0.51	0.58	0.58	0.49
Number of offers for units bought above p-avg						
or sold below p-avg in the previous period	14	-	14	13	20	9
(C) rel. freq. of bids > p-own & asks < p-own	0.19	-	0.14	0.08	0.30	0.22
Number of offers for units						
that were not traded in the previous period	22	-	17	32	21	17
(D) rel. freq. of bids < p-own & asks > p-own	0.48	-	0.41	0.66	0.43	0.41
Aggressive offer frequency ratio (c)/(C)	0.83	-	0.37	0.26	1.66	1.04
Moderate offer frequency ratio (d)/(D)	0.88	-	0.81	1.14	0.73	0.84

Panel VII.B.17

Double Auction Market Session Results Session: M2841 Number of traders: 8 File: DAMM2841.SES

Market clearing quantity q* : 7
Market clearing price p* : 559

	Average of Periods	Period 1	Period 2	Period 3	Period 4	Period 5
Deviation of traded quantity from q*	-0.40	-1	-1	0	0	0
Efficiency	0.99	0.97	0.99	1.00	1.00	1.00
Average contract price (deviation from p*)	-1.94	-2.00	-2.83	-2.00	-1.43	-1.43
Minimum contract price (deviation from p*)	-5.20	-10.00	-4.00	-4.00	-4.00	-4.00
Maximum contract price (deviation from p*)	2.20	9.00	0.00	0.00	0.00	2.00
Market clearing price concordance coefficient	0.89	0.78	0.89	0.92	0.94	0.92
Rel. freq. of falling price paths	0.20	0	1	0	0	0
Rel. freq. of rising price paths	0.40	0	0	1	1	0
Rel. freq. of hanging price paths	0.00	0	0	0	0	0
Rel. freq. of standing price paths	0.40	1	0	0	0	1
Rel. freq. of unknown price paths	0.00	0	0	0	0	0
Rel. freq. of contract proposals by buyers	0.64	0.83	0.67	0.71	0.29	0.71
... rel. freq. of these with price < p*	0.84	0.40	1.00	1.00	1.00	0.80
... rel. freq. of these with price = p*	0.04	0.00	0.00	0.00	0.00	0.20
... rel. freq. of these with price > p*	0.12	0.60	0.00	0.00	0.00	0.00
... avg. buyer proposed contract price - p*	-2.54	-1.60	-4.00	-2.40	-2.50	-2.20
Rel. freq. of contract proposals by sellers	0.36	0.17	0.33	0.29	0.71	0.29
... rel. freq. of these with price < p*	0.58	1.00	0.50	0.50	0.40	0.50
... rel. freq. of these with price = p*	0.32	0.00	0.50	0.50	0.60	0.00
... rel. freq. of these with price > p*	0.10	0.00	0.00	0.00	0.00	0.50
... avg. seller proposed contract price - p*	-1.20	-4.00	-0.50	-1.00	-1.00	0.50
Rel. freq. of proposer's curses - all traders	0.28	0.67	0.17	0.14	0.29	0.14
Rel. freq. of proposer's curses - all buyers	0.12	0.60	0.00	0.00	0.00	0.00
Rel. freq. of proposer's curses - all sellers	0.58	1.00	0.50	0.50	0.40	0.50
Rel. freq. of proposer's curses - buyer 1	0.00	.	0.00	0.00	.	0.00
Rel. freq. of proposer's curses - buyer 2	0.00	0.00	0.00	0.00	0.00	0.00
Rel. freq. of proposer's curses - buyer 3	0.10	0.50	0.00	0.00	0.00	0.00
Rel. freq. of proposer's curses - buyer 4	0.25	1.00	0.00	0.00	.	0.00
Rel. freq. of proposer's curses - seller 5	0.00	.	0.00	.	.	.
Rel. freq. of proposer's curses - seller 6	0.50	.	.	0.50	0.50	.
Rel. freq. of proposer's curses - seller 7	1.00	.	.	.	1.00	1.00
Rel. freq. of proposer's curses - seller 8	0.50	1.00	1.00	.	0.00	0.00
Number of market ask and market bid matches	6.40	6.00	6.00	7.00	7.00	6.00
Number of contracts per market ask-bid match	1.03	1.00	1.00	1.00	1.00	1.17
Avg. rank of traded red.value at trading time	2.08	2.00	2.00	1.57	2.14	2.67
Avg. rank of traded unit cost at trading time	2.04	2.00	2.00	1.86	1.86	2.50
Number of submitted bids per contract	6.10	6.33	7.33	5.00	7.43	4.43
... rel. freq. of bids < p* - 200	0.00	0.00	0.00	0.00	0.00	0.00
... rel. freq. of bids < p*	0.93	0.92	0.98	0.97	0.94	0.84
... rel. freq. of bids = p*	0.04	0.00	0.02	0.03	0.06	0.10
... rel. freq. of bids > p*	0.03	0.08	0.00	0.00	0.00	0.06
Number of submitted asks per contract	5.06	5.00	5.00	4.71	5.43	5.14
... rel. freq. of asks < p*	0.28	0.27	0.43	0.27	0.21	0.19
... rel. freq. of asks = p*	0.15	0.00	0.10	0.27	0.29	0.11
... rel. freq. of asks > p*	0.57	0.73	0.47	0.45	0.50	0.69
... rel. freq. of asks > p* + 200	0.00	0.00	0.00	0.00	0.00	0.00
Number of offers for all units in periods > 1	75	.	74	68	90	67
(a) rel. freq. of bids > p-max & asks < p-min	0.01	.	0.00	0.00	0.00	0.03
(b) rel. freq. of bids > p-avg & asks < p-avg	0.13	.	0.15	0.13	0.11	0.13
(c) rel. freq. of bids > p-own & asks < p-own	0.20	.	0.26	0.26	0.13	0.13
(d) rel. freq. of bids < p-own & asks > p-own	0.50	.	0.45	0.50	0.60	0.46
Number of offers for units bought above p-avg or sold below p-avg in the previous period	15	.	13	10	19	18
(C) rel. freq. of bids > p-own & asks < p-own	0.03	.	0.00	0.00	0.00	0.11
Number of offers for units that were not traded in the previous period	26	.	24	33	25	22
(D) rel. freq. of bids < p-own & asks > p-own	0.26	.	0.33	0.36	0.24	0.09
Aggressive offer frequency ratio (c)/(C)	0.21	.	0.00	0.00	0.00	0.83
Moderate offer frequency ratio (d)/(D)	0.52	.	0.75	0.73	0.40	0.20

Panel VII.B.18

Double Auction Market Session Results	Session: M2A11	Number of traders: 10	File: DAMM2A11.SES			
		Average	Period	Period	Period	Period	Period

		Average of Periods	Period 1	Period 2	Period 3	Period 4	Period 5
Market clearing quantity q* :	9						
Market clearing price p* :	797						
Deviation of traded quantity from q*		0.00	0	0	0	0	0
Efficiency		1.00	1.00	1.00	1.00	1.00	1.00
Average contract price (deviation from p*)		-0.91	-2.11	-1.67	-0.22	-0.11	-0.44
Minimum contract price (deviation from p*)		-3.80	-8.00	-6.00	-2.00	-2.00	-1.00
Maximum contract price (deviation from p*)		1.40	3.00	2.00	1.00	1.00	0.00
Market clearing price concordance coefficient		0.91	0.81	0.85	0.94	0.96	0.97
Rel. freq. of falling price paths		0.00	0	0	0	0	0
Rel. freq. of rising price paths		1.00	1	1	1	1	1
Rel. freq. of hanging price paths		0.00	0	0	0	0	0
Rel. freq. of standing price paths		0.00	0	0	0	0	0
Rel. freq. of unknown price paths		0.00	0	0	0	0	0
Rel. freq. of contract proposals by buyers		0.13	0.11	0.00	0.22	0.11	0.22
... rel. freq. of these with price < p*		0.50	0.00	.	1.00	0.00	1.00
... rel. freq. of these with price = p*		0.50	1.00	.	0.00	1.00	0.00
... rel. freq. of these with price > p*		0.00	0.00	.	0.00	0.00	0.00
... avg. buyer proposed contract price - p*		-0.75	0.00	.	-2.00	0.00	-1.00
Rel. freq. of contract proposals by sellers		0.87	0.89	1.00	0.78	0.89	0.78
... rel. freq. of these with price < p*		0.37	0.50	0.67	0.14	0.25	0.29
... rel. freq. of these with price = p*		0.40	0.25	0.11	0.43	0.50	0.71
... rel. freq. of these with price > p*		0.23	0.25	0.22	0.43	0.25	0.00
... avg. seller proposed contract price - p*		-0.83	-2.38	-1.67	0.29	-0.13	-0.29
Rel. freq. of proposer's curses - all traders		0.33	0.44	0.67	0.11	0.22	0.22
Rel. freq. of proposer's curses - all buyers		0.00	0.00	.	0.00	0.00	0.00
Rel. freq. of proposer's curses - all sellers		0.37	0.50	0.67	0.14	0.25	0.29
Rel. freq. of proposer's curses - buyer 1		.	.	.	.	.	.
Rel. freq. of proposer's curses - buyer 2		0.00	0.00	.	.	.	0.00
Rel. freq. of proposer's curses - buyer 3		0.00	.	.	0.00	0.00	.
Rel. freq. of proposer's curses - buyer 4		0.00	.	.	0.00	.	.
Rel. freq. of proposer's curses - buyer 5		0.00	.	.	.	.	0.00
Rel. freq. of proposer's curses - seller 6		0.33	0.00	1.00	0.00	.	.
Rel. freq. of proposer's curses - seller 7		0.40	0.50	0.50	0.00	0.50	0.50
Rel. freq. of proposer's curses - seller 8		0.30	0.50	0.50	0.50	0.00	0.00
Rel. freq. of proposer's curses - seller 9		0.30	1.00	0.50	0.00	0.00	0.00
Rel. freq. of proposer's curses - seller 10		0.60	0.50	1.00	0.00	0.50	1.00
Number of market ask and market bid matches		8.40	9.00	7.00	9.00	9.00	8.00
Number of contracts per market ask-bid match		1.07	1.00	1.29	1.00	1.00	1.13
Avg. rank of traded red.value at trading time		2.27	2.33	2.29	2.44	1.89	2.38
Avg. rank of traded unit cost at trading time		2.22	1.89	2.71	2.11	2.11	2.25
Number of submitted bids per contract		5.00	5.78	4.67	5.67	4.78	4.11
... rel. freq. of bids < p* - 200		0.01	0.02	0.00	0.02	0.00	0.00
... rel. freq. of bids < p*		0.84	0.88	0.83	0.84	0.81	0.84
... rel. freq. of bids = p*		0.11	0.08	0.10	0.10	0.14	0.16
... rel. freq. of bids > p*		0.04	0.04	0.07	0.06	0.05	0.00
Number of submitted asks per contract		5.67	6.89	5.22	5.78	5.22	5.22
... rel. freq. of asks < p*		0.10	0.06	0.21	0.12	0.04	0.09
... rel. freq. of asks = p*		0.12	0.08	0.06	0.10	0.13	0.21
... rel. freq. of asks > p*		0.78	0.85	0.72	0.79	0.83	0.70
... rel. freq. of asks > p* + 200		0.00	0.00	0.00	0.00	0.00	0.00
Number of offers for all units in periods > 1		91	.	89	103	90	84
(a) rel. freq. of bids > p-max & asks < p-min		0.00	.	0.00	0.00	0.00	0.00
(b) rel. freq. of bids > p-avg & asks < p-avg		0.11	.	0.21	0.14	0.04	0.05
(c) rel. freq. of bids > p-own & asks < p-own		0.16	.	0.19	0.19	0.09	0.15
(d) rel. freq. of bids < p-own & asks > p-own		0.51	.	0.52	0.52	0.49	0.50
Number of offers for units bought above p-avg							
or sold below p-avg in the previous period		17	.	23	25	11	8
(C) rel. freq. of bids > p-own & asks < p-own		0.02	.	0.04	0.04	0.00	0.00
Number of offers for units							
that were not traded in the previous period		27	.	28	29	26	23
(D) rel. freq. of bids < p-own & asks > p-own		0.09	.	0.21	0.07	0.00	0.09
Aggressive offer frequency ratio (c)/(C)		0.11	.	0.23	0.21	0.00	0.00
Moderate offer frequency ratio (d)/(D)		0.18	.	0.41	0.13	0.00	0.17

VII.C. Zero-Intelligence Trader Simulation Data Panels

The panels contained in this appendix display the results of simulation runs of the alternating double auction market populated with zero-intelligence trader agents. Each simulation consisted of 10000 rounds. A simulation round corresponded to an experimental session, i.e. consisted of five market periods.

In each offer cycle of a simulated market, each zero-intelligence trading agent submits an offer that is randomly drawn from the range of feasible offers with uniform probability. The range of feasible offers of such an agent is always limited by the *no-loss* rule, i.e. an offer that if accepted would lead to a loss for the agent is non-feasible. The range of offers is further limited by the technical price range, i.e. the range of positive integers in which all offers must lie. The technical price range in the **baseline simulations** (panels VII.C.0.0 to VII.C.0.2a) was set to [200,...,700], with $p^* = 500$ in all baseline simulations. Five combinations of market type and number of traders were used for the baseline simulations: market type M0 with 8 players, market type M1 with 8 players, market type M1 with 12 players, market type M2 with 8 players, and market type M2 with 10 players. The results of the corresponding zero-intelligence trader simulations are reported in the panels VII.C.0.0, VII.C.0.1a, VII.C.0.1b, VII.C.0.2a, and VII.C.0.2b.

The technical price range in the **full price range simulations** was set to [1,2,...,9999], with $p^* = 500$ in the M0 simulation (panel VII.C.1.0) and p^* equal to the experimental value in the other simulations (panels VII.C.1.1 to VII.C.1.18). The panels of the full price range simulations are sorted as the panels in the previous appendix (and as in all preceding tables): the panels corresponding to the inexperienced sessions are displayed first, ordered from the market type 1 to the market type 2 sessions and from sessions with less players to those with more players. Then the panels of the experienced sessions follow in a similar manner.

The zero-intelligence trading agents were further bounded by the *no-crossing* and the *ask-bid-spread-reduction* rules, just as the human counterparts in the experiment. The former rule disallows asks lower than the current market bid and bids higher than the current market ask, given that a corresponding market offer of the other market side is standing. The latter rule requires any new ask to be smaller than the current market ask and any new bid to be greater than the current market bid, given that a corresponding market offer of the own market side is standing.)

A zero-intelligence trading agent was not allowed to submit an offer, if the lower bound of the range of his feasible offers was greater than the upper bound of that range. The same market period termination criterion as in the experiment was used: the market period ended, when a number z of offer cycles passed without any new offers. The value of z was set to 3 in all simulations.

The results are ordered in the same manner as the experimental results reported in the panels of the previous appendix. The abbreviations used are the same as in the previous appendix. As before, prices and offers are normalized on the market clearing price, i.e. reported as deviation from p*. The reported values are - in general - averages of the corresponding parameter over the 10000 rounds of the simulations. In some cases, however, an additional line titled *median value in the simulation* has been inserted. A value reported in such a line is the median of the simulation distribution of the parameter reported in the previous line. Note that the values reported in the column *Average of Periods* are averages of the five period averages over the 10000 rounds, not averages of the corresponding parameter over 50000 rounds. In this column, in lines titled *median value in the simulation*, the median of the five period averages is reported. This median can be decisively different than the median of the values themselves, since the distribution of the five period averages is generally more *tightly* clustered around the average (or median) value. (The notes concerning the lines containing median values are analogously true for the lines titled *lower quartile bound in the simulation* and *upper quartile bound in the simulation*. Here, the boundary values for the lower or upper quartile of the simulation distribution of the corresponding parameters are reported.)

At the end of each simulation result panel, the simulation distribution of price paths is reported. The relative (and cumulative) frequency distribution of simulation rounds in which n (or more) periods have price paths of the same type are displayed, where n = 0...5. The table is to be read in the following manner: The entry in the column *falling - f(x)* of the line *n = 4* corresponds to the relative frequency with which simulation rounds had falling price paths in four of the five market periods. The entry in the column *falling - F(x)* of the same line corresponds to the cumulative frequency with which simulation rounds had falling price paths in four or more of the five periods. (Obviously, the line *n = 0* must contain the value 1.0 in all *F(x)* columns, since the cumulative frequency with which simulation rounds had price paths of one type in zero or more of the five market periods must be one.)

Panel VII.C.0.0

Double Auction Market Simulation Results   File: DAMM0800.SIM

Number of simulation rounds	: 10000		Periods per round :			5
Market type	: MO		Number of traders :			8
Market rules	: AltDAM		Technical price range : [300 ... 700]			
No-Crossing rule	: YES		Ask-Bid-Spread-Reduction rule :			YES
Trader type	: ZI					

Market clearing quantity q*	: 10					
Market clearing price p*	: 500					

	Average of Periods	Period 1	Period 2	Period 3	Period 4	Period 5
Deviation of traded quantity from q*	0.40	0.40	0.40	0.40	0.41	0.40
Efficiency	0.99	0.99	0.99	0.99	0.99	0.99
... median value in the simulation	1.00	1.00	1.00	1.00	1.00	1.00
Average contract price (deviation from p*)	0.01	-0.00	-0.00	-0.02	0.05	0.01
... median value in the simulation	0.01	0.01	0.01	-0.08	0.11	0.01
Minimum contract price (deviation from p*)	-13.22	-13.22	-13.24	-13.20	-13.19	-13.25
... median value in the simulation	-13.99	-13.99	-13.99	-13.99	-13.99	-13.99
Maximum contract price (deviation from p*)	13.20	13.19	13.18	13.18	13.23	13.23
... median value in the simulation	14.01	14.01	14.01	14.01	14.01	14.01
Market clearing price concordance coefficient	0.51	0.51	0.51	0.51	0.51	0.51
... median value in the simulation	0.52	0.51	0.51	0.51	0.51	0.51
Rel. freq. of falling price paths	0.32	0.31	0.31	0.32	0.33	0.32
Rel. freq. of rising price paths	0.32	0.32	0.31	0.32	0.31	0.31
Rel. freq. of hanging price paths	0.18	0.18	0.18	0.18	0.18	0.19
Rel. freq. of standing price paths	0.18	0.19	0.19	0.18	0.19	0.18
Rel. freq. of unknown price paths	0.00	0.00	0.00	0.00	0.00	0.00
Rel. freq. of contract proposals by buyers	0.50	0.50	0.50	0.50	0.50	0.50
... lower quartile bound in the simulation	0.45	0.37	0.37	0.37	0.37	0.37
... median value in the simulation	0.51	0.51	0.51	0.51	0.51	0.51
... upper quartile bound in the simulation	0.57	0.64	0.64	0.64	0.64	0.64
... rel. freq. of these with price < p*	0.36	0.37	0.36	0.37	0.36	0.36
... rel. freq. of these with price = p*	0.09	0.08	0.09	0.09	0.09	0.09
... rel. freq. of these with price > p*	0.55	0.55	0.55	0.55	0.55	0.55
... avg. buyer proposed contract price - p*	1.86	1.82	1.88	1.84	1.89	1.87
Rel. freq. of contract proposals by sellers	0.50	0.50	0.50	0.50	0.50	0.50
... lower quartile bound in the simulation	0.45	0.37	0.37	0.37	0.37	0.37
... median value in the simulation	0.51	0.51	0.51	0.51	0.51	0.51
... upper quartile bound in the simulation	0.57	0.64	0.64	0.64	0.64	0.64
... rel. freq. of these with price < p*	0.55	0.55	0.55	0.55	0.55	0.55
... rel. freq. of these with price = p*	0.09	0.09	0.09	0.09	0.09	0.09
... rel. freq. of these with price > p*	0.36	0.36	0.36	0.36	0.36	0.36
... avg. seller proposed contract price - p*	-1.85	-1.85	-1.84	-1.90	-1.82	-1.81
Rel. freq. of proposer's curses - all traders	0.57	0.57	0.57	0.57	0.57	0.57
... median value in the simulation	0.57	0.61	0.61	0.61	0.61	0.61
Rel. freq. of proposer's curses - all buyers	0.55	0.55	0.55	0.55	0.55	0.55
Rel. freq. of proposer's curses - all sellers	0.55	0.55	0.55	0.55	0.55	0.55
Rel. freq. of proposer's curses - buyer 1	0.67	0.67	0.67	0.67	0.68	0.68
Rel. freq. of proposer's curses - buyer 2	0.48	0.48	0.49	0.49	0.48	0.48
Rel. freq. of proposer's curses - buyer 3	0.55	0.55	0.55	0.54	0.55	0.55
Rel. freq. of proposer's curses - buyer 4	0.57	0.57	0.57	0.57	0.56	0.57
Rel. freq. of proposer's curses - seller 5	0.68	0.67	0.68	0.68	0.67	0.68
Rel. freq. of proposer's curses - seller 6	0.48	0.48	0.49	0.48	0.48	0.49
Rel. freq. of proposer's curses - seller 7	0.55	0.55	0.55	0.55	0.56	0.55
Rel. freq. of proposer's curses - seller 8	0.57	0.57	0.57	0.57	0.56	0.56

Panel VII.C.0.0 (continued)

	Average of Periods	Period 1	Period 2	Period 3	Period 4	Period 5
Number of market ask and market bid matches	9.64	9.64	9.64	9.63	9.63	9.63
Number of contracts per market ask-bid match	1.08	1.08	1.08	1.08	1.08	1.08
Avg. rank of traded red.value at trading time	1.82	1.82	1.82	1.82	1.82	1.82
... median value in the simulation	1.82	1.82	1.82	1.82	1.82	1.82
Avg. rank of traded unit cost at trading time	1.82	1.82	1.82	1.82	1.82	1.82
... median value in the simulation	1.83	1.84	1.82	1.82	1.82	1.82
Number of submitted bids per contract	9.82	9.82	9.83	9.81	9.81	9.81
... median value in the simulation	9.82	9.82	9.82	9.82	9.81	9.81
... rel. freq. of bids < p* - 200	0.00	0.00	0.00	0.00	0.00	0.00
... rel. freq. of bids < p*	0.84	0.84	0.84	0.84	0.84	0.84
... rel. freq. of bids = p*	0.02	0.02	0.02	0.02	0.02	0.02
... rel. freq. of bids > p*	0.14	0.14	0.14	0.14	0.14	0.14
Number of submitted asks per contract	9.81	9.81	9.81	9.83	9.80	9.82
... median value in the simulation	9.82	9.81	9.81	9.82	9.82	9.81
... rel. freq. of asks < p*	0.14	0.14	0.14	0.14	0.14	0.14
... rel. freq. of asks = p*	0.02	0.02	0.02	0.02	0.02	0.02
... rel. freq. of asks > p*	0.84	0.84	0.84	0.84	0.84	0.84
... rel. freq. of asks > p* + 200	0.00	0.00	0.00	0.00	0.00	0.00
Number of offers for all units in periods > 1	204	·	204	204	204	204
(a) rel. freq. of bids > p-max & asks < p-min	0.02	·	0.02	0.02	0.02	0.02
(b) rel. freq. of bids > p-avg & asks < p-avg	0.15	·	0.15	0.15	0.15	0.15
(c) rel. freq. of bids > p-own & asks < p-own	0.11	·	0.11	0.11	0.11	0.11
(d) rel. freq. of bids < p-own & asks > p-own	0.84	·	0.84	0.84	0.84	0.84
Number of offers for units bought above p-avg or sold below p-avg in the previous period	78	·	78	78	78	78
(C) rel. freq. of bids > p-own & asks < p-own	0.06	·	0.06	0.06	0.06	0.07
Number of offers for units that were not traded in the previous period	33	·	33	33	33	33
(D) rel. freq. of bids < p-own & asks > p-own	0.87	·	0.87	0.87	0.87	0.87
Aggressive offer frequency ratio (c)/(C)	0.58	·	0.57	0.58	0.58	0.59
... median value in the simulation	0.57	·	0.54	0.55	0.55	0.56
Moderate offer frequency ratio (d)/(D)	1.04	·	1.04	1.04	1.04	1.03
... median value in the simulation	1.04	·	1.04	1.04	1.04	1.04

Relative frequencies f(x) and cumulative distributions F(x) of the simulation rounds
(5 subsequent periods) in which any n periods have price paths of the same type.

	falling		rising		hanging		standing		unknown	
n	f(x)	F(x)	f(x)	F(x)	f(x)	F(x)	f(x)	F(x)	f(x)	F(x)
5	0.00	0.00	0.00	0.00	0.00	0.00	0.00	0.00	0.00	0.00
4	0.04	0.04	0.03	0.03	0.01	0.01	0.00	0.01	0.00	0.00
3	0.15	0.18	0.15	0.18	0.04	0.05	0.04	0.05	0.00	0.00
2	0.33	0.51	0.32	0.51	0.18	0.23	0.18	0.23	0.00	0.00
1	0.34	0.85	0.34	0.84	0.40	0.63	0.41	0.64	0.00	0.00
0	0.15	1.00	0.15	1.00	0.37	1.00	0.36	1.00	1.00	1.00

Panel VII.C.0.1a

Double Auction Market Simulation Results   File: DAMM1800.SIM

Number of simulation rounds	: 10000	Periods per round :	5
Market type	: M1	Number of traders :	8
Market rules	: AltDAM	Technical price range :	[300 ... 700]
No-Crossing rule	: YES	Ask-Bid-Spread-Reduction rule :	YES
Trader type	: ZI		

Market clearing quantity q*   :   10
Market clearing price   p*   :   500

	Average of Periods	Period 1	Period 2	Period 3	Period 4	Period 5
Deviation of traded quantity from q*	0.27	0.27	0.27	0.27	0.27	0.27
Efficiency	0.99	0.99	0.99	0.99	0.99	0.99
... median value in the simulation	1.00	1.00	1.00	1.00	1.00	1.00
Average contract price (deviation from p*)	-1.86	-1.90	-1.80	-1.86	-1.87	-1.87
... median value in the simulation	-1.80	-1.88	-1.78	-1.88	-1.80	-1.80
Minimum contract price (deviation from p*)	-19.56	-19.58	-19.45	-19.54	-19.62	-19.59
... median value in the simulation	-20.99	-20.99	-20.99	-20.99	-20.99	-20.99
Maximum contract price (deviation from p*)	13.05	13.00	13.10	13.07	13.02	13.05
... median value in the simulation	14.01	14.01	14.01	14.01	14.01	14.01
Market clearing price concordance coefficient	0.52	0.52	0.52	0.52	0.52	0.52
... median value in the simulation	0.53	0.52	0.53	0.53	0.52	0.52
Rel. freq. of falling price paths	0.22	0.23	0.23	0.22	0.22	0.22
Rel. freq. of rising price paths	0.40	0.41	0.40	0.40	0.40	0.40
Rel. freq. of hanging price paths	0.26	0.25	0.25	0.26	0.26	0.26
Rel. freq. of standing price paths	0.12	0.11	0.12	0.12	.0.12	0.12
Rel. freq. of unknown price paths	0.00	0.00	0.00	0.00	0.00	0.00
Rel. freq. of contract proposals by buyers	0.51	0.51	0.51	0.51	0.51	0.50
... lower quartile bound in the simulation	0.45	0.41	0.37	0.41	0.41	0.41
... median value in the simulation	0.51	0.51	0.51	0.51	0.51	0.51
... upper quartile bound in the simulation	0.57	0.64	0.64	0.64	0.64	0.64
... rel. freq. of these with price < p*	0.40	0.40	0.40	0.40	0.40	0.40
... rel. freq. of these with price = p*	0.08	0.08	0.08	0.08	0.08	0.08
... rel. freq. of these with price > p*	0.52	0.52	0.53	0.52	0.52	0.52
... avg. buyer proposed contract price - p*	0.23	0.18	0.27	0.26	0.18	0.25
Rel. freq. of contract proposals by sellers	0.49	0.49	0.49	0.49	0.49	0.50
... lower quartile bound in the simulation	0.44	0.37	0.37	0.37	0.37	0.37
... median value in the simulation	0.50	0.51	0.51	0.51	0.51	0.51
... upper quartile bound in the simulation	0.56	0.61	0.64	0.61	0.61	0.61
... rel. freq. of these with price < p*	0.58	0.58	0.57	0.58	0.58	0.57
... rel. freq. of these with price = p*	0.08	0.08	0.08	0.08	0.08	0.08
... rel. freq. of these with price > p*	0.34	0.34	0.34	0.34	0.34	0.35
... avg. seller proposed contract price - p*	-4.02	-4.07	-3.97	-4.05	-4.01	-4.01
Rel. freq. of proposer's curses - all traders	0.57	0.57	0.57	0.57	0.57	0.57
... median value in the simulation	0.57	0.61	0.61	0.61	0.61	0.61
Rel. freq. of proposer's curses - all buyers	0.52	0.52	0.53	0.52	0.52	0.52
Rel. freq. of proposer's curses - all sellers	0.58	0.58	0.57	0.58	0.58	0.57
Rel. freq. of proposer's curses - buyer 1	0.64	0.64	0.65	0.65	0.63	0.64
Rel. freq. of proposer's curses - buyer 2	0.46	0.47	0.47	0.47	0.46	0.46
Rel. freq. of proposer's curses - buyer 3	0.52	0.52	0.52	0.52	0.52	0.52
Rel. freq. of proposer's curses - buyer 4	0.53	0.52	0.53	0.53	0.53	0.53
Rel. freq. of proposer's curses - seller 5	0.69	0.69	0.69	0.69	0.69	0.68
Rel. freq. of proposer's curses - seller 6	0.50	0.50	0.50	0.50	0.50	0.50
Rel. freq. of proposer's curses - seller 7	0.60	0.61	0.60	0.60	0.60	0.60
Rel. freq. of proposer's curses - seller 8	0.60	0.60	0.60	0.60	0.60	0.60

Panel VII.C.0.1a (continued)

	Average of Periods	Period 1	Period 2	Period 3	Period 4	Period 5
Number of market ask and market bid matches	9.50	9.50	9.48	9.50	9.51	9.51
Number of contracts per market ask-bid match	1.08	1.08	1.08	1.08	1.08	1.08
Avg. rank of traded red.value at trading time	1.85	1.84	1.84	1.84	1.85	1.85
... median value in the simulation	1.85	1.84	1.84	1.84	1.86	1.84
Avg. rank of traded unit cost at trading time	1.80	1.80	1.80	1.80	1.80	1.80
... median value in the simulation	1.81	1.81	1.81	1.81	1.81	1.81
Number of submitted bids per contract	9.70	9.70	9.68	9.70	9.70	9.71
... median value in the simulation	9.70	9.71	9.71	9.71	9.71	9.71
... rel. freq. of bids < p^* - 200	0.00	0.00	0.00	0.00	0.00	0.00
... rel. freq. of bids < p^*	0.84	0.84	0.84	0.84	0.84	0.84
... rel. freq. of bids = p^*	0.02	0.02	0.02	0.02	0.02	0.02
... rel. freq. of bids > p^*	0.13	0.13	0.13	0.13	0.13	0.13
Number of submitted asks per contract	9.71	9.70	9.71	9.72	9.72	9.72
... median value in the simulation	9.72	9.71	9.71	9.73	9.71	9.71
... rel. freq. of asks < p^*	0.16	0.16	0.15	0.16	0.16	0.16
... rel. freq. of asks = p^*	0.02	0.02	0.02	0.02	0.02	0.02
... rel. freq. of asks > p^*	0.82	0.82	0.83	0.82	0.82	0.82
... rel. freq. of asks > p^* + 200	0.00	0.00	0.00	0.00	0.00	0.00
Number of offers for all units in periods > 1	199	-	199	199	199	199
(a) rel. freq. of bids > p-max & asks < p-min	0.02	-	0.02	0.02	0.02	0.02
(b) rel. freq. of bids > p-avg & asks < p-avg	0.16	-	0.16	0.16	0.16	0.16
(c) rel. freq. of bids > p-own & asks < p-own	0.12	-	0.12	0.12	0.12	0.12
(d) rel. freq. of bids < p-own & asks > p-own	0.84	-	0.84	0.84	0.84	0.84
Number of offers for units bought above p-avg or sold below p-avg in the previous period	76	-	76	76	76	76
(C) rel. freq. of bids > p-own & asks < p-own	0.07	-	0.07	0.07	0.07	0.07
Number of offers for units that were not traded in the previous period	35	-	35	35	35	35
(D) rel. freq. of bids < p-own & asks > p-own	0.87	-	0.87	0.87	0.87	0.87
Aggressive offer frequency ratio (c)/(C)	0.60	-	0.60	0.60	0.60	0.60
... median value in the simulation	0.60	-	0.57	0.57	0.57	0.57
Moderate offer frequency ratio (d)/(D)	1.04	-	1.04	1.04	1.04	1.04
... median value in the simulation	1.05	-	1.05	1.05	1.05	1.05

Relative frequencies f(x) and cumulative distributions F(x) of the simulation rounds (5 subsequent periods) in which any n periods have price paths of the same type.

n	falling f(x)	F(x)	rising f(x)	F(x)	hanging f(x)	F(x)	standing f(x)	F(x)	unknown f(x)	F(x)
5	0.00	0.00	0.01	0.01	0.00	0.00	0.00	0.00	0.00	0.00
4	0.01	0.01	0.08	0.09	0.02	0.02	0.00	0.00	0.00	0.00
3	0.07	0.08	0.23	0.32	0.09	0.11	0.01	0.01	0.00	0.00
2	0.23	0.31	0.35	0.67	0.27	0.38	0.09	0.11	0.00	0.00
1	0.41	0.72	0.26	0.92	0.39	0.77	0.36	0.46	0.00	0.00
0	0.28	1.00	0.08	1.00	0.23	1.00	0.54	1.00	1.00	1.00

Panel VII.C.0.1b

Double Auction Market Simulation Results   File: DAMM1C00.SIM

Number of simulation rounds	:	10000		Periods per round :			5
Market type	:	M1		Number of traders :			12
Market rules	:	AltDAM		Technical price range :	[300	...	700]
No-Crossing rule	:	YES		Ask-Bid-Spread-Reduction rule :			YES
Trader type	:	ZI					

Market clearing quantity q*	:	14	
Market clearing price p*	:	500	

	Average of Periods	Period 1	Period 2	Period 3	Period 4	Period 5
Deviation of traded quantity from q*	0.67	0.68	0.66	0.67	0.66	0.67
Efficiency	0.99	0.99	0.99	0.99	0.99	0.99
... median value in the simulation	0.99	0.99	0.99	0.99	0.99	0.99
Average contract price (deviation from p*)	-2.51	-2.48	-2.50	-2.53	-2.52	-2.54
... median value in the simulation	-2.59	-2.49	-2.49	-2.59	-2.55	-2.59
Minimum contract price (deviation from p*)	-30.05	-30.01	-30.05	-29.99	-30.08	-30.13
... median value in the simulation	-31.99	-31.99	-31.99	-31.99	-31.99	-31.99
Maximum contract price (deviation from p*)	20.21	20.23	20.18	20.22	20.22	20.21
... median value in the simulation	21.01	22.01	21.01	22.01	22.01	21.01
Market clearing price concordance coefficient	0.53	0.53	0.53	0.53	0.53	0.53
... median value in the simulation	0.53	0.53	0.53	0.53	0.53	0.53
Rel. freq. of falling price paths	0.19	0.19	0.19	0.18	0.19	0.19
Rel. freq. of rising price paths	0.40	0.40	0.40	0.41	0.41	0.40
Rel. freq. of hanging price paths	0.29	0.29	0.29	0.30	0.28	0.30
Rel. freq. of standing price paths	0.11	0.12	0.12	0.11	0.11	0.11
Rel. freq. of unknown price paths	0.00	0.00	0.00	0.00	0.00	0.00
Rel. freq. of contract proposals by buyers	0.50	0.50	0.51	0.50	0.51	0.50
... lower quartile bound in the simulation	0.46	0.41	0.41	0.41	0.41	0.41
... median value in the simulation	0.51	0.51	0.51	0.51	0.51	0.51
... upper quartile bound in the simulation	0.56	0.63	0.63	0.63	0.63	0.61
... rel. freq. of these with price < p*	0.43	0.42	0.43	0.43	0.43	0.43
... rel. freq. of these with price = p*	0.05	0.05	0.05	0.05	0.05	0.05
... rel. freq. of these with price > p*	0.52	0.53	0.52	0.52	0.52	0.52
... avg. buyer proposed contract price - p*	-0.09	-0.08	-0.03	-0.11	-0.12	-0.12
Rel. freq. of contract proposals by sellers	0.50	0.50	0.49	0.50	0.49	0.50
... lower quartile bound in the simulation	0.45	0.38	0.38	0.38	0.38	0.40
... median value in the simulation	0.50	0.51	0.51	0.51	0.51	0.51
... upper quartile bound in the simulation	0.55	0.61	0.61	0.61	0.61	0.61
... rel. freq. of these with price < p*	0.57	0.57	0.57	0.57	0.57	0.57
... rel. freq. of these with price = p*	0.05	0.05	0.05	0.05	0.05	0.05
... rel. freq. of these with price > p*	0.38	0.38	0.38	0.38	0.38	0.38
... avg. seller proposed contract price - p*	-5.01	-4.98	-5.01	-5.04	-4.99	-5.01
Rel. freq. of proposer's curses - all traders	0.56	0.56	0.56	0.56	0.56	0.56
... median value in the simulation	0.57	0.58	0.58	0.58	0.58	0.58
Rel. freq. of proposer's curses - all buyers	0.52	0.53	0.52	0.52	0.52	0.52
Rel. freq. of proposer's curses - all sellers	0.57	0.57	0.57	0.57	0.57	0.57
Rel. freq. of proposer's curses - buyer 1	0.63	0.63	0.63	0.63	0.63	0.63
Rel. freq. of proposer's curses - buyer 2	0.49	0.49	0.49	0.49	0.49	0.49
Rel. freq. of proposer's curses - buyer 3	0.53	0.52	0.53	0.53	0.53	0.53
Rel. freq. of proposer's curses - buyer 4	0.55	0.56	0.55	0.55	0.54	0.55
Rel. freq. of proposer's curses - buyer 5	0.53	0.53	0.54	0.53	0.54	0.54
Rel. freq. of proposer's curses - buyer 6	0.51	0.52	0.52	0.51	0.51	0.51
Rel. freq. of proposer's curses - seller 7	0.66	0.67	0.67	0.66	0.66	0.66
Rel. freq. of proposer's curses - seller 8	0.51	0.51	0.51	0.52	0.51	0.51
Rel. freq. of proposer's curses - seller 9	0.58	0.58	0.58	0.58	0.58	0.58
Rel. freq. of proposer's curses - seller 10	0.62	0.62	0.62	0.62	0.62	0.61
Rel. freq. of proposer's curses - seller 11	0.61	0.62	0.61	0.61	0.61	0.61
Rel. freq. of proposer's curses - seller 12	0.58	0.57	0.58	0.58	0.57	0.58

Panel VII.C.0.1b (continued)

	Average of Periods	Period 1	Period 2	Period 3	Period 4	Period 5
Number of market ask and market bid matches	13.29	13.27	13.29	13.29	13.30	13.27
Number of contracts per market ask-bid match	1.10	1.11	1.10	1.10	1.10	1.11
Avg. rank of traded red.value at trading time	2.43	2.43	2.43	2.43	2.43	2.44
... median value in the simulation	2.44	2.44	2.43	2.43	2.46	2.44
Avg. rank of traded unit cost at trading time	2.35	2.36	2.35	2.35	2.36	2.36
... median value in the simulation	2.36	2.36	2.36	2.36	2.36	2.36
Number of submitted bids per contract	11.81	11.79	11.83	11.81	11.83	11.80
... median value in the simulation	11.82	11.81	11.87	11.82	11.86	11.81
... rel. freq. of bids < p* - 200	0.00	0.00	0.00	0.00	0.00	0.00
... rel. freq. of bids < p*	0.85	0.85	0.85	0.85	0.85	0.85
... rel. freq. of bids = p*	0.02	0.02	0.02	0.02	0.02	0.02
... rel. freq. of bids > p*	0.14	0.14	0.14	0.14	0.14	0.14
Number of submitted asks per contract	11.80	11.78	11.80	11.81	11.82	11.77
... median value in the simulation	11.81	11.81	11.81	11.86	11.82	11.81
... rel. freq. of asks < p*	0.16	0.16	0.16	0.16	0.16	0.16
... rel. freq. of asks = p*	0.01	0.01	0.01	0.01	0.01	0.01
... rel. freq. of asks > p*	0.83	0.83	0.83	0.83	0.83	0.83
... rel. freq. of asks > p* + 200	0.00	0.00	0.00	0.00	0.00	0.00
Number of offers for all units in periods > 1	346	-	346	346	346	345
(a) rel. freq. of bids > p-max & asks < p-min	0.01	-	0.01	0.01	0.01	0.01
(b) rel. freq. of bids > p-avg & asks < p-avg	0.16	-	0.16	0.16	0.16	0.16
(c) rel. freq. of bids > p-own & asks < p-own	0.11	-	0.11	0.11	0.11	0.11
(d) rel. freq. of bids > p-own & asks > p-own	0.85	-	0.85	0.85	0.85	0.85
Number of offers for units bought above p-avg or sold below p-avg in the previous period	130	-	131	130	130	130
(C) rel. freq. of bids > p-own & asks < p-own	0.07	-	0.07	0.07	0.07	0.07
Number of offers for units that were not traded in the previous period	70	-	69	70	70	70
(D) rel. freq. of bids < p-own & asks > p-own	0.87	-	0.88	0.87	0.87	0.88
Aggressive offer frequency ratio (c)/(C)	0.61	-	0.61	0.61	0.61	0.61
... median value in the simulation	0.61	-	0.59	0.59	0.59	0.59
Moderate offer frequency ratio (d)/(D)	1.03	-	1.03	1.03	1.03	1.03
... median value in the simulation	1.04	-	1.04	1.04	1.04	1.04

Relative frequencies f(x) and cumulative distributions F(x) of the simulation rounds
(5 subsequent periods) in which any n periods have price paths of the same type.

	falling		rising		hanging		standing		unknown	
n	f(x)	F(x)	f(x)	F(x)	f(x)	F(x)	f(x)	F(x)	f(x)	F(x)
5	0.00	0.00	0.01	0.01	0.00	0.00	0.00	0.00	0.00	0.00
4	0.00	0.00	0.08	0.09	0.03	0.03	0.00	0.00	0.00	0.00
3	0.05	0.05	0.24	0.33	0.13	0.15	0.01	0.01	0.00	0.00
2	0.19	0.24	0.34	0.67	0.31	0.46	0.09	0.10	0.00	0.00
1	0.41	0.65	0.25	0.92	0.36	0.82	0.36	0.46	0.00	0.00
0	0.35	1.00	0.08	1.00	0.18	1.00	0.54	1.00	1.00	1.00

Panel VII.C.0.2a

Double Auction Market Simulation Results   File: DAMM2800.SIM

Number of simulation rounds	:	10000	Periods per round :	5
Market type	:	M2	Number of traders :	8
Market rules	:	AltDAM	Technical price range :	[300 ... 700]
No-Crossing rule	:	YES	Ask-Bid-Spread-Reduction rule :	YES
Trader type	:	ZI		

Market clearing quantity q*	:	7
Market clearing price p*	:	500

	Average of Periods	Period 1	Period 2	Period 3	Period 4	Period 5
Deviation of traded quantity from q*	0.23	0.23	0.22	0.23	0.23	0.22
Efficiency	0.98	0.98	0.98	0.98	0.98	0.98
... median value in the simulation	0.99	0.99	0.99	0.99	0.99	0.99
Average contract price (deviation from p*)	2.20	2.24	2.07	2.26	2.29	2.14
... median value in the simulation	2.01	2.29	2.01	2.29	2.29	2.01
Minimum contract price (deviation from p*)	-15.36	-15.32	-15.43	-15.40	-15.23	-15.42
... median value in the simulation	-15.99	-15.99	-15.99	-15.99	-15.99	-15.99
Maximum contract price (deviation from p*)	22.70	22.75	22.53	22.77	22.69	22.77
... median value in the simulation	24.01	24.01	24.01	24.01	24.01	24.01
Market clearing price concordance coefficient	0.55	0.54	0.55	0.54	0.55	0.54
... median value in the simulation	0.55	0.55	0.55	0.55	0.55	0.55
Rel. freq. of falling price paths	0.40	0.40	0.40	0.40	0.40	0.40
Rel. freq. of rising price paths	0.25	0.25	0.25	0.24	0.24	0.25
Rel. freq. of hanging price paths	0.12	0.12	0.13	0.12	0.12	0.12
Rel. freq. of standing price paths	0.23	0.22	0.23	0.23	0.24	0.22
Rel. freq. of unknown price paths	0.00	0.00	0.00	0.00	0.00	0.00
Rel. freq. of contract proposals by buyers	0.49	0.49	0.49	0.50	0.50	0.49
... lower quartile bound in the simulation	0.43	0.29	0.29	0.29	0.29	0.29
... median value in the simulation	0.50	0.51	0.51	0.51	0.51	0.51
... upper quartile bound in the simulation	0.57	0.63	0.63	0.72	0.72	0.72
... rel. freq. of these with price < p*	0.37	0.37	0.37	0.37	0.37	0.37
... rel. freq. of these with price = p*	0.08	0.08	0.08	0.08	0.08	0.08
... rel. freq. of these with price > p*	0.55	0.55	0.54	0.55	0.55	0.55
... avg. buyer proposed contract price - p*	4.52	4.55	4.37	4.62	4.56	4.53
Rel. freq. of contract proposals by sellers	0.51	0.51	0.51	0.50	0.50	0.51
... lower quartile bound in the simulation	0.44	0.38	0.38	0.29	0.29	0.29
... median value in the simulation	0.51	0.51	0.51	0.51	0.51	0.51
... upper quartile bound in the simulation	0.58	0.72	0.72	0.72	0.72	0.72
... rel. freq. of these with price < p*	0.51	0.51	0.51	0.51	0.51	0.51
... rel. freq. of these with price = p*	0.08	0.08	0.08	0.08	0.08	0.08
... rel. freq. of these with price > p*	0.41	0.41	0.40	0.41	0.41	0.41
... avg. seller proposed contract price - p*	-0.10	-0.02	-0.22	-0.12	0.02	-0.17
Rel. freq. of proposer's curses - all traders	0.55	0.55	0.55	0.55	0.55	0.55
... median value in the simulation	0.55	0.58	0.58	0.58	0.58	0.58
Rel. freq. of proposer's curses - all buyers	0.55	0.55	0.54	0.55	0.55	0.55
Rel. freq. of proposer's curses - all sellers	0.51	0.51	0.51	0.51	0.51	0.51
Rel. freq. of proposer's curses - buyer 1	0.71	0.70	0.71	0.71	0.71	0.71
Rel. freq. of proposer's curses - buyer 2	0.65	0.65	0.64	0.66	0.65	0.65
Rel. freq. of proposer's curses - buyer 3	0.37	0.37	0.36	0.37	0.37	0.37
Rel. freq. of proposer's curses - buyer 4	0.53	0.54	0.53	0.53	0.53	0.53
Rel. freq. of proposer's curses - seller 5	0.67	0.67	0.68	0.67	0.66	0.66
Rel. freq. of proposer's curses - seller 6	0.61	0.62	0.61	0.62	0.61	0.61
Rel. freq. of proposer's curses - seller 7	0.35	0.35	0.35	0.35	0.34	0.35
Rel. freq. of proposer's curses - seller 8	0.45	0.44	0.45	0.44	0.44	0.46

Panel VII.C.0.2a (continued)

	Average of Periods	Period 1	Period 2	Period 3	Period 4	Period 5
Number of market ask and market bid matches	6.76	6.76	6.76	6.77	6.75	6.76
Number of contracts per market ask-bid match	1.07	1.07	1.07	1.07	1.07	1.07
Avg. rank of traded red.value at trading time	1.71	1.71	1.71	1.71	1.71	1.71
... median value in the simulation	1.71	1.72	1.72	1.72	1.72	1.72
Avg. rank of traded unit cost at trading time	1.77	1.77	1.77	1.78	1.77	1.77
... median value in the simulation	1.78	1.76	1.76	1.76	1.76	1.76
Number of submitted bids per contract	11.02	11.04	11.04	11.01	10.99	11.04
... median value in the simulation	11.03	11.01	11.01	11.01	11.01	11.01
... rel. freq. of bids < p* - 200	0.00	0.00	0.00	0.00	0.00	0.00
... rel. freq. of bids < p*	0.85	0.85	0.85	0.85	0.85	0.85
... rel. freq. of bids = p*	0.02	0.02	0.02	0.02	0.02	0.02
... rel. freq. of bids > p*	0.13	0.13	0.13	0.13	0.13	0.13
Number of submitted asks per contract	11.11	11.12	11.11	11.10	11.11	11.12
... median value in the simulation	11.12	11.15	11.15	11.15	11.15	11.15
... rel. freq. of asks < p*	0.12	0.12	0.12	0.12	0.12	0.12
... rel. freq. of asks = p*	0.02	0.02	0.02	0.02	0.02	0.02
... rel. freq. of asks > p*	0.86	0.86	0.86	0.86	0.87	0.86
... rel. freq. of asks > p* + 200	0.00	0.00	0.00	0.00	0.00	0.00
Number of offers for all units in periods > 1	160	-	160	160	160	160
(a) rel. freq. of bids > p-max & asks < p-min	0.03	-	0.03	0.03	0.03	0.03
(b) rel. freq. of bids > p-avg & asks < p-avg	0.14	-	0.14	0.14	0.14	0.14
(c) rel. freq. of bids > p-own & asks < p-own	0.14	-	0.14	0.14	0.14	0.14
(d) rel. freq. of bids < p-own & asks > p-own	0.81	-	0.81	0.81	0.81	0.82
Number of offers for units bought above p-avg or sold below p-avg in the previous period	50	-	50	49	50	50
(C) rel. freq. of bids > p-own & asks < p-own	0.09	-	0.09	0.09	0.09	0.09
Number of offers for units that were not traded in the previous period	53	-	53	53	53	53
(D) rel. freq. of bids < p-own & asks > p-own	0.82	-	0.82	0.82	0.82	0.82
Aggressive offer frequency ratio (c)/(C)	0.64	-	0.63	0.64	0.64	0.64
... median value in the simulation	0.63	-	0.59	0.59	0.59	0.59
Moderate offer frequency ratio (d)/(D)	1.01	-	1.01	1.01	1.01	1.01
... median value in the simulation	1.02	-	1.02	1.02	1.02	1.02

Relative frequencies f(x) and cumulative distributions F(x) of the simulation rounds (5 subsequent periods) in which any n periods have price paths of the same type.

	falling		rising		hanging		standing		unknown	
n	f(x)	F(x)	f(x)	F(x)	f(x)	F(x)	f(x)	F(x)	f(x)	F(x)
5	0.01	0.01	0.00	0.00	0.00	0.00	0.00	0.00	0.00	0.00
4	0.07	0.08	0.02	0.02	0.00	0.00	0.01	0.01	0.00	0.00
3	0.23	0.31	0.08	0.10	0.02	0.02	0.07	0.08	0.00	0.00
2	0.36	0.67	0.26	0.36	0.10	0.12	0.24	0.32	0.00	0.00
1	0.26	0.92	0.40	0.76	0.36	0.48	0.40	0.73	0.00	0.00
0	0.08	1.00	0.24	1.00	0.52	1.00	0.28	1.00	1.00	1.00

Panel VII.C.0.2b

Double Auction Market Simulation Results   File: DAMM2A00.SIM

Number of simulation rounds	: 10000		Periods per round :		5
Market type	: M2		Number of traders :		10
Market rules	: AltDAM		Technical price range :	[300 ...	700]
No-Crossing rule	: YES		Ask-Bid-Spread-Reduction rule :		YES
Trader type	: ZI				

Market clearing quantity q* :   9
Market clearing price   p* : 500

	Average of Periods	Period 1	Period 2	Period 3	Period 4	Period 5
Deviation of traded quantity from q*	0.45	0.45	0.45	0.45	0.45	0.44
Efficiency	0.98	0.98	0.98	0.98	0.98	0.98
... median value in the simulation	0.99	0.99	0.99	0.99	0.99	0.99
Average contract price (deviation from p*)	1.47	1.45	1.49	1.47	1.43	1.49
... median value in the simulation	1.46	1.45	1.51	1.45	1.40	1.46
Minimum contract price (deviation from p*)	-11.37	-11.40	-11.36	-11.36	-11.36	-11.35
... median value in the simulation	-11.99	-11.99	-11.99	-11.99	-11.99	-11.99
Maximum contract price (deviation from p*)	16.78	16.80	16.84	16.70	16.69	16.88
... median value in the simulation	18.01	18.01	18.01	18.01	18.01	18.01
Market clearing price concordance coefficient	0.52	0.52	0.52	0.52	0.53	0.52
... median value in the simulation	0.53	0.53	0.53	0.53	0.53	0.53
Rel. freq. of falling price paths	0.39	0.39	0.40	0.40	0.39	0.39
Rel. freq. of rising price paths	0.23	0.23	0.22	0.23	0.23	0.22
Rel. freq. of hanging price paths	0.13	0.13	0.13	0.12	0.13	0.13
Rel. freq. of standing price paths	0.25	0.25	0.25	0.25	0.26	0.26
Rel. freq. of unknown price paths	0.00	0.00	0.00	0.00	0.00	0.00
Rel. freq. of contract proposals by buyers	0.49	0.49	0.49	0.49	0.49	0.49
... lower quartile bound in the simulation	0.44	0.34	0.34	0.34	0.34	0.34
... median value in the simulation	0.50	0.51	0.51	0.51	0.51	0.51
... upper quartile bound in the simulation	0.56	0.67	0.67	0.67	0.67	0.67
... rel. freq. of these with price < p*	0.36	0.37	0.36	0.36	0.36	0.36
... rel. freq. of these with price = p*	0.08	0.08	0.08	0.08	0.08	0.08
... rel. freq. of these with price > p*	0.56	0.55	0.56	0.56	0.56	0.56
... avg. buyer proposed contract price - p*	3.27	3.20	3.30	3.26	3.26	3.33
Rel. freq. of contract proposals by sellers	0.51	0.51	0.51	0.51	0.51	0.51
... lower quartile bound in the simulation	0.45	0.34	0.34	0.34	0.34	0.34
... median value in the simulation	0.51	0.51	0.51	0.51	0.51	0.51
... upper quartile bound in the simulation	0.57	0.67	0.67	0.67	0.67	0.67
... rel. freq. of these with price < p*	0.52	0.52	0.52	0.52	0.52	0.52
... rel. freq. of these with price = p*	0.08	0.08	0.08	0.08	0.08	0.08
... rel. freq. of these with price > p*	0.40	0.40	0.40	0.40	0.40	0.40
... avg. seller proposed contract price - p*	-0.29	-0.24	-0.29	-0.31	-0.34	-0.24
Rel. freq. of proposer's curses - all traders	0.56	0.55	0.56	0.56	0.56	0.56
... median value in the simulation	0.56	0.56	0.56	0.56	0.56	0.56
Rel. freq. of proposer's curses - all buyers	0.56	0.55	0.56	0.56	0.56	0.56
Rel. freq. of proposer's curses - all sellers	0.52	0.52	0.52	0.52	0.52	0.52
Rel. freq. of proposer's curses - buyer 1	0.71	0.71	0.71	0.71	0.71	0.71
Rel. freq. of proposer's curses - buyer 2	0.68	0.67	0.68	0.68	0.68	0.68
Rel. freq. of proposer's curses - buyer 3	0.62	0.61	0.62	0.62	0.62	0.62
Rel. freq. of proposer's curses - buyer 4	0.37	0.37	0.37	0.37	0.37	0.37
Rel. freq. of proposer's curses - buyer 5	0.48	0.48	0.47	0.48	0.47	0.49
Rel. freq. of proposer's curses - seller 6	0.67	0.67	0.66	0.67	0.67	0.67
Rel. freq. of proposer's curses - seller 7	0.64	0.63	0.64	0.64	0.64	0.63
Rel. freq. of proposer's curses - seller 8	0.57	0.57	0.57	0.57	0.58	0.58
Rel. freq. of proposer's curses - seller 9	0.35	0.35	0.35	0.35	0.35	0.35
Rel. freq. of proposer's curses - seller 10	0.42	0.41	0.42	0.42	0.43	0.43

Panel VII.C.0.2b (continued)

	Average of Periods	Period 1	Period 2	Period 3	Period 4	Period 5
Number of market ask and market bid matches	8.68	8.69	8.68	8.68	8.69	8.68
Number of contracts per market ask-bid match	1.09	1.09	1.09	1.09	1.09	1.09
Avg. rank of traded red.value at trading time	1.98	1.98	1.98	1.98	1.98	1.97
... median value in the simulation	1.98	2.01	2.01	2.01	2.01	2.01
Avg. rank of traded unit cost at trading time	2.06	2.06	2.06	2.06	2.06	2.07
... median value in the simulation	2.07	2.07	2.07	2.07	2.07	2.08
Number of submitted bids per contract	12.48	12.49	12.47	12.48	12.47	12.47
... median value in the simulation	12.49	12.56	12.56	12.51	12.51	12.51
... rel. freq. of bids < $p^* - 200$	0.00	0.00	0.00	0.00	0.00	0.00
... rel. freq. of bids < p^*	0.87	0.87	0.87	0.87	0.87	0.87
... rel. freq. of bids = p^*	0.02	0.02	0.02	0.02	0.02	0.02
... rel. freq. of bids > p^*	0.12	0.12	0.12	0.12	0.12	0.12
Number of submitted asks per contract	12.49	12.50	12.48	12.50	12.49	12.50
... median value in the simulation	12.52	12.56	12.56	12.56	12.56	12.56
... rel. freq. of asks < p^*	0.10	0.11	0.10	0.10	0.11	0.10
... rel. freq. of asks = p^*	0.02	0.02	0.02	0.02	0.02	0.02
... rel. freq. of asks > p^*	0.88	0.87	0.88	0.88	0.87	0.88
... rel. freq. of asks > $p^* + 200$	0.00	0.00	0.00	0.00	0.00	0.00
Number of offers for all units in periods > 1	236	-	236	236	236	236
(a) rel. freq. of bids > p-max & asks < p-min	0.02	-	0.02	0.02	0.02	0.02
(b) rel. freq. of bids > p-avg & asks < p-avg	0.12	-	0.12	0.12	0.12	0.12
(c) rel. freq. of bids > p-own & asks < p-own	0.12	-	0.12	0.12	0.12	0.12
(d) rel. freq. of bids < p-own & asks > p-own	0.83	-	0.84	0.83	0.83	0.84
Number of offers for units bought above p-avg or sold below p-avg in the previous period	74	-	74	74	74	74
(C) rel. freq. of bids > p-own & asks < p-own	0.07	-	0.07	0.07	0.07	0.07
Number of offers for units that were not traded in the previous period	75	-	75	75	75	75
(D) rel. freq. of bids < p-own & asks > p-own	0.83	-	0.83	0.83	0.83	0.83
Aggressive offer frequency ratio (c)/(C)	0.59	-	0.59	0.59	0.59	0.59
... median value in the simulation	0.58	-	0.54	0.55	0.55	0.54
Moderate offer frequency ratio (d)/(D)	0.99	-	0.99	0.99	0.99	0.99
... median value in the simulation	1.00	-	1.00	1.00	1.00	1.00

Relative frequencies f(x) and cumulative distributions F(x) of the simulation rounds (5 subsequent periods) in which any n periods have price paths of the same type.

	falling		rising		hanging		standing		unknown	
n	f(x)	F(x)	f(x)	F(x)	f(x)	F(x)	f(x)	F(x)	f(x)	F(x)
5	0.01	0.01	0.00	0.00	0.00	0.00	0.00	0.00	0.00	0.00
4	0.08	0.09	0.01	0.01	0.00	0.00	0.02	0.02	0.00	0.00
3	0.22	0.30	0.07	0.08	0.02	0.02	0.09	0.10	0.00	0.00
2	0.35	0.65	0.24	0.31	0.11	0.13	0.28	0.38	0.00	0.00
1	0.27	0.92	0.41	0.73	0.36	0.49	0.39	0.77	0.00	0.00
0	0.08	1.00	0.27	1.00	0.51	1.00	0.23	1.00	1.00	1.00

Panel VII.C.1.0

Double Auction Market Simulation Results   File: DAMM0810.SIM

Number of simulation rounds	:	10000		Periods per round :				5
Market type	:	MO		Number of traders :				8
Market rules	:	AltDAM		Technical price range : [1 ... 9999]				
No-Crossing rule	:	YES		Ask-Bid-Spread-Reduction rule :				YES
Trader type	:	ZI						

			Average of Periods	Period 1	Period 2	Period 3	Period 4	Period 5
Market clearing quantity q*	:	10						
Market clearing price p*	:	500						

	Average of Periods	Period 1	Period 2	Period 3	Period 4	Period 5
Deviation of traded quantity from q*	0.45	0.45	0.46	0.45	0.45	0.46
Efficiency	0.99	0.99	0.99	0.99	0.99	0.99
... median value in the simulation	0.99	1.00	1.00	1.00	1.00	1.00
Average contract price (deviation from p*)	6.24	6.25	6.21	6.27	6.23	6.22
... median value in the simulation	6.64	6.71	6.64	6.71	6.64	6.64
Minimum contract price (deviation from p*)	-5.61	-5.65	-5.57	-5.58	-5.60	-5.63
... median value in the simulation	-2.99	-2.99	-2.99	-2.99	-2.99	-2.99
Maximum contract price (deviation from p*)	17.04	17.05	17.02	17.06	17.03	17.03
... median value in the simulation	18.01	18.01	18.01	18.01	18.01	18.01
Market clearing price concordance coefficient	0.43	0.43	0.43	0.43	0.43	0.43
... median value in the simulation	0.44	0.43	0.43	0.43	0.43	0.43
Rel. freq. of falling price paths	0.81	0.81	0.81	0.81	0.81	0.81
Rel. freq. of rising price paths	0.01	0.01	0.01	0.01	0.01	0.01
Rel. freq. of hanging price paths	0.00	0.00	0.00	0.00	0.00	0.00
Rel. freq. of standing price paths	0.18	0.18	0.18	0.18	0.18	0.18
Rel. freq. of unknown price paths	0.00	0.00	0.00	0.00	0.00	0.00
Rel. freq. of contract proposals by buyers	0.78	0.78	0.78	0.78	0.78	0.78
... lower quartile bound in the simulation	0.74	0.71	0.71	0.71	0.71	0.71
... median value in the simulation	0.79	0.81	0.81	0.81	0.81	0.81
... upper quartile bound in the simulation	0.83	0.91	0.91	0.91	0.91	0.91
... rel. freq. of these with price < p*	0.10	0.10	0.10	0.10	0.10	0.10
... rel. freq. of these with price = p*	0.10	0.10	0.10	0.10	0.10	0.10
... rel. freq. of these with price > p*	0.80	0.80	0.80	0.80	0.80	0.80
... avg. buyer proposed contract price · p*	6.97	6.95	6.97	6.99	6.96	6.99
Rel. freq. of contract proposals by sellers	0.22	0.22	0.22	0.22	0.22	0.22
... lower quartile bound in the simulation	0.18	0.11	0.11	0.11	0.11	0.11
... median value in the simulation	0.22	0.20	0.20	0.20	0.20	0.20
... upper quartile bound in the simulation	0.27	0.31	0.31	0.31	0.31	0.31
... rel. freq. of these with price < p*	0.28	0.28	0.28	0.28	0.27	0.28
... rel. freq. of these with price = p*	0.06	0.06	0.06	0.06	0.06	0.06
... rel. freq. of these with price > p*	0.67	0.67	0.66	0.67	0.67	0.66
... avg. seller proposed contract price · p*	3.87	3.87	3.85	3.91	3.93	3.77
Rel. freq. of proposer's curses · all traders	0.69	0.69	0.69	0.69	0.69	0.69
... median value in the simulation	0.70	0.71	0.71	0.71	0.71	0.71
Rel. freq. of proposer's curses · all buyers	0.80	0.80	0.80	0.80	0.80	0.80
Rel. freq. of proposer's curses · all sellers	0.28	0.28	0.28	0.28	0.27	0.28
Rel. freq. of proposer's curses · buyer 1	0.94	0.95	0.94	0.94	0.94	0.95
Rel. freq. of proposer's curses · buyer 2	0.68	0.68	0.68	0.69	0.68	0.68
Rel. freq. of proposer's curses · buyer 3	0.83	0.83	0.82	0.83	0.83	0.83
Rel. freq. of proposer's curses · buyer 4	0.93	0.93	0.93	0.93	0.93	0.93
Rel. freq. of proposer's curses · seller 5	0.40	0.39	0.40	0.40	0.39	0.40
Rel. freq. of proposer's curses · seller 6	0.26	0.26	0.26	0.25	0.25	0.26
Rel. freq. of proposer's curses · seller 7	0.26	0.25	0.26	0.26	0.25	0.26
Rel. freq. of proposer's curses · seller 8	0.24	0.22	0.25	0.23	0.23	0.24

Panel VII.C.1.0 (continued)

	Average of Periods	Period 1	Period 2	Period 3	Period 4	Period 5
Number of market ask and market bid matches	10.04	10.04,	10.04	10.04	10.04	10.05
Number of contracts per market ask-bid match	1.04	1.04	1.04	1.04	1.04	1.04
Avg. rank of traded red.value at trading time	1.30	1.30	1.30	1.30	1.31	1.31
... median value in the simulation	1.30	1.28	1.28	1.28	1.28	1.28
Avg. rank of traded unit cost at trading time	2.05	2.05	2.05	2.05	2.05	2.05
... median value in the simulation	2.06	2.10	2.08	2.10	2.10	2.10
Number of submitted bids per contract	13.47	13.48	13.46	13.46	13.46	13.47
... median value in the simulation	13.01	13.01	13.01	13.01	13.01	13.01
... rel. freq. of bids < p* - 200	0.18	0.18	0.18	0.18	0.18	0.18
... rel. freq. of bids < p*	0.73	0.73	0.73	0.73	0.73	0.73
... rel. freq. of bids = p*	0.03	0.03	0.03	0.03	0.03	0.03
... rel. freq. of bids > p*	0.24	0.24	0.24	0.24	0.24	0.24
Number of submitted asks per contract	17.35	17.36	17.35	17.39	17.34	17.32
... median value in the simulation	13.01	13.01	13.01	13.01	13.01	13.01
... rel. freq. of asks < p*	0.02	0.02	0.02	0.02	0.02	0.02
... rel. freq. of asks = p*	0.01	0.01	0.01	0.01	0.01	0.01
... rel. freq. of asks > p*	0.97	0.97	0.97	0.97	0.97	0.97
... rel. freq. of asks > p* + 200	0.43	0.43	0.43	0.43	0.43	0.43
Number of offers for all units in periods > 1	322	-	322	322	322	322
(a) rel. freq. of bids > p-max & asks < p-min	0.01	-	0.01	0.01	0.01	0.01
(b) rel. freq. of bids > p-avg & asks < p-avg	0.11	-	0.11	0.11	0.11	0.11
(c) rel. freq. of bids > p-own & asks < p-own	0.06	-	0.06	0.06	0.06	0.06
(d) rel. freq. of bids < p-own & asks > p-own	0.88	-	0.88	0.88	0.88	0.88
Number of offers for units bought above p-avg or sold below p-avg in the previous period	122	-	122	123	122	123
(C) rel. freq. of bids > p-own & asks < p-own	0.04	-	0.04	0.04	0.04	0.04
Number of offers for units that were not traded in the previous period	46	-	46	46	46	46
(D) rel. freq. of bids < p-own & asks > p-own	0.90	-	0.90	0.90	0.90	0.90
Aggressive offer frequency ratio (c)/(C)	0.59	-	0.59	0.58	0.59	0.59
... median value in the simulation	0.58	-	0.54	0.53	0.53	0.53
Moderate offer frequency ratio (d)/(D)	1.03	-	1.02	1.03	1.03	1.02
... median value in the simulation	1.03	-	1.03	1.03	1.03	1.03

Relative frequencies f(x) and cumulative distributions F(x) of the simulation rounds (5 subsequent periods) in which any n periods have price paths of the same type.

	falling		rising		hanging		standing		unknown	
n	f(x)	F(x)	f(x)	F(x)	f(x)	F(x)	f(x)	F(x)	f(x)	F(x)
5	0.35	0.35	0.00	0.00	0.00	0.00	0.00	0.00	0.00	0.00
4	0.40	0.75	0.00	0.00	0.00	0.00	0.00	0.00	0.00	0.00
3	0.19	0.95	0.00	0.00	0.00	0.00	0.04	0.05	0.00	0.00
2	0.05	0.99	0.00	0.00	0.00	0.00	0.18	0.23	0.00	0.00
1	0.01	1.00	0.03	0.04	0.02	0.02	0.40	0.62	0.00	0.00
0	0.00	1.00	0.96	1.00	0.98	1.00	0.38	1.00	1.00	1.00

Panel VII.C.1.1

Double Auction Market Simulation Results   File: DAMM1810.SIM

Number of simulation rounds	:	10000
Market type	:	M1
Market rules	:	AltDAM
No-Crossing rule	:	YES
Trader type	:	ZI

Periods per round :	5
Number of traders :	8
Technical price range : [1 ... 9999]	
Ask-Bid-Spread-Reduction rule :	YES

Market clearing quantity q*   :   10
Market clearing price   p*   :   261

	Average of Periods	Period 1	Period 2	Period 3	Period 4	Period 5
Deviation of traded quantity from q*	0.35	0.34	0.35	0.35	0.35	0.35
Efficiency	0.99	0.99	0.99	0.99	0.99	0.99
... median value in the simulation	1.00	0.99	0.99	0.99	0.99	0.99
Average contract price (deviation from p*)	6.96	6.99	6.95	6.97	6.95	6.95
... median value in the simulation	7.51	7.55	7.51	7.51	7.51	7.51
Minimum contract price (deviation from p*)	-4.95	-4.88	-4.95	-4.94	-4.98	-4.97
... median value in the simulation	-1.99	-1.99	-1.99	-1.99	-1.99	-1.99
Maximum contract price (deviation from p*)	17.27	17.30	17.26	17.27	17.27	17.27
... median value in the simulation	18.01	18.01	18.01	18.01	18.01	18.01
Market clearing price concordance coefficient	0.53	0.53	0.53	0.53	0.53	0.53
... median value in the simulation	0.53	0.52	0.53	0.53	0.53	0.53
Rel. freq. of falling price paths	0.86	0.86	0.86	0.86	0.86	0.86
Rel. freq. of rising price paths	0.00	0.01	0.01	0.00	0.00	0.00
Rel. freq. of hanging price paths	0.00	0.00	0.00	0.00	0.00	0.00
Rel. freq. of standing price paths	0.13	0.13	0.13	0.13	0.13	0.13
Rel. freq. of unknown price paths	0.00	0.00	0.00	0.00	0.00	0.00
Rel. freq. of contract proposals by buyers	0.83	0.83	0.83	0.83	0.83	0.83
... lower quartile bound in the simulation	0.79	0.73	0.73	0.73	0.73	0.73
... median value in the simulation	0.84	0.82	0.82	0.82	0.82	0.82
... upper quartile bound in the simulation	0.89	0.91	0.91	0.91	0.91	0.91
... rel. freq. of these with price < p*	0.07	0.07	0.07	0.07	0.07	0.07
... rel. freq. of these with price = p*	0.10	0.10	0.10	0.10	0.10	0.10
... rel. freq. of these with price > p*	0.83	0.83	0.83	0.83	0.83	0.82
... avg. buyer proposed contract price - p*	7.43	7.44	7.42	7.44	7.43	7.40
Rel. freq. of contract proposals by sellers	0.17	0.17	0.17	0.17	0.17	0.17
... lower quartile bound in the simulation	0.13	0.10	0.10	0.10	0.10	0.10
... median value in the simulation	0.17	0.19	0.19	0.19	0.19	0.19
... upper quartile bound in the simulation	0.22	0.28	0.28	0.28	0.28	0.28
... rel. freq. of these with price < p*	0.21	0.21	0.22	0.21	0.22	0.21
... rel. freq. of these with price = p*	0.05	0.05	0.05	0.05	0.05	0.05
... rel. freq. of these with price > p*	0.73	0.74	0.73	0.74	0.73	0.74
... avg. seller proposed contract price - p*	4.87	4.92	4.86	4.94	4.74	4.86
Rel. freq. of proposer's curses - all traders	0.73	0.73	0.73	0.73	0.73	0.72
... median value in the simulation	0.73	0.73	0.73	0.73	0.73	0.73
Rel. freq. of proposer's curses - all buyers	0.83	0.83	0.83	0.83	0.83	0.82
Rel. freq. of proposer's curses - all sellers	0.21	0.21	0.22	0.21	0.22	0.21
Rel. freq. of proposer's curses - buyer 1	0.96	0.97	0.96	0.96	0.96	0.96
Rel. freq. of proposer's curses - buyer 2	0.71	0.71	0.71	0.71	0.70	0.70
Rel. freq. of proposer's curses - buyer 3	0.88	0.88	0.88	0.87	0.88	0.87
Rel. freq. of proposer's curses - buyer 4	0.96	0.96	0.96	0.96	0.96	0.96
Rel. freq. of proposer's curses - seller 5	0.31	0.31	0.31	0.30	0.31	0.31
Rel. freq. of proposer's curses - seller 6	0.20	0.19	0.19	0.20	0.20	0.20
Rel. freq. of proposer's curses - seller 7	0.20	0.21	0.20	0.20	0.22	0.19
Rel. freq. of proposer's curses - seller 8	0.19	0.18	0.19	0.19	0.19	0.20

Panel VII.C.1.1 (continued)

	Average of Periods	Period 1	Period 2	Period 3	Period 4	Period 5
Number of market ask and market bid matches	10.03	10.03	10.03	10.03	10.02	10.04
Number of contracts per market ask-bid match	1.03	1.03	1.03	1.03	1.03	1.03
Avg. rank of traded red.value at trading time	1.22	1.22	1.22	1.22	1.22	1.22
... median value in the simulation	1.22	1.19	1.19	1.19	1.19	1.19
Avg. rank of traded unit cost at trading time	2.05	2.05	2.05	2.05	2.05	2.05
... median value in the simulation	2.06	2.10	2.10	2.10	2.10	2.10
Number of submitted bids per contract	12.69	12.70	12.69	12.66	12.68	12.71
... median value in the simulation	12.71	12.81	12.73	12.71	12.71	12.73
... rel. freq. of bids < p* - 200	0.07	0.07	0.07	0.07	0.07	0.07
... rel. freq. of bids < p*	0.69	0.69	0.69	0.69	0.69	0.69
... rel. freq. of bids = p*	0.03	0.03	0.03	0.03	0.03	0.03
... rel. freq. of bids > p*	0.28	0.28	0.28	0.28	0.28	0.28
Number of submitted asks per contract	17.81	17.82	17.84	17.77	17.79	17.84
... median value in the simulation	13.01	13.01	13.01	13.01	13.01	13.01
... rel. freq. of asks < p*	0.01	0.01	0.01	0.01	0.01	0.01
... rel. freq. of asks = p*	0.01	0.01	0.01	0.01	0.01	0.01
... rel. freq. of asks > p*	0.98	0.98	0.98	0.98	0.98	0.98
... rel. freq. of asks > p* + 200	0.43	0.43	0.43	0.43	0.43	0.43
Number of offers for all units in periods > 1	315	·	315	315	315	316
(a) rel. freq. of bids > p-max & asks < p-min	0.01	·	0.01	0.01	0.01	0.01
(b) rel. freq. of bids > p-avg & asks < p-avg	0.11	·	0.11	0.11	0.11	0.11
(c) rel. freq. of bids > p-own & asks < p-own	0.06	·	0.06	0.06	0.06	0.06
(d) rel. freq. of bids < p-own & asks > p-own	0.87	·	0.87	0.87	0.87	0.87
Number of offers for units bought above p-avg or sold below p-avg in the previous period	117	·	117	117	117	117
(C) rel. freq. of bids > p-own & asks < p-own	0.04	·	0.04	0.04	0.04	0.04
Number of offers for units that were not traded in the previous period	48	·	48	48	48	48
(D) rel. freq. of bids < p-own & asks > p-own	0.90	·	0.90	0.90	0.90	0.90
Aggressive offer frequency ratio (c)/(C)	0.60	·	0.61	0.60	0.60	0.60
... median value in the simulation	0.59	·	0.55	0.55	0.55	0.55
Moderate offer frequency ratio (d)/(D)	1.03	·	1.03	1.03	1.03	1.03
... median value in the simulation	1.03	·	1.03	1.03	1.03	1.03

Relative frequencies f(x) and cumulative distributions F(x) of the simulation rounds (5 subsequent periods) in which any n periods have price paths of the same type.

	falling		rising		hanging		standing		unknown	
n	f(x)	F(x)	f(x)	F(x)	f(x)	F(x)	f(x)	F(x)	f(x)	F(x)
5	0.48	0.48	0.00	0.00	0.00	0.00	0.00	0.00	0.00	0.00
4	0.37	0.86	0.00	0.00	0.00	0.00	0.00	0.00	0.00	0.00
3	0.12	0.98	0.00	0.00	0.00	0.00	0.02	0.02	0.00	0.00
2	0.02	1.00	0.00	0.00	0.00	0.00	0.11	0.13	0.00	0.00
1	0.00	1.00	0.02	0.02	0.01	0.01	0.36	0.50	0.00	0.00
0	0.00	1.00	0.98	1.00	0.99	1.00	0.50	1.00	1.00	1.00

Panel VII.C.1.2

Double Auction Market Simulation Results   File: DAMM1820.SIM

Number of simulation rounds	:	10000
Market type	:	M1
Market rules	:	AltDAM
No-Crossing rule	:	YES
Trader type	:	ZI

Periods per round :		5
Number of traders :		8
Technical price range :	[1 ...	9999]
Ask-Bid-Spread-Reduction rule :		YES

Market clearing quantity q*	:	10
Market clearing price p*	:	390

	Average of Periods	Period 1	Period 2	Period 3	Period 4	Period 5
Deviation of traded quantity from q*	0.34	0.34	0.34	0.34	0.34	0.35
Efficiency	0.99	0.99	0.99	0.99	0.99	0.99
... median value in the simulation	1.00	0.99	0.99	0.99	0.99	0.99
Average contract price (deviation from p*)	6.37	6.37	6.37	6.36	6.39	6.37
... median value in the simulation	7.01	7.01	7.01	6.91	7.01	7.01
Minimum contract price (deviation from p*)	-6.68	-6.66	-6.65	-6.75	-6.69	-6.67
... median value in the simulation	-2.99	-2.99	-2.99	-2.99	-2.99	-2.99
Maximum contract price (deviation from p*)	17.13	17.13	17.11	17.14	17.15	17.14
... median value in the simulation	18.01	18.01	18.01	18.01	18.01	18.01
Market clearing price concordance coefficient	0.53	0.53	0.53	0.53	0.53	0.53
... median value in the simulation	0.54	0.53	0.53	0.53	0.53	0.53
Rel. freq. of falling price paths	0.82	0.82	0.82	0.82	0.82	0.82
Rel. freq. of rising price paths	0.01	0.01	0.01	0.01	0.01	0.01
Rel. freq. of hanging price paths	0.01	0.01	0.01	0.01	0.01	0.01
Rel. freq. of standing price paths	0.16	0.16	0.16	0.16	0.16	0.16
Rel. freq. of unknown price paths	0.00	0.00	0.00	0.00	0.00	0.00
Rel. freq. of contract proposals by buyers	0.80	0.80	0.80	0.80	0.80	0.80
... lower quartile bound in the simulation	0.76	0.71	0.71	0.71	0.71	0.71
... median value in the simulation	0.81	0.81	0.81	0.81	0.81	0.81
... upper quartile bound in the simulation	0.85	0.91	0.91	0.91	0.91	0.91
... rel. freq. of these with price < p*	0.09	0.09	0.09	0.09	0.09	0.09
... rel. freq. of these with price = p*	0.10	0.10	0.10	0.10	0.10	0.10
... rel. freq. of these with price > p*	0.81	0.81	0.81	0.81	0.81	0.81
... avg. buyer proposed contract price - p*	7.09	7.10	7.09	7.10	7.08	7.10
Rel. freq. of contract proposals by sellers	0.20	0.20	0.20	0.20	0.20	0.20
... lower quartile bound in the simulation	0.16	0.11	0.11	0.10	0.11	0.11
... median value in the simulation	0.20	0.20	0.20	0.20	0.20	0.20
... upper quartile bound in the simulation	0.25	0.31	0.31	0.31	0.31	0.31
... rel. freq. of these with price < p*	0.26	0.26	0.25	0.26	0.25	0.25
... rel. freq. of these with price = p*	0.05	0.05	0.05	0.05	0.05	0.05
... rel. freq. of these with price > p*	0.69	0.69	0.69	0.69	0.69	0.69
... avg. seller proposed contract price - p*	3.76	3.70	3.82	3.68	3.83	3.77
Rel. freq. of proposer's curses - all traders	0.71	0.71	0.71	0.71	0.71	0.71
... median value in the simulation	0.71	0.71	0.71	0.71	0.71	0.71
Rel. freq. of proposer's curses - all buyers	0.81	0.81	0.81	0.81	0.81	0.81
Rel. freq. of proposer's curses - all sellers	0.26	0.26	0.25	0.26	0.25	0.25
Rel. freq. of proposer's curses - buyer 1	0.95	0.95	0.95	0.95	0.95	0.95
Rel. freq. of proposer's curses - buyer 2	0.69	0.69	0.69	0.69	0.69	0.69
Rel. freq. of proposer's curses - buyer 3	0.86	0.86	0.86	0.86	0.86	0.86
Rel. freq. of proposer's curses - buyer 4	0.94	0.94	0.94	0.94	0.94	0.94
Rel. freq. of proposer's curses - seller 5	0.36	0.37	0.36	0.37	0.36	0.36
Rel. freq. of proposer's curses - seller 6	0.23	0.24	0.23	0.24	0.23	0.23
Rel. freq. of proposer's curses - seller 7	0.24	0.23	0.24	0.24	0.24	0.24
Rel. freq. of proposer's curses - seller 8	0.23	0.23	0.23	0.23	0.23	0.24

Panel VII.C.1.2 (continued)

	Average of Periods	Period 1	Period 2	Period 3	Period 4	Period 5
Number of market ask and market bid matches	9.97	9.95	9.97	9.97	9.97	9.98
Number of contracts per market ask-bid match	1.04	1.04	1.04	1.04	1.04	1.04
Avg. rank of traded red.value at trading time	1.27	1.27	1.28	1.27	1.28	1.27
... median value in the simulation	1.27	1.26	1.26	1.26	1.26	1.23
Avg. rank of traded unit cost at trading time	2.03	2.03	2.03	2.04	2.03	2.04
... median value in the simulation	2.04	2.06	2.06	2.06	2.06	2.06
Number of submitted bids per contract	13.19	13.17	13.19	13.21	13.20	13.20
... median value in the simulation	13.01	13.01	13.01	13.01	13.01	13.01
... rel. freq. of bids < p* - 200	0.14	0.14	0.14	0.14	0.14	0.14
... rel. freq. of bids < p*	0.72	0.72	0.72	0.72	0.72	0.72
... rel. freq. of bids = p*	0.03	0.03	0.03	0.03	0.03	0.03
... rel. freq. of bids > p*	0.26	0.26	0.26	0.26	0.26	0.26
Number of submitted asks per contract	17.60	17.61	17.59	17.61	17.63	17.59
... median value in the simulation	13.01	13.01	13.01	13.01	13.01	13.01
... rel. freq. of asks < p*	0.02	0.02	0.02	0.02	0.02	0.02
... rel. freq. of asks = p*	0.01	0.01	0.01	0.01	0.01	0.01
... rel. freq. of asks > p*	0.97	0.97	0.97	0.97	0.97	0.97
... rel. freq. of asks > p* + 200	0.43	0.43	0.43	0.43	0.43	0.43
Number of offers for all units in periods > 1	318	-	318	318	318	318
(a) rel. freq. of bids > p-max & asks < p-min	0.01	-	0.01	0.01	0.01	0.01
(b) rel. freq. of bids > p-avg & asks < p-avg	0.11	-	0.11	0.11	0.11	0.11
(c) rel. freq. of bids > p-own & asks < p-own	0.06	-	0.06	0.06	0.06	0.06
(d) rel. freq. of bids < p-own & asks > p-own	0.88	-	0.88	0.88	0.88	0.88
Number of offers for units bought above p-avg or sold below p-avg in the previous period	118	-	118	118	119	118
(C) rel. freq. of bids > p-own & asks < p-own	0.04	-	0.04	0.04	0.04	0.04
Number of offers for units that were not traded in the previous period	50	-	50	49	50	49
(D) rel. freq. of bids < p-own & asks > p-own	0.90	-	0.90	0.90	0.90	0.90
Aggressive offer frequency ratio (c)/(C)	0.59	-	0.60	0.59	0.59	0.60
... median value in the simulation	0.58	-	0.55	0.54	0.53	0.55
Moderate offer frequency ratio (d)/(D)	1.02	-	1.02	1.02	1.03	1.02
... median value in the simulation	1.03	-	1.03	1.03	1.03	1.03

Relative frequencies f(x) and cumulative distributions F(x) of the simulation rounds (5 subsequent periods) in which any n periods have price paths of the same type.

	falling		rising		hanging		standing		unknown	
n	f(x)	F(x)	f(x)	F(x)	f(x)	F(x)	f(x)	F(x)	f(x)	F(x)
5	0.37	0.37	0.00	0.00	0.00	0.00	0.00	0.00	0.00	0.00
4	0.40	0.78	0.00	0.00	0.00	0.00	0.00	0.00	0.00	0.00
3	0.18	0.96	0.00	0.00	0.00	0.00	0.03	0.03	0.00	0.00
2	0.04	1.00	0.00	0.00	0.00	0.00	0.16	0.19	0.00	0.00
1	0.00	1.00	0.05	0.06	0.03	0.03	0.40	0.58	0.00	0.00
0	0.00	1.00	0.94	1.00	0.97	1.00	0.42	1.00	1.00	1.00

Panel VII.C.1.3

```
Double Auction Market Simulation Results                              File: DAMM1830.SIM

Number of simulation rounds       :  10000          Periods per round :          5
Market type                       :     M1          Number of traders :          8
Market rules                      : AltDAM       Technical price range : [    1 ... 9999]
No-Crossing rule                  :    YES    Ask-Bid-Spread-Reduction rule :        YES
Trader type                       :     ZI

Market clearing quantity q*       :     10
Market clearing price    p*       :    893
```

	Average of Periods	Period 1	Period 2	Period 3	Period 4	Period 5
Deviation of traded quantity from q*	0.33	0.32	0.32	0.33	0.34	0.33
Efficiency	0.99	0.99	0.99	0.99	0.99	0.99
... median value in the simulation	1.00	0.99	0.99	0.99	0.99	0.99
Average contract price (deviation from p*)	4.61	4.61	4.58	4.63	4.59	4.63
... median value in the simulation	5.10	5.10	5.10	5.10	5.10	5.10
Minimum contract price (deviation from p*)	-11.35	-11.36	-11.43	-11.30	-11.34	-11.32
... median value in the simulation	-9.99	-9.99	-10.99	-9.99	-9.99	-9.99
Maximum contract price (deviation from p*)	16.70	16.69	16.71	16.71	16.70	16.70
... median value in the simulation	18.01	18.01	18.01	18.01	18.01	18.01
Market clearing price concordance coefficient	0.53	0.53	0.53	0.53	0.53	0.53
... median value in the simulation	0.53	0.53	0.53	0.53	0.53	0.53
Rel. freq. of falling price paths	0.70	0.71	0.69	0.70	0.71	0.70
Rel. freq. of rising price paths	0.05	0.05	0.05	0.05	0.05	0.05
Rel. freq. of hanging price paths	0.04	0.04	0.04	0.03	0.04	0.04
Rel. freq. of standing price paths	0.21	0.21	0.22	0.21	0.21	0.21
Rel. freq. of unknown price paths	0.00	0.00	0.00	0.00	0.00	0.00
Rel. freq. of contract proposals by buyers	0.72	0.73	0.72	0.73	0.72	0.73
... lower quartile bound in the simulation	0.68	0.61	0.61	0.61	0.61	0.61
... median value in the simulation	0.73	0.73	0.73	0.73	0.73	0.73
... upper quartile bound in the simulation	0.79	0.91	0.91	0.91	0.82	0.91
... rel. freq. of these with price < p*	0.13	0.14	0.13	0.13	0.13	0.13
... rel. freq. of these with price = p*	0.10	0.10	0.10	0.10	0.10	0.10
... rel. freq. of these with price > p*	0.77	0.77	0.77	0.77	0.77	0.77
... avg. buyer proposed contract price - p*	6.08	6.06	6.06	6.09	6.10	6.10
Rel. freq. of contract proposals by sellers	0.28	0.27	0.28	0.27	0.28	0.27
... lower quartile bound in the simulation	0.23	0.11	0.11	0.11	0.19	0.11
... median value in the simulation	0.28	0.28	0.28	0.28	0.28	0.28
... upper quartile bound in the simulation	0.33	0.40	0.40	0.40	0.40	0.40
... rel. freq. of these with price < p*	0.37	0.36	0.36	0.37	0.37	0.37
... rel. freq. of these with price = p*	0.06	0.06	0.06	0.06	0.06	0.06
... rel. freq. of these with price > p*	0.57	0.58	0.58	0.57	0.57	0.57
... avg. seller proposed contract price - p*	1.00	1.12	0.94	0.99	0.92	1.03
Rel. freq. of proposer's curses - all traders	0.67	0.67	0.67	0.67	0.67	0.67
... median value in the simulation	0.67	0.71	0.71	0.71	0.71	0.71
Rel. freq. of proposer's curses - all buyers	0.77	0.77	0.77	0.77	0.77	0.77
Rel. freq. of proposer's curses - all sellers	0.37	0.36	0.36	0.37	0.37	0.37
Rel. freq. of proposer's curses - buyer 1	0.91	0.91	0.91	0.91	0.91	0.90
Rel. freq. of proposer's curses - buyer 2	0.66	0.66	0.65	0.65	0.66	0.66
Rel. freq. of proposer's curses - buyer 3	0.81	0.81	0.81	0.81	0.81	0.81
Rel. freq. of proposer's curses - buyer 4	0.87	0.87	0.88	0.87	0.88	0.88
Rel. freq. of proposer's curses - seller 5	0.48	0.48	0.48	0.49	0.48	0.49
Rel. freq. of proposer's curses - seller 6	0.34	0.34	0.34	0.34	0.34	0.33
Rel. freq. of proposer's curses - seller 7	0.35	0.35	0.36	0.34	0.35	0.35
Rel. freq. of proposer's curses - seller 8	0.34	0.33	0.33	0.34	0.34	0.34

Panel VII.C.1.3 (continued)

	Average of Periods	Period 1	Period 2	Period 3	Period 4	Period 5
Number of market ask and market bid matches	9.84	9.82	9.83	9.84	9.85	9.84
Number of contracts per market ask-bid match	1.05	1.05	1.05	1.05	1.05	1.05
Avg. rank of traded red.value at trading time	1.42	1.42	1.41	1.42	1.42	1.42
... median value in the simulation	1.42	1.41	1.39	1.41	1.41	1.41
Avg. rank of traded unit cost at trading time	1.99	1.99	1.99	2.00	1.99	1.99
... median value in the simulation	2.00	2.01	2.01	2.01	2.01	2.01
Number of submitted bids per contract	14.15	14.14	14.15	14.16	14.13	14.16
... median value in the simulation	13.01	13.01	13.01	13.01	13.01	13.01
... rel. freq. of bids < p* - 200	0.24	0.24	0.24	0.24	0.24	0.24
... rel. freq. of bids < p*	0.77	0.77	0.77	0.77	0.77	0.77
... rel. freq. of bids = p*	0.02	0.02	0.02	0.02	0.02	0.02
... rel. freq. of bids > p*	0.20	0.20	0.20	0.20	0.20	0.20
Number of submitted asks per contract	17.14	17.11	17.14	17.15	17.12	17.17
... median value in the simulation	13.01	13.01	13.01	13.01	13.01	13.01
... rel. freq. of asks < p*	0.03	0.03	0.03	0.03	0.03	0.03
... rel. freq. of asks = p*	0.01	0.01	0.01	0.01	0.01	0.01
... rel. freq. of asks > p*	0.96	0.96	0.96	0.96	0.96	0.96
... rel. freq. of asks > p* + 200	0.43	0.43	0.43	0.43	0.43	0.43
Number of offers for all units in periods > 1	323	·	322	323	323	323
(a) rel. freq. of bids > p-max & asks < p-min	0.01	·	0.01	0.01	0.01	0.01
(b) rel. freq. of bids > p-avg & asks < p-avg	0.10	·	0.10	0.10	0.10	0.10
(c) rel. freq. of bids > p-own & asks < p-own	0.07	·	0.07	0.07	0.07	0.07
(d) rel. freq. of bids < p-own & asks > p-own	0.89	·	0.89	0.89	0.89	0.89
Number of offers for units bought above p-avg or sold below p-avg in the previous period	121	·	120	121	121	121
(C) rel. freq. of bids > p-own & asks < p-own	0.04	·	0.04	0.04	0.04	0.04
Number of offers for units that were not traded in the previous period	52	·	52	53	52	52
(D) rel. freq. of bids < p-own & asks > p-own	0.91	·	0.91	0.91	0.91	0.91
Aggressive offer frequency ratio (c)/(C)	0.57	·	0.57	0.57	0.57	0.56
... median value in the simulation	0.56	·	0.53	0.52	0.52	0.51
Moderate offer frequency ratio (d)/(D)	1.03	·	1.03	1.03	1.02	1.03
... median value in the simulation	1.03	·	1.03	1.03	1.03	1.03

Relative frequencies f(x) and cumulative distributions F(x) of the simulation rounds (5 subsequent periods) in which any n periods have price paths of the same type.

	falling		rising		hanging		standing		unknown	
n	f(x)	F(x)	f(x)	F(x)	f(x)	F(x)	f(x)	F(x)	f(x)	F(x)
5	0.17	0.17	0.00	0.00	0.00	0.00	0.00	0.00	0.00	0.00
4	0.36	0.53	0.00	0.00	0.00	0.00	0.01	0.01	0.00	0.00
3	0.31	0.84	0.00	0.00	0.00	0.00	0.06	0.07	0.00	0.00
2	0.13	0.97	0.02	0.02	0.01	0.01	0.22	0.29	0.00	0.00
1	0.03	1.00	0.21	0.23	0.16	0.17	0.40	0.69	0.00	0.00
0	0.00	1.00	0.77	1.00	0.83	1.00	0.31	1.00	1.00	1.00

Panel VII.C.1.4

Double Auction Market Simulation Results   File: DAMM1840.SIM

Number of simulation rounds	: 10000	Periods per round :			5
Market type	: M1	Number of traders :			8
Market rules	: AltDAM	Technical price range : [1 ... 9999]			
No-Crossing rule	: YES	Ask-Bid-Spread-Reduction rule :			YES
Trader type	: ZI				

Market clearing quantity q*	:	10
Market clearing price p*	:	211

	Average of Periods	Period 1	Period 2	Period 3	Period 4	Period 5
Deviation of traded quantity from q*	0.35	0.35	0.35	0.35	0.35	0.35
Efficiency	0.99	0.99	0.99	0.99	0.99	0.99
... median value in the simulation	1.00	0.99	0.99	0.99	0.99	0.99
Average contract price (deviation from p*)	7.25	7.25	7.25	7.28	7.25	7.22
... median value in the simulation	7.70	7.73	7.73	7.73	7.73	7.70
Minimum contract price (deviation from p*)	-4.00	-3.98	-4.02	-3.93	-4.00	-4.07
... median value in the simulation	-1.99	-1.99	-1.99	-1.99	-1.99	-1.99
Maximum contract price (deviation from p*)	17.34	17.33	17.35	17.35	17.32	17.33
... median value in the simulation	18.01	18.01	18.01	18.01	18.01	18.01
Market clearing price concordance coefficient	0.53	0.53	0.53	0.53	0.53	0.53
... median value in the simulation	0.53	0.52	0.52	0.52	0.52	0.52
Rel. freq. of falling price paths	0.88	0.88	0.88	0.88	0.88	0.88
Rel. freq. of rising price paths	0.00	0.00	0.00	0.00	0.00	0.00
Rel. freq. of hanging price paths	0.00	0.00	0.00	0.00	0.00	0.00
Rel. freq. of standing price paths	0.11	0.12	0.11	0.11	0.11	0.11
Rel. freq. of unknown price paths	0.00	0.00	0.00	0.00	0.00	0.00
Rel. freq. of contract proposals by buyers	0.84	0.85	0.84	0.85	0.84	0.84
... lower quartile bound in the simulation	0.81	0.81	0.81	0.81	0.81	0.81
... median value in the simulation	0.85	0.91	0.91	0.91	0.91	0.91
... upper quartile bound in the simulation	0.89	0.91	0.91	0.91	0.91	0.91
... rel. freq. of these with price < p*	0.07	0.07	0.07	0.07	0.06	0.07
... rel. freq. of these with price = p*	0.10	0.10	0.10	0.10	0.10	0.10
... rel. freq. of these with price > p*	0.83	0.83	0.83	0.83	0.83	0.83
... avg. buyer proposed contract price - p*	7.59	7.59	7.61	7.60	7.58	7.57
Rel. freq. of contract proposals by sellers	0.16	0.15	0.16	0.15	0.16	0.16
... lower quartile bound in the simulation	0.12	0.10	0.10	0.10	0.10	0.10
... median value in the simulation	0.16	0.11	0.11	0.11	0.11	0.11
... upper quartile bound in the simulation	0.20	0.20	0.20	0.20	0.20	0.20
... rel. freq. of these with price < p*	0.19	0.19	0.19	0.18	0.19	0.19
... rel. freq. of these with price = p*	0.05	0.05	0.05	0.05	0.05	0.05
... rel. freq. of these with price > p*	0.76	0.77	0.76	0.77	0.77	0.76
... avg. seller proposed contract price - p*	5.56	5.67	5.44	5.58	5.57	5.56
Rel. freq. of proposer's curses - all traders	0.74	0.74	0.74	0.74	0.74	0.74
... median value in the simulation	0.75	0.73	0.73	0.73	0.73	0.73
Rel. freq. of proposer's curses - all buyers	0.83	0.83	0.83	0.83	0.83	0.83
Rel. freq. of proposer's curses - all sellers	0.19	0.19	0.19	0.18	0.19	0.19
Rel. freq. of proposer's curses - buyer 1	0.97	0.97	0.97	0.97	0.97	0.97
Rel. freq. of proposer's curses - buyer 2	0.71	0.71	0.71	0.71	0.71	0.71
Rel. freq. of proposer's curses - buyer 3	0.89	0.89	0.89	0.89	0.89	0.89
Rel. freq. of proposer's curses - buyer 4	0.97	0.97	0.97	0.97	0.97	0.97
Rel. freq. of proposer's curses - seller 5	0.27	0.28	0.28	0.27	0.27	0.27
Rel. freq. of proposer's curses - seller 6	0.17	0.18	0.17	0.16	0.17	0.17
Rel. freq. of proposer's curses - seller 7	0.17	0.18	0.17	0.16	0.18	0.19
Rel. freq. of proposer's curses - seller 8	0.17	0.16	0.17	0.16	0.17	0.17

Panel VII.C.1.4 (continued)

	Average of Periods	Period 1	Period 2	Period 3	Period 4	Period 5
Number of market ask and market bid matches	10.06	10.06	10.05	10.07	10.06	10.06
Number of contracts per market ask-bid match	1.03	1.03	1.03	1.03	1.03	1.03
Avg. rank of traded red.value at trading time	1.19	1.19	1.19	1.19	1.19	1.19
... median value in the simulation	1.19	1.17	1.17	1.16	1.16	1.17
Avg. rank of traded unit cost at trading time	2.05	2.05	2.05	2.05	2.05	2.05
... median value in the simulation	2.06	2.10	2.10	2.10	2.10	2.10
Number of submitted bids per contract	12.43	12.42	12.43	12.44	12.41	12.43
... median value in the simulation	12.44	12.46	12.46	12.51	12.46	12.46
... rel. freq. of bids < p* - 200	0.01	0.01	0.01	0.01	0.01	0.01
... rel. freq. of bids < p*	0.67	0.67	0.67	0.67	0.67	0.67
... rel. freq. of bids = p*	0.03	0.03	0.03	0.03	0.03	0.03
... rel. freq. of bids > p*	0.30	0.30	0.30	0.30	0.30	0.30
Number of submitted asks per contract	17.95	17.96	17.94	17.97	17.93	17.95
... median value in the simulation	13.01	13.01	13.01	13.01	13.01	13.01
... rel. freq. of asks < p*	0.01	0.01	0.01	0.01	0.01	0.01
... rel. freq. of asks = p*	0.01	0.01	0.01	0.01	0.01	0.01
... rel. freq. of asks > p*	0.98	0.98	0.98	0.98	0.98	0.98
... rel. freq. of asks > p* + 200	0.43	0.43	0.43	0.43	0.43	0.43
Number of offers for all units in periods > 1	314	-	314	314	314	314
(a) rel. freq. of bids > p-max & asks < p-min	0.01	-	0.01	0.01	0.01	0.01
(b) rel. freq. of bids > p-avg & asks < p-avg	0.11	-	0.11	0.11	0.11	0.11
(c) rel. freq. of bids > p-own & asks < p-own	0.06	-	0.06	0.06	0.06	0.06
(d) rel. freq. of bids < p-own & asks > p-own	0.87	-	0.87	0.87	0.87	0.87
Number of offers for units bought above p-avg or sold below p-avg in the previous period	116	-	117	117	116	116
(C) rel. freq. of bids > p-own & asks < p-own	0.04	-	0.04	0.03	0.04	0.04
Number of offers for units that were not traded in the previous period	47	-	47	48	47	47
(D) rel. freq. of bids < p-own & asks > p-own	0.89	-	0.89	0.89	0.89	0.89
Aggressive offer frequency ratio (c)/(C)	0.61	-	0.60	0.60	0.61	0.60
... median value in the simulation	0.60	-	0.55	0.55	0.56	0.55
Moderate offer frequency ratio (d)/(D)	1.03	-	1.02	1.03	1.03	1.03
... median value in the simulation	1.03	-	1.03	1.03	1.03	1.03

Relative frequencies f(x) and cumulative distributions F(x) of the simulation rounds
(5 subsequent periods) in which any n periods have price paths of the same type.

	falling		rising		hanging		standing		unknown	
n	f(x)	F(x)	f(x)	F(x)	f(x)	F(x)	f(x)	F(x)	f(x)	F(x)
5	0.53	0.53	0.00	0.00	0.00	0.00	0.00	0.00	0.00	0.00
4	0.36	0.89	0.00	0.00	0.00	0.00	0.00	0.00	0.00	0.00
3	0.09	0.99	0.00	0.00	0.00	0.00	0.01	0.01	0.00	0.00
2	0.01	1.00	0.00	0.00	0.00	0.00	0.09	0.10	0.00	0.00
1	0.00	1.00	0.02	0.02	0.01	0.01	0.36	0.46	0.00	0.00
0	0.00	1.00	0.98	1.00	0.99	1.00	0.54	1.00	1.00	1.00

Panel VII.C.1.5

Double Auction Market Simulation Results   File: DAMM1C10.SIM

Number of simulation rounds	:	10000	Periods per round :		5
Market type	:	M1	Number of traders :		12
Market rules	:	AltDAM	Technical price range : [	1 ...	9999]
No-Crossing rule	:	YES	Ask-Bid-Spread-Reduction rule :		YES
Trader type	:	ZI			

Market clearing quantity q*   :   14
Market clearing price   p*   :   763

	Average of Periods	Period 1	Period 2	Period 3	Period 4	Period 5
Deviation of traded quantity from q*	0.68	0.68	0.67	0.68	0.69	0.70
Efficiency	0.99	0.99	0.99	0.99	0.99	0.99
... median value in the simulation	0.99	0.99	0.99	0.99	0.99	0.99
Average contract price (deviation from p*)	7.28	7.30	7.29	7.29	7.25	7.25
... median value in the simulation	7.80	7.82	7.82	7.80	7.80	7.80
Minimum contract price (deviation from p*)	-17.09	-16.92	-17.29	-17.08	-17.11	-17.03
... median value in the simulation	-15.99	-15.99	-15.99	-15.99	-15.99	-15.99
Maximum contract price (deviation from p*)	24.58	24.57	24.60	24.56	24.58	24.60
... median value in the simulation	25.01	25.01	25.01	25.01	25.01	25.01
Market clearing price concordance coefficient	0.54	0.54	0.54	0.54	0.54	0.54
... median value in the simulation	0.54	0.54	0.54	0.54	0.54	0.54
Rel. freq. of falling price paths	0.74	0.74	0.74	0.74	0.74	0.74
Rel. freq. of rising price paths	0.02	0.02	0.02	0.02	0.02	0.02
Rel. freq. of hanging price paths	0.02	0.02	0.01	0.01	0.02	0.02
Rel. freq. of standing price paths	0.22	0.23	0.23	0.23	0.22	0.22
Rel. freq. of unknown price paths	0.00	0.00	0.00	0.00	0.00	0.00
Rel. freq. of contract proposals by buyers	0.71	0.71	0.71	0.71	0.70	0.71
... lower quartile bound in the simulation	0.67	0.61	0.61	0.61	0.61	0.61
... median value in the simulation	0.71	0.72	0.72	0.72	0.72	0.74
... upper quartile bound in the simulation	0.76	0.81	0.81	0.81	0.81	0.81
... rel. freq. of these with price < p*	0.14	0.14	0.14	0.14	0.14	0.15
... rel. freq. of these with price = p*	0.07	0.07	0.07	0.07	0.07	0.07
... rel. freq. of these with price > p*	0.79	0.79	0.79	0.79	0.79	0.79
... avg. buyer proposed contract price · p*	8.82	8.81	8.87	8.77	8.84	8.80
Rel. freq. of contract proposals by sellers	0.29	0.29	0.29	0.29	0.30	0.29
... lower quartile bound in the simulation	0.25	0.20	0.20	0.20	0.20	0.20
... median value in the simulation	0.30	0.29	0.29	0.29	0.29	0.27
... upper quartile bound in the simulation	0.34	0.40	0.40	0.40	0.40	0.40
... rel. freq. of these with price < p*	0.30	0.30	0.30	0.30	0.30	0.30
... rel. freq. of these with price = p*	0.04	0.04	0.04	0.04	0.04	0.04
... rel. freq. of these with price > p*	0.66	0.66	0.66	0.66	0.66	0.66
... avg. seller proposed contract price · p*	3.98	4.06	3.95	4.07	3.89	3.94
Rel. freq. of proposer's curses - all traders	0.66	0.65	0.66	0.65	0.66	0.66
... median value in the simulation	0.66	0.67	0.67	0.67	0.67	0.67
Rel. freq. of proposer's curses - all buyers	0.79	0.79	0.79	0.79	0.79	0.79
Rel. freq. of proposer's curses - all sellers	0.30	0.30	0.30	0.30	0.30	0.30
Rel. freq. of proposer's curses · buyer 1	0.91	0.91	0.91	0.90	0.91	0.91
Rel. freq. of proposer's curses · buyer 2	0.68	0.67	0.68	0.68	0.68	0.68
Rel. freq. of proposer's curses · buyer 3	0.77	0.77	0.77	0.77	0.77	0.76
Rel. freq. of proposer's curses · buyer 4	0.89	0.89	0.89	0.88	0.89	0.88
Rel. freq. of proposer's curses · buyer 5	0.90	0.90	0.90	0.90	0.89	0.89
Rel. freq. of proposer's curses · buyer 6	0.89	0.89	0.89	0.89	0.89	0.89
Rel. freq. of proposer's curses · seller 7	0.39	0.40	0.40	0.39	0.39	0.40
Rel. freq. of proposer's curses · seller 8	0.30	0.31	0.29	0.29	0.30	0.30
Rel. freq. of proposer's curses · seller 9	0.31	0.31	0.32	0.31	0.32	0.32
Rel. freq. of proposer's curses · seller 10	0.30	0.30	0.30	0.31	0.32	0.30
Rel. freq. of proposer's curses · seller 11	0.30	0.29	0.29	0.29	0.30	0.30
Rel. freq. of proposer's curses · seller 12	0.27	0.26	0.28	0.27	0.28	0.28

Panel VII.C.1.5 (continued)

	Average of Periods	Period 1	Period 2	Period 3	Period 4	Period 5
Number of market ask and market bid matches	13.74	13.73	13.74	13.74	13.73	13.77
Number of contracts per market ask-bid match	1.07	1.07	1.07	1.07	1.07	1.07
Avg. rank of traded red.value at trading time	1.66	1.66	1.66	1.66	1.66	1.67
... median value in the simulation	1.66	1.65	1.65	1.65	1.65	1.65
Avg. rank of traded unit cost at trading time	2.68	2.68	2.68	2.68	2.67	2.68
... median value in the simulation	2.69	2.70	2.68	2.68	2.69	2.70
Number of submitted bids per contract	17.34	17.33	17.37	17.33	17.32	17.34
... median value in the simulation	13.01	13.01	13.01	13.01	13.01	13.01
... rel. freq. of bids < p* - 200	0.25	0.25	0.25	0.25	0.25	0.25
... rel. freq. of bids < p*	0.76	0.76	0.76	0.76	0.76	0.76
... rel. freq. of bids = p*	0.02	0.02	0.02	0.02	0.02	0.02
... rel. freq. of bids > p*	0.22	0.22	0.22	0.22	0.22	0.22
Number of submitted asks per contract	21.15	21.13	21.19	21.15	21.13	21.17
... median value in the simulation	13.01	13.01	13.01	13.01	13.01	13.01
... rel. freq. of asks < p*	0.03	0.03	0.03	0.03	0.03	0.03
... rel. freq. of asks = p*	0.01	0.01	0.01	0.01	0.01	0.01
... rel. freq. of asks > p*	0.97	0.97	0.97	0.97	0.97	0.97
... rel. freq. of asks > p* + 200	0.45	0.45	0.45	0.45	0.45	0.45
Number of offers for all units in periods > 1	565	-	565	564	564	565
(a) rel. freq. of bids > p-max & asks < p-min	0.01	-	0.01	0.01	0.01	0.01
(b) rel. freq. of bids > p-avg & asks < p-avg	0.10	-	0.10	0.10	0.11	0.10
(c) rel. freq. of bids > p-own & asks < p-own	0.07	-	0.07	0.07	0.06	0.07
(d) rel. freq. of bids < p-own & asks > p-own	0.89	-	0.89	0.89	0.89	0.89
Number of offers for units bought above p-avg or sold below p-avg in the previous period	204	-	204	204	204	205
(C) rel. freq. of bids > p-own & asks < p-own	0.04	-	0.04	0.04	0.04	0.04
Number of offers for units that were not traded in the previous period	109	-	109	109	109	109
(D) rel. freq. of bids < p-own & asks > p-own	0.91	-	0.91	0.91	0.91	0.91
Aggressive offer frequency ratio (c)/(C)	0.59	-	0.59	0.59	0.59	0.59
... median value in the simulation	0.59	-	0.56	0.56	0.56	0.55
Moderate offer frequency ratio (d)/(D)	1.02	-	1.02	1.02	1.02	1.02
... median value in the simulation	1.03	-	1.03	1.03	1.03	1.03

Relative frequencies f(x) and cumulative distributions F(x) of the simulation rounds
(5 subsequent periods) in which any n periods have price paths of the same type.

	falling		rising		hanging		standing		unknown	
n	f(x)	F(x)	f(x)	F(x)	f(x)	F(x)	f(x)	F(x)	f(x)	F(x)
5	0.22	0.22	0.00	0.00	0.00	0.00	0.00	0.00	0.00	0.00
4	0.40	0.62	0.00	0.00	0.00	0.00	0.01	0.01	0.00	0.00
3	0.27	0.89	0.00	0.00	0.00	0.00	0.07	0.08	0.00	0.00
2	0.09	0.98	0.00	0.00	0.00	0.00	0.24	0.31	0.00	0.00
1	0.02	1.00	0.08	0.09	0.07	0.08	0.41	0.72	0.00	0.00
0	0.00	1.00	0.91	1.00	0.92	1.00	0.28	1.00	1.00	1.00

Panel VII.C.1.6

Double Auction Market Simulation Results   File: DAMM2810.SIM

Number of simulation rounds	:	10000
Market type	:	M2
Market rules	:	AltDAM
No-Crossing rule	:	YES
Trader type	:	ZI

Periods per round	:	5
Number of traders	:	8
Technical price range	: [	1 ... 9999]
Ask-Bid-Spread-Reduction rule	:	YES

Market clearing quantity q*	:	7
Market clearing price p*	:	351

	Average of Periods	Period 1	Period 2	Period 3	Period 4	Period 5
Deviation of traded quantity from q*	0.26	0.26	0.25	0.26	0.26	0.25
Efficiency	0.97	0.97	0.97	0.97	0.97	0.97
... median value in the simulation	0.98	0.97	0.98	0.97	0.98	0.98
Average contract price (deviation from p*)	13.27	13.31	13.31	13.23	13.19	13.33
... median value in the simulation	14.01	14.01	14.01	13.86	13.86	14.01
Minimum contract price (deviation from p*)	-5.47	-5.42	-5.39	-5.55	-5.59	-5.41
... median value in the simulation	-3.99	-3.99	-3.99	-3.99	-4.99	-3.99
Maximum contract price (deviation from p*)	33.06	33.07	33.14	33.04	33.01	33.05
... median value in the simulation	35.01	35.01	35.01	35.01	35.01	35.01
Market clearing price concordance coefficient	0.36	0.36	0.36	0.36	0.36	0.36
... median value in the simulation	0.36	0.34	0.34	0.34	0.34	0.34
Rel. freq. of falling price paths	0.86	0.87	0.86	0.86	0.86	0.86
Rel. freq. of rising price paths	0.00	0.00	0.00	0.01	0.00	0.00
Rel. freq. of hanging price paths	0.00	0.00	0.00	0.00	0.00	0.00
Rel. freq. of standing price paths	0.13	0.12	0.13	0.13	0.13	0.13
Rel. freq. of unknown price paths	0.00	0.00	0.00	0.00	0.00	0.00
Rel. freq. of contract proposals by buyers	0.74	0.74	0.74	0.74	0.74	0.74
... lower quartile bound in the simulation	0.69	0.58	0.58	0.58	0.58	0.58
... median value in the simulation	0.75	0.76	0.76	0.72	0.72	0.76
... upper quartile bound in the simulation	0.81	0.86	0.86	0.86	0.86	0.86
... rel. freq. of these with price < p*	0.10	0.10	0.10	0.10	0.10	0.10
... rel. freq. of these with price = p*	0.12	0.12	0.12	0.12	0.12	0.12
... rel. freq. of these with price > p*	0.78	0.78	0.78	0.78	0.78	0.78
... avg. buyer proposed contract price - p*	14.29	14.30	14.36	14.30	14.18	14.30
Rel. freq. of contract proposals by sellers	0.26	0.26	0.26	0.26	0.26	0.26
... lower quartile bound in the simulation	0.20	0.15	0.15	0.15	0.15	0.15
... median value in the simulation	0.26	0.26	0.26	0.29	0.29	0.26
... upper quartile bound in the simulation	0.32	0.43	0.43	0.43	0.43	0.43
... rel. freq. of these with price < p*	0.22	0.22	0.22	0.23	0.22	0.22
... rel. freq. of these with price = p*	0.05	0.05	0.05	0.05	0.05	0.05
... rel. freq. of these with price > p*	0.72	0.73	0.72	0.72	0.72	0.73
... avg. seller proposed contract price - p*	10.68	10.74	10.68	10.56	10.63	10.76
Rel. freq. of proposer's curses - all traders	0.64	0.64	0.64	0.64	0.64	0.64
... median value in the simulation	0.65	0.63	0.72	0.72	0.63	0.72
Rel. freq. of proposer's curses - all buyers	0.78	0.78	0.78	0.78	0.78	0.78
Rel. freq. of proposer's curses - all sellers	0.22	0.22	0.22	0.23	0.22	0.22
Rel. freq. of proposer's curses - buyer 1	0.96	0.96	0.96	0.96	0.96	0.96
Rel. freq. of proposer's curses - buyer 2	0.94	0.94	0.94	0.94	0.94	0.94
Rel. freq. of proposer's curses - buyer 3	0.53	0.53	0.53	0.53	0.53	0.53
Rel. freq. of proposer's curses - buyer 4	0.82	0.82	0.82	0.82	0.82	0.82
Rel. freq. of proposer's curses - seller 5	0.34	0.34	0.34	0.34	0.34	0.34
Rel. freq. of proposer's curses - seller 6	0.27	0.26	0.27	0.28	0.27	0.26
Rel. freq. of proposer's curses - seller 7	0.15	0.14	0.15	0.15	0.15	0.15
Rel. freq. of proposer's curses - seller 8	0.16	0.15	0.16	0.16	0.16	0.16

Panel VII.C.1.6 (continued)

	Average of Periods	Period 1	Period 2	Period 3	Period 4	Period 5
Number of market ask and market bid matches	7.04	7.05	7.03	7.04	7.04	7.03
Number of contracts per market ask-bid match	1.03	1.03	1.03	1.03	1.03	1.03
Avg. rank of traded red.value at trading time	1.22	1.22	1.22	1.22	1.23	1.22
... median value in the simulation	1.22	1.15	1.15	1.15	1.15	1.15
Avg. rank of traded unit cost at trading time	2.04	2.04	2.04	2.04	2.03	2.04
... median value in the simulation	2.05	2.08	2.01	2.09	2.01	2.07
Number of submitted bids per contract	13.86	13.88	13.87	13.86	13.84	13.85
... median value in the simulation	13.01	13.01	13.01	13.01	13.01	13.01
... rel. freq. of bids < p* - 200	0.14	0.14	0.14	0.14	0.14	0.14
... rel. freq. of bids < p*	0.73	0.73	0.73	0.73	0.73	0.73
... rel. freq. of bids = p*	0.02	0.02	0.02	0.02	0.02	0.02
... rel. freq. of bids > p*	0.25	0.25	0.25	0.25	0.25	0.25
Number of submitted asks per contract	20.13	20.15	20.15	20.12	20.08	20.13
... median value in the simulation	13.01	13.01	13.01	13.01	13.01	13.01
... rel. freq. of asks < p*	0.02	0.02	0.02	0.02	0.02	0.02
... rel. freq. of asks = p*	0.01	0.01	0.01	0.01	0.01	0.01
... rel. freq. of asks > p*	0.97	0.97	0.97	0.97	0.97	0.97
... rel. freq. of asks > p* + 200	0.44	0.44	0.44	0.44	0.44	0.44
Number of offers for all units in periods > 1	247	-	247	247	246	246
(a) rel. freq. of bids > p-max & asks < p-min	0.01	-	0.01	0.01	0.01	0.01
(b) rel. freq. of bids > p-avg & asks < p-avg	0.12	-	0.12	0.12	0.12	0.12
(c) rel. freq. of bids > p-own & asks < p-own	0.09	-	0.09	0.09	0.09	0.09
(d) rel. freq. of bids < p-own & asks > p-own	0.87	-	0.87	0.87	0.87	0.87
Number of offers for units bought above p-avg or sold below p-avg in the previous period	77	-	77	77	77	77
(C) rel. freq. of bids > p-own & asks < p-own	0.06	-	0.06	0.06	0.06	0.06
Number of offers for units that were not traded in the previous period	76	-	76	76	75	75
(D) rel. freq. of bids < p-own & asks > p-own	0.89	-	0.89	0.89	0.89	0.89
Aggressive offer frequency ratio (c)/(C)	0.70	-	0.71	0.70	0.71	0.70
... median value in the simulation	0.70	-	0.67	0.65	0.66	0.65
Moderate offer frequency ratio (d)/(D)	1.02	-	1.03	1.02	1.03	1.02
... median value in the simulation	1.03	-	1.03	1.03	1.03	1.03

Relative frequencies f(x) and cumulative distributions F(x) of the simulation rounds
(5 subsequent periods) in which any n periods have price paths of the same type.

| | falling | | rising | | hanging | | standing | | unknown | |
n	f(x)	F(x)	f(x)	F(x)	f(x)	F(x)	f(x)	F(x)	f(x)	F(x)
5	0.48	0.48	0.00	0.00	0.00	0.00	0.00	0.00	0.00	0.00
4	0.37	0.86	0.00	0.00	0.00	0.00	0.00	0.00	0.00	0.00
3	0.12	0.98	0.00	0.00	0.00	0.00	0.02	0.02	0.00	0.00
2	0.02	1.00	0.00	0.00	0.00	0.00	0.11	0.13	0.00	0.00
1	0.00	1.00	0.02	0.02	0.02	0.02	0.37	0.50	0.00	0.00
0	0.00	1.00	0.98	1.00	0.98	1.00	0.50	1.00	1.00	1.00

Panel VII.C.1.7

Double Auction Market Simulation Results   File: DAMM2820.SIM

Number of simulation rounds	: 10000
Market type	: M2
Market rules	: AltDAM
No-Crossing rule	: YES
Trader type	: ZI

Periods per round :   5
Number of traders :   8
Technical price range : [1 ... 9999]
Ask-Bid-Spread-Reduction rule :   YES

Market clearing quantity q*	: 7
Market clearing price p*	: 739

	Average of Periods	Period 1	Period 2	Period 3	Period 4	Period 5
Deviation of traded quantity from q*	0.25	0.26	0.25	0.25	0.25	0.25
Efficiency	0.97	0.97	0.97	0.97	0.97	0.97
... median value in the simulation	0.98	0.99	0.99	0.99	0.99	0.99
Average contract price (deviation from p*)	11.30	11.29	11.36	11.33	11.25	11.28
... median value in the simulation	12.01	12.01	12.15	12.01	12.01	12.01
Minimum contract price (deviation from p*)	-8.07	-8.04	-8.03	-8.07	-8.10	-8.12
... median value in the simulation	-5.99	--5.99	-5.99	-5.99	-5.99	-5.99
Maximum contract price (deviation from p*)	32.02	32.01	32.04	32.05	32.03	31.98
... median value in the simulation	34.01	34.01	34.01	34.01	34.01	34.01
Market clearing price concordance coefficient	0.39	0.39	0.39	0.39	0.39	0.39
... median value in the simulation	0.39	0.38	0.37	0.38	0.38	0.38
Rel. freq. of falling price paths	0.78	0.78	0.78	0.78	0.78	0.78
Rel. freq. of rising price paths	0.02	0.02	0.02	0.02	0.02	0.02
Rel. freq. of hanging price paths	0.01	0.01	0.01	0.01	0.01	0.01
Rel. freq. of standing price paths	0.19	0.19	0.19	0.19	0.19	0.19
Rel. freq. of unknown price paths	0.00	0.00	0.00	0.00	0.00	0.00
Rel. freq. of contract proposals by buyers	0.68	0.68	0.69	0.68	0.68	0.68
... lower quartile bound in the simulation	0.63	0.58	0.58	0.58	0.58	0.58
... median value in the simulation	0.69	0.72	0.72	0.72	0.72	0.72
... upper quartile bound in the simulation	0.75	0.86	0.86	0.86	0.86	0.86
... rel. freq. of these with price < p*	0.14	0.14	0.14	0.14	0.14	0.13
... rel. freq. of these with price = p*	0.11	0.11	0.11	0.11	0.11	0.11
... rel. freq. of these with price > p*	0.75	0.75	0.75	0.75	0.75	0.75
... avg. buyer proposed contract price - p*	13.07	13.06	13.12	13.19	12.96	13.04
Rel. freq. of contract proposals by sellers	0.32	0.32	0.31	0.32	0.32	0.32
... lower quartile bound in the simulation	0.26	0.15	0.15	0.15	0.15	0.15
... median value in the simulation	0.32	0.29	0.29	0.29	0.29	0.29
... upper quartile bound in the simulation	0.38	0.43	0.43	0.43	0.43	0.43
... rel. freq. of these with price < p*	0.30	0.30	0.30	0.30	0.30	0.30
... rel. freq. of these with price = p*	0.06	0.05	0.06	0.06	0.06	0.06
... rel. freq. of these with price > p*	0.64	0.65	0.65	0.64	0.64	0.65
... avg. seller proposed contract price - p*	7.82	7.90	7.85	7.64	7.90	7.80
Rel. freq. of proposer's curses - all traders	0.62	0.62	0.62	0.62	0.61	0.62
... median value in the simulation	0.62	0.63	0.63	0.63	0.63	0.63
Rel. freq. of proposer's curses - all buyers	0.75	0.75	0.75	0.75	0.75	0.75
Rel. freq. of proposer's curses - all sellers	0.30	0.30	0.30	0.30	0.30	0.30
Rel. freq. of proposer's curses - buyer 1	0.93	0.93	0.93	0.93	0.93	0.93
Rel. freq. of proposer's curses - buyer 2	0.90	0.90	0.90	0.90	0.90	0.91
Rel. freq. of proposer's curses - buyer 3	0.51	0.51	0.51	0.51	0.50	0.51
Rel. freq. of proposer's curses - buyer 4	0.78	0.77	0.78	0.77	0.78	0.78
Rel. freq. of proposer's curses - seller 5	0.44	0.43	0.44	0.44	0.43	0.44
Rel. freq. of proposer's curses - seller 6	0.36	0.35	0.35	0.36	0.37	0.36
Rel. freq. of proposer's curses - seller 7	0.20	0.20	0.20	0.20	0.20	0.20
Rel. freq. of proposer's curses - seller 8	0.22	0.22	0.22	0.21	0.21	0.21

Panel VII.C.1.7 (continued)

	Average of Periods	Period 1	Period 2	Period 3	Period 4	Period 5
Number of market ask and market bid matches	6.97	6.98	6.96	6.97	6.97	6.96
Number of contracts per market ask-bid match	1.04	1.04	1.04	1.04	1.04	1.04
Avg. rank of traded red.value at trading time	1.33	1.33	1.33	1.32	1.32	1.33
... median value in the simulation	1.32	1.29	1.29	1.29	1.29	1.29
Avg. rank of traded unit cost at trading time	2.00	2.00	2.00	2.00	2.00	2.00
... median value in the simulation	2.01	2.01	2.01	2.01	2.01	2.01
Number of submitted bids per contract	15.00	15.01	15.01	15.01	14.99	14.99
... median value in the simulation	13.01	13.01	13.01	13.01	13.01	13.01
... rel. freq. of bids < p^* - 200	0.24	0.24	0.24	0.24	0.24	0.24
... rel. freq. of bids < p^*	0.78	0.78	0.78	0.78	0.78	0.78
... rel. freq. of bids = p^*	0.02	0.02	0.02	0.02	0.02	0.02
... rel. freq. of bids > p^*	0.20	0.20	0.20	0.20	0.20	0.20
Number of submitted asks per contract	19.69	19.72	19.69	19.70	19.70	19.67
... median value in the simulation	13.01	13.01	13.01	13.01	13.01	13.01
... rel. freq. of asks < p^*	0.03	0.03	0.03	0.03	0.03	0.03
... rel. freq. of asks = p^*	0.01	0.01	0.01	0.01	0.01	0.01
... rel. freq. of asks > p^*	0.96	0.96	0.96	0.96	0.96	0.96
... rel. freq. of asks > p^* + 200	0.45	0.45	0.45	0.45	0.45	0.45
Number of offers for all units in periods > 1	251	-	252	252	251	251
(a) rel. freq. of bids > p-max & asks < p-min	0.01	-	0.01	0.01	0.01	0.01
(b) rel. freq. of bids > p-avg & asks < p-avg	0.12	-	0.12	0.12	0.12	0.12
(c) rel. freq. of bids > p-own & asks < p-own	0.09	-	0.09	0.09	0.09	0.09
(d) rel. freq. of bids < p-own & asks > p-own	0.87	-	0.87	0.87	0.87	0.87
Number of offers for units bought above p-avg or sold below p-avg in the previous period	79	-	79	79	79	79
(C) rel. freq. of bids > p-own & asks < p-own	0.06	-	0.06	0.06	0.06	0.06
Number of offers for units that were not traded in the previous period	79	-	79	79	79	79
(D) rel. freq. of bids < p-own & asks > p-own	0.89	-	0.89	0.89	0.89	0.89
Aggressive offer frequency ratio (c)/(C)	0.69	-	0.69	0.69	0.69	0.69
... median value in the simulation	0.68	-	0.65	0.63	0.64	0.64
Moderate offer frequency ratio (d)/(D)	1.02	-	1.02	1.02	1.02	1.02
... median value in the simulation	1.03	-	1.03	1.03	1.03	1.03

Relative frequencies f(x) and cumulative distributions F(x) of the simulation rounds (5 subsequent periods) in which any n periods have price paths of the same type.

	falling		rising		hanging		standing		unknown	
n	f(x)	F(x)	f(x)	F(x)	f(x)	F(x)	f(x)	F(x)	f(x)	F(x)
5	0.29	0.29	0.00	0.00	0.00	0.00	0.00	0.00	0.00	0.00
4	0.41	0.70	0.00	0.00	0.00	0.00	0.01	0.01	0.00	0.00
3	0.23	0.93	0.00	0.00	0.00	0.00	0.04	0.05	0.00	0.00
2	0.06	0.99	0.00	0.00	0.00	0.00	0.19	0.24	0.00	0.00
1	0.01	1.00	0.09	0.09	0.06	0.06	0.42	0.66	0.00	0.00
0	0.00	1.00	0.91	1.00	0.94	1.00	0.34	1.00	1.00	1.00

Panel VII.C.1.8

Double Auction Market Simulation Results   File: DAMM2830.SIM

Number of simulation rounds	:	10000
Market type	:	M2
Market rules	:	AltDAM
No-Crossing rule	:	YES
Trader type	:	ZI

Periods per round :	5
Number of traders :	8
Technical price range : [1 ... 9999]
Ask-Bid-Spread-Reduction rule :	YES

Market clearing quantity q*   :   7
Market clearing price   p*   :   281

	Average of Periods	Period 1	Period 2	Period 3	Period 4	Period 5
Deviation of traded quantity from q*	0.26	0.26	0.26	0.27	0.27	0.26
Efficiency	0.97	0.97	0.97	0.97	0.97	0.97
... median value in the simulation	0.98	0.98	0.97	0.97	0.97	0.97
Average contract price (deviation from p*)	13.75	13.76	13.79	13.73	13.72	13.76
... median value in the simulation	14.29	14.29	14.38	14.29	14.29	14.29
Minimum contract price (deviation from p*)	-4.77	-4.74	-4.72	-4.80	-4.83	-4.74
... median value in the simulation	-2.99	-2.99	-2.99	-2.99	-3.99	-2.99
Maximum contract price (deviation from p*)	33.27	33.28	33.29	33.21	33.29	33.28
... median value in the simulation	35.01	35.01	35.01	35.01	35.01	35.01
Market clearing price concordance coefficient	0.35	0.35	0.35	0.35	0.35	0.35
... median value in the simulation	0.35	0.34	0.33	0.34	0.34	0.34
Rel. freq. of falling price paths	0.88	0.87	0.88	0.88	0.88	0.88
Rel. freq. of rising price paths	0.00	0.00	0.00	0.00	0.00	0.00
Rel. freq. of hanging price paths	0.00	0.00	0.00	0.00	0.00	0.00
Rel. freq. of standing price paths	0.11	0.12	0.11	0.11	0.12	0.12
Rel. freq. of unknown price paths	0.00	0.00	0.00	0.00	0.00	0.00
Rel. freq. of contract proposals by buyers	0.76	0.76	0.75	0.75	0.76	0.76
... lower quartile bound in the simulation	0.71	0.63	0.63	0.63	0.63	0.63
... median value in the simulation	0.76	0.76	0.76	0.76	0.76	0.76
... upper quartile bound in the simulation	0.82	0.88	0.88	0.88	0.88	0.88
... rel. freq. of these with price < p*	0.09	0.09	0.09	0.09	0.10	0.09
... rel. freq. of these with price = p*	0.12	0.12	0.12	0.12	0.12	0.12
... rel. freq. of these with price > p*	0.78	0.79	0.78	0.78	0.78	0.79
... avg. buyer proposed contract price - p*	14.54	14.60	14.55	14.51	14.51	14.55
Rel. freq. of contract proposals by sellers	0.24	0.24	0.25	0.25	0.24	0.24
... lower quartile bound in the simulation	0.19	0.13	0.13	0.13	0.13	0.13
... median value in the simulation	0.25	0.26	0.26	0.26	0.26	0.26
... upper quartile bound in the simulation	0.31	0.38	0.38	0.38	0.38	0.38
... rel. freq. of these with price < p*	0.20	0.20	0.20	0.20	0.20	0.20
... rel. freq. of these with price = p*	0.05	0.05	0.05	0.05	0.05	0.05
... rel. freq. of these with price > p*	0.75	0.75	0.75	0.75	0.75	0.75
... avg. seller proposed contract price - p*	11.57	11.51	11.53	11.71	11.61	11.51
Rel. freq. of proposer's curses - all traders	0.65	0.65	0.65	0.65	0.65	0.65
... median value in the simulation	0.66	0.72	0.72	0.72	0.72	0.72
Rel. freq. of proposer's curses - all buyers	0.78	0.79	0.78	0.78	0.78	0.79
Rel. freq. of proposer's curses - all sellers	0.20	0.20	0.20	0.20	0.20	0.20
Rel. freq. of proposer's curses - buyer 1	0.97	0.97	0.97	0.97	0.97	0.97
Rel. freq. of proposer's curses - buyer 2	0.95	0.95	0.95	0.95	0.95	0.95
Rel. freq. of proposer's curses - buyer 3	0.54	0.54	0.53	0.54	0.54	0.53
Rel. freq. of proposer's curses - buyer 4	0.83	0.83	0.83	0.83	0.82	0.83
Rel. freq. of proposer's curses - seller 5	0.31	0.32	0.32	0.31	0.31	0.31
Rel. freq. of proposer's curses - seller 6	0.25	0.24	0.24	0.25	0.25	0.24
Rel. freq. of proposer's curses - seller 7	0.12	0.12	0.12	0.13	0.13	0.12
Rel. freq. of proposer's curses - seller 8	0.14	0.13	0.15	0.13	0.13	0.14

Panel VII.C.1.8 (continued)

	Average of Periods	Period 1	Period 2	Period 3	Period 4	Period 5
Number of market ask and market bid matches	7.06	7.06	7.06	7.06	7.06	7.05
Number of contracts per market ask-bid match	1.03	1.03	1.03	1.03	1.03	1.03
Avg. rank of traded red.value at trading time	1.19	1.19	1.19	1.19	1.20	1.20
... median value in the simulation	1.19	1.15	1.15	1.15	1.15	1.15
Avg. rank of traded unit cost at trading time	2.05	2.05	2.05	2.05	2.06	2.05
... median value in the simulation	2.06	2.09	2.09	2.09	2.13	2.09
Number of submitted bids per contract	13.52	13.54	13.53	13.51	13.51	13.53
... median value in the simulation	13.01	13.01	13.01	13.01	13.01	13.01
... rel. freq. of bids < p^* - 200	0.09	0.09	0.09	0.09	0.09	0.09
... rel. freq. of bids < p^*	0.72	0.72	0.72	0.72	0.72	0.72
... rel. freq. of bids = p^*	0.02	0.02	0.02	0.02	0.02	0.02
... rel. freq. of bids > p^*	0.26	0.26	0.26	0.26	0.26	0.26
Number of submitted asks per contract	20.26	20.29	20.26	20.25	20.25	20.26
... median value in the simulation	13.01	13.01	13.01	13.01	13.01	13.01
... rel. freq. of asks < p^*	0.02	0.02	0.02	0.02	0.02	0.02
... rel. freq. of asks = p^*	0.01	0.01	0.01	0.01	0.01	0.01
... rel. freq. of asks > p^*	0.97	0.97	0.97	0.97	0.97	0.97
... rel. freq. of asks > p^* + 200	0.44	0.44	0.44	0.44	0.44	0.44
Number of offers for all units in periods > 1	245	-	245	245	245	245
(a) rel. freq. of bids > p-max & asks < p-min	0.01	-	0.01	0.01	0.01	0.01
(b) rel. freq. of bids > p-avg & asks < p-avg	0.12	-	0.12	0.12	0.12	0.12
(c) rel. freq. of bids > p-own & asks < p-own	0.09	-	0.09	0.09	0.09	0.09
(d) rel. freq. of bids < p-own & asks > p-own	0.87	-	0.87	0.87	0.87	0.87
Number of offers for units bought above p-avg or sold below p-avg in the previous period	77	-	77	77	77	77
(C) rel. freq. of bids > p-own & asks < p-own	0.06	-	0.06	0.06	0.06	0.06
Number of offers for units that were not traded in the previous period	75	-	75	75	75	74
(D) rel. freq. of bids < p-own & asks > p-own	0.89	-	0.89	0.89	0.89	0.89
Aggressive offer frequency ratio (c)/(C)	0.70	-	0.70	0.70	0.71	0.70
... median value in the simulation	0.70	-	0.66	0.65	0.66	0.66
Moderate offer frequency ratio (d)/(D)	1.03	-	1.03	1.03	1.03	1.03
... median value in the simulation	1.03	-	1.03	1.03	1.03	1.03

Relative frequencies f(x) and cumulative distributions F(x) of the simulation rounds (5 subsequent periods) in which any n periods have price paths of the same type.

	falling		rising		hanging		standing		unknown	
n	f(x)	F(x)	f(x)	F(x)	f(x)	F(x)	f(x)	F(x)	f(x)	F(x)
5	0.53	0.53	0.00	0.00	0.00	0.00	0.00	0.00	0.00	0.00
4	0.36	0.88	0.00	0.00	0.00	0.00	0.00	0.00	0.00	0.00
3	0.10	0.98	0.00	0.00	0.00	0.00	0.01	0.01	0.00	0.00
2	0.02	1.00	0.00	0.00	0.00	0.00	0.09	0.11	0.00	0.00
1	0.00	1.00	0.02	0.02	0.01	0.01	0.35	0.45	0.00	0.00
0	0.00	1.00	0.98	1.00	0.99	1.00	0.55	1.00	1.00	1.00

Panel VII.C.1.9

Double Auction Market Simulation Results   File: DAMM2840.SIM

Number of simulation rounds	: 10000	Periods per round :	5
Market type	: M2	Number of traders :	8
Market rules	: AltDAM	Technical price range : [1 ... 9999]	
No-Crossing rule	: YES	Ask-Bid-Spread-Reduction rule :	YES
Trader type	: ZI		

Market clearing quantity q*   :   7
Market clearing price   p*   :   836

	Average of Periods	Period 1	Period 2	Period 3	Period 4	Period 5
Deviation of traded quantity from q*	0.25	0.25	0.25	0.25	0.26	0.25
Efficiency	0.97	0.97	0.97	0.97	0.97	0.97
... median value in the simulation	0.98	0.99	0.99	0.99	0.99	0.99
Average contract price (deviation from p*)	10.93	10.91	10.97	10.96	10.92	10.90
... median value in the simulation	11.63	11.63	11.72	11.72	11.58	11.63
Minimum contract price (deviation from p*)	-8.64	-8.61	-8.62	-8.60	-8.75	-8.64
... median value in the simulation	-6.99	-6.99	-6.99	-6.99	-6.99	-6.99
Maximum contract price (deviation from p*)	31.87	31.87	31.87	31.87	31.92	31.82
... median value in the simulation	34.01	34.01	34.01	34.01	34.01	34.01
Market clearing price concordance coefficient	0.39	0.39	0.39	0.39	0.39	0.39
... median value in the simulation	0.40	0.38	0.38	0.38	0.38	0.39
Rel. freq. of falling price paths	0.77	0.76	0.77	0.77	0.77	0.77
Rel. freq. of rising price paths	0.02	0.02	0.02	0.02	0.02	0.02
Rel. freq. of hanging price paths	0.02	0.02	0.01	0.02	0.02	0.02
Rel. freq. of standing price paths	0.20	0.20	0.20	0.19	0.20	0.20
Rel. freq. of unknown price paths	0.00	0.00	0.00	0.00	0.00	0.00
Rel. freq. of contract proposals by buyers	0.67	0.67	0.68	0.67	0.67	0.67
... lower quartile bound in the simulation	0.62	0.58	0.58	0.58	0.58	0.58
... median value in the simulation	0.68	0.72	0.72	0.72	0.72	0.72
... upper quartile bound in the simulation	0.75	0.86	0.86	0.86	0.86	0.86
... rel. freq. of these with price < p*	0.14	0.15	0.14	0.14	0.14	0.14
... rel. freq. of these with price = p*	0.11	0.11	0.11	0.11	0.11	0.11
... rel. freq. of these with price > p*	0.75	0.75	0.75	0.75	0.75	0.75
... avg. buyer proposed contract price - p*	12.93	12.81	12.98	13.00	12.95	12.92
Rel. freq. of contract proposals by sellers	0.33	0.33	0.32	0.33	0.33	0.33
... lower quartile bound in the simulation	0.26	0.15	0.15	0.15	0.15	0.15
... median value in the simulation	0.33	0.29	0.29	0.29	0.29	0.29
... upper quartile bound in the simulation	0.39	0.43	0.43	0.43	0.43	0.43
... rel. freq. of these with price < p*	0.32	0.32	0.31	0.31	0.32	0.31
... rel. freq. of these with price = p*	0.06	0.06	0.06	0.06	0.05	0.06
... rel. freq. of these with price > p*	0.63	0.63	0.62	0.63	0.63	0.63
... avg. seller proposed contract price - p*	7.23	7.37	7.14	7.21	7.26	7.17
Rel. freq. of proposer's curses - all traders	0.61	0.61	0.62	0.62	0.61	0.61
... median value in the simulation	0.62	0.63	0.63	0.63	0.63	0.63
Rel. freq. of proposer's curses - all buyers	0.75	0.75	0.75	0.75	0.75	0.75
Rel. freq. of proposer's curses - all sellers	0.32	0.32	0.31	0.31	0.32	0.31
Rel. freq. of proposer's curses - buyer 1	0.93	0.92	0.93	0.93	0.93	0.93
Rel. freq. of proposer's curses - buyer 2	0.89	0.89	0.90	0.90	0.89	0.89
Rel. freq. of proposer's curses - buyer 3	0.51	0.51	0.51	0.51	0.50	0.51
Rel. freq. of proposer's curses - buyer 4	0.77	0.77	0.77	0.77	0.76	0.77
Rel. freq. of proposer's curses - seller 5	0.46	0.46	0.46	0.45	0.46	0.46
Rel. freq. of proposer's curses - seller 6	0.37	0.38	0.37	0.38	0.38	0.37
Rel. freq. of proposer's curses - seller 7	0.21	0.22	0.22	0.21	0.22	0.22
Rel. freq. of proposer's curses - seller 8	0.23	0.22	0.24	0.24	0.23	0.23

Panel VII.C.1.9 (continued)

	Average of Periods	Period 1	Period 2	Period 3	Period 4	Period 5
Number of market ask and market bid matches	6.96	6.96	6.96	6.96	6.97	6.95
Number of contracts per market ask-bid match	1.04	1.04	1.04	1.04	1.04	1.04
Avg. rank of traded red.value at trading time	1.34	1.34	1.34	1.34	1.34	1.34
... median value in the simulation	1.34	1.29	1.29	1.29	1.29	1.29
Avg. rank of traded unit cost at trading time	1.99	1.99	1.99	1.99	1.99	1.99
... median value in the simulation	2.00	2.01	2.01	2.01	2.01	2.01
Number of submitted bids per contract	15.19	15.20	15.19	15.19	15.20	15.19
... median value in the simulation	13.01	13.01	13.01	13.01	13.01	13.01
... rel. freq. of bids < p* - 200	0.25	0.25	0.25	0.25	0.25	0.25
... rel. freq. of bids < p*	0.79	0.79	0.79	0.79	0.79	0.79
... rel. freq. of bids = p*	0.02	0.02	0.02	0.02	0.02	0.02
... rel. freq. of bids > p*	0.19	0.19	0.19	0.19	0.19	0.19
Number of submitted asks per contract	19.61	19.62	19.61	19.60	19.63	19.58
... median value in the simulation	13.01	13.01	13.01	13.01	13.01	13.01
... rel. freq. of asks < p*	0.03	0.03	0.03	0.03	0.03	0.03
... rel. freq. of asks = p*	0.01	0.01	0.01	0.01	0.01	0.01
... rel. freq. of asks > p*	0.96	0.96	0.96	0.96	0.96	0.96
... rel. freq. of asks > p* + 200	0.45	0.45	0.45	0.45	0.45	0.45
Number of offers for all units in periods > 1	252	·	252	252	253	252
(a) rel. freq. of bids > p-max & asks < p-min	0.01	·	0.01	0.01	0.01	0.01
(b) rel. freq. of bids > p-avg & asks < p-avg	0.11	·	0.11	0.11	0.11	0.11
(c) rel. freq. of bids > p-own & asks < p-own	0.09	·	0.09	0.09	0.09	0.09
(d) rel. freq. of bids > p-own & asks > p-own	0.87	·	0.87	0.87	0.87	0.87
Number of offers for units bought above p-avg or sold below p-avg in the previous period	79	·	79	79	79	79
(C) rel. freq. of bids > p-own & asks < p-own	0.06	·	0.06	0.06	0.06	0.06
Number of offers for units that were not traded in the previous period	79	·	80	79	79	79
(D) rel. freq. of bids < p-own & asks > p-own	0.89	·	0.89	0.89	0.89	0.89
Aggressive offer frequency ratio (c)/(C)	0.68	·	0.68	0.68	0.68	0.68
... median value in the simulation	0.67	·	0.63	0.63	0.63	0.62
Moderate offer frequency ratio (d)/(D)	1.02	·	1.02	1.02	1.02	1.02
... median value in the simulation	1.03	·	1.03	1.03	1.03	1.03

Relative frequencies f(x) and cumulative distributions F(x) of the simulation rounds
(5 subsequent periods) in which any n periods have price paths of the same type.

	falling		rising		hanging		standing		unknown	
n	f(x)	F(x)	f(x)	F(x)	f(x)	F(x)	f(x)	F(x)	f(x)	F(x)
5	0.26	0.26	0.00	0.00	0.00	0.00	0.00	0.00	0.00	0.00
4	0.41	0.67	0.00	0.00	0.00	0.00	0.01	0.01	0.00	0.00
3	0.25	0.92	0.00	0.00	0.00	0.00	0.05	0.05	0.00	0.00
2	0.07	0.99	0.00	0.00	0.00	0.00	0.20	0.25	0.00	0.00
1	0.01	1.00	0.10	0.11	0.07	0.08	0.42	0.67	0.00	0.00
0	0.00	1.00	0.89	1.00	0.92	1.00	0.33	1.00	1.00	1.00

Panel VII.C.1.10

Double Auction Market Simulation Results   File: DAMM2A10.SIM

Number of simulation rounds	: 10000		Periods per round :		5
Market type	: M2		Number of traders :		10
Market rules	: AltDAM		Technical price range : [	1 ...	9999]
No-Crossing rule	: YES		Ask-Bid-Spread-Reduction rule :		YES
Trader type	: ZI				

Market clearing quantity q*   :   9
Market clearing price   p*   : 342

	Average of Periods	Period 1	Period 2	Period 3	Period 4	Period 5
Deviation of traded quantity from q*	0.53	0.54	0.54	0.53	0.54	0.53
Efficiency	0.97	0.97	0.97	0.97	0.97	0.97
... median value in the simulation	0.98	0.98	0.98	0.98	0.98	0.98
Average contract price (deviation from p*)	8.78	8.77	8.78	8.80	8.78	8.76
... median value in the simulation	9.12	9.12	9.12	9.21	9.12	9.12
Minimum contract price (deviation from p*)	-4.44	-4.43	-4.46	-4.38	-4.43	-4.49
... median value in the simulation	-2.99	-2.99	-2.99	-2.99	-2.99	-2.99
Maximum contract price (deviation from p*)	22.50	22.49	22.49	22.53	22.48	22.49
... median value in the simulation	24.01	24.01	24.01	24.01	24.01	24.01
Market clearing price concordance coefficient	0.35	0.35	0.35	0.35	0.35	0.35
... median value in the simulation	0.35	0.35	0.34	0.34	0.34	0.34
Rel. freq. of falling price paths	0.86	0.86	0.86	0.85	0.86	0.86
Rel. freq. of rising price paths	0.00	0.00	0.00	0.00	0.00	0.00
Rel. freq. of hanging price paths	0.00	0.00	0.00	0.00	0.00	0.00
Rel. freq. of standing price paths	0.14	0.14	0.13	0.14	0.13	0.14
Rel. freq. of unknown price paths	0.00	0.00	0.00	0.00	0.00	0.00
Rel. freq. of contract proposals by buyers	0.76	0.76	0.75	0.76	0.76	0.75
... lower quartile bound in the simulation	0.71	0.67	0.67	0.67	0.67	0.67
... median value in the simulation	0.77	0.78	0.78	0.78	0.78	0.78
... upper quartile bound in the simulation	0.81	0.89	0.89	0.89	0.89	0.89
... rel. freq. of these with price < p*	0.11	0.11	0.11	0.11	0.11	0.11
... rel. freq. of these with price = p*	0.10	0.10	0.10	0.10	0.10	0.10
... rel. freq. of these with price > p*	0.79	0.79	0.79	0.79	0.79	0.79
... avg. buyer proposed contract price - p*	9.42	9.39	9.42	9.44	9.44	9.41
Rel. freq. of contract proposals by sellers	0.24	0.24	0.25	0.24	0.24	0.25
... lower quartile bound in the simulation	0.20	0.12	0.12	0.12	0.12	0.12
... median value in the simulation	0.25	0.23	0.23	0.23	0.23	0.23
... upper quartile bound in the simulation	0.30	0.34	0.34	0.34	0.34	0.34
... rel. freq. of these with price < p*	0.22	0.22	0.22	0.21	0.22	0.22
... rel. freq. of these with price = p*	0.05	0.06	0.05	0.06	0.05	0.05
... rel. freq. of these with price > p*	0.73	0.73	0.73	0.73	0.73	0.73
... avg. seller proposed contract price - p*	7.05	7.06	7.15	7.06	6.98	7.00
Rel. freq. of proposer's curses - all traders	0.66	0.66	0.66	0.66	0.66	0.66
... median value in the simulation	0.67	0.67	0.67	0.67	0.67	0.67
Rel. freq. of proposer's curses - all buyers	0.79	0.79	0.79	0.79	0.79	0.79
Rel. freq. of proposer's curses - all sellers	0.22	0.22	0.22	0.21	0.22	0.22
Rel. freq. of proposer's curses - buyer 1	0.95	0.95	0.95	0.96	0.96	0.95
Rel. freq. of proposer's curses - buyer 2	0.96	0.95	0.95	0.96	0.96	0.96
Rel. freq. of proposer's curses - buyer 3	0.93	0.93	0.94	0.93	0.93	0.93
Rel. freq. of proposer's curses - buyer 4	0.56	0.56	0.55	0.55	0.55	0.55
Rel. freq. of proposer's curses - buyer 5	0.73	0.73	0.73	0.73	0.73	0.73
Rel. freq. of proposer's curses - seller 6	0.33	0.33	0.34	0.32	0.34	0.34
Rel. freq. of proposer's curses - seller 7	0.29	0.30	0.31	0.29	0.29	0.29
Rel. freq. of proposer's curses - seller 8	0.23	0.22	0.22	0.23	0.23	0.23
Rel. freq. of proposer's curses - seller 9	0.14	0.14	0.14	0.14	0.15	0.14
Rel. freq. of proposer's curses - seller 10	0.14	0.14	0.14	0.13	0.15	0.14

Panel VII.C.1.10 (continued)

	Average of Periods	Period 1	Period 2	Period 3	Period 4	Period 5
Number of market ask and market bid matches	9.14	9.14	9.14	9.14	9.14	9.14
Number of contracts per market ask-bid match	1.04	1.04	1.04	1.04	1.04	1.04
Avg. rank of traded red.value at trading time	1.31	1.31	1.31	1.31	1.32	1.31
... median value in the simulation	1.31	1.28	1.28	1.28	1.28	1.28
Avg. rank of traded unit cost at trading time	2.41	2.41	2.40	2.41	2.41	2.41
... median value in the simulation	2.42	2.44	2.43	2.44	2.45	2.44
Number of submitted bids per contract	15.72	15.73	15.73	15.72	15.72	15.71
... median value in the simulation	13.01	13.01	13.01	13.01	13.01	13.01
... rel. freq. of bids < p^* - 200	0.14	0.14	0.14	0.14	0.14	0.14
... rel. freq. of bids < p^*	0.75	0.75	0.75	0.75	0.75	0.75
... rel. freq. of bids = p^*	0.02	0.02	0.02	0.02	0.02	0.02
... rel. freq. of bids > p^*	0.23	0.23	0.23	0.23	0.23	0.23
Number of submitted asks per contract	22.21	22.23	22.21	22.24	22.20	22.19
... median value in the simulation	13.01	13.01	13.01	13.01	13.01	13.01
... rel. freq. of asks < p^*	0.02	0.02	0.02	0.02	0.02	0.02
... rel. freq. of asks = p^*	0.01	0.01	0.01	0.01	0.01	0.01
... rel. freq. of asks > p^*	0.98	0.98	0.98	0.98	0.98	0.98
... rel. freq. of asks > p^* + 200	0.44	0.44	0.44	0.44	0.44	0.44
Number of offers for all units in periods > 1	361	-	362	362	361	361
(a) rel. freq. of bids > p-max & asks < p-min	0.01	-	0.01	0.01	0.01	0.01
(b) rel. freq. of bids > p-avg & asks < p-avg	0.11	-	0.11	0.11	0.11	0.11
(c) rel. freq. of bids > p-own & asks < p-own	0.07	-	0.07	0.07	0.07	0.07
(d) rel. freq. of bids > p-own & asks > p-own	0.88	-	0.88	0.88	0.88	0.88
Number of offers for units bought above p-avg or sold below p-avg in the previous period	113	-	113	114	113	113
(C) rel. freq. of bids > p-own & asks < p-own	0.05	-	0.05	0.05	0.05	0.05
Number of offers for units that were not traded in the previous period	105	-	105	105	105	105
(D) rel. freq. of bids < p-own & asks > p-own	0.89	-	0.89	0.89	0.89	0.89
Aggressive offer frequency ratio (c)/(C)	0.65	-	0.65	0.64	0.64	0.65
... median value in the simulation	0.64	-	0.61	0.59	0.60	0.61
Moderate offer frequency ratio (d)/(D)	1.02	-	1.01	1.02	1.02	1.02
... median value in the simulation	1.02	-	1.02	1.02	1.02	1.02

Relative frequencies f(x) and cumulative distributions F(x) of the simulation rounds (5 subsequent periods) in which any n periods have price paths of the same type.

	falling		rising		hanging		standing		unknown	
n	f(x)	F(x)	f(x)	F(x)	f(x)	F(x)	f(x)	F(x)	f(x)	F(x)
5	0.47	0.47	0.00	0.00	0.00	0.00	0.00	0.00	0.00	0.00
4	0.38	0.85	0.00	0.00	0.00	0.00	0.00	0.00	0.00	0.00
3	0.13	0.98	0.00	0.00	0.00	0.00	0.02	0.02	0.00	0.00
2	0.02	1.00	0.00	0.00	0.00	0.00	0.12	0.14	0.00	0.00
1	0.00	1.00	0.01	0.01	0.01	0.01	0.38	0.52	0.00	0.00
0	0.00	1.00	0.99	1.00	0.99	1.00	0.48	1.00	1.00	1.00

Panel VII.C.1.11

Double Auction Market Simulation Results   File: DAMM1811.SIM

Number of simulation rounds	: 10000	Periods per round :			5
Market type	: M1	Number of traders :			8
Market rules	: AltDAM	Technical price range : [1 ... 9999]			
No-Crossing rule	: YES	Ask-Bid-Spread-Reduction rule :			YES
Trader type	: ZI				

Market clearing quantity q*   :   10
Market clearing price   p*   :   829

	Average of Periods	Period 1	Period 2	Period 3	Period 4	Period 5
Deviation of traded quantity from q*	0.33	0.32	0.32	0.32	0.33	0.33
Efficiency	0.99	0.99	0.99	0.99	0.99	0.99
... median value in the simulation	1.00	0.99	0.99	0.99	0.99	0.99
Average contract price (deviation from p*)	4.80	4.81	4.77	4.80	4.77	4.84
... median value in the simulation	5.37	5.26	5.28	5.28	5.28	5.37
Minimum contract price (deviation from p*)	-10.91	-10.88	-10.99	-10.98	-10.92	-10.78
... median value in the simulation	-8.99	-9.99	-9.99	-9.99	-8.99	-8.99
Maximum contract price (deviation from p*)	16.74	16.72	16.76	16.76	16.73	16.77
... median value in the simulation	18.01	18.01	18.01	18.01	18.01	18.01
Market clearing price concordance coefficient	0.53	0.53	0.53	0.53	0.53	0.53
... median value in the simulation	0.54	0.53	0.53	0.53	0.53	0.53
Rel. freq. of falling price paths	0.72	0.71	0.72	0.72	0.72	0.72
Rel. freq. of rising price paths	0.04	0.04	0.04	0.04	0.05	0.04
Rel. freq. of hanging price paths	0.03	0.03	0.04	0.03	0.03	0.03
Rel. freq. of standing price paths	0.21	0.21	0.20	0.21	0.20	0.21
Rel. freq. of unknown price paths	0.00	0.00	0.00	0.00	0.00	0.00
Rel. freq. of contract proposals by buyers	0.73	0.73	0.73	0.73	0.73	0.73
... lower quartile bound in the simulation	0.69	0.61	0.61	0.61	0.61	0.61
... median value in the simulation	0.74	0.73	0.73	0.73	0.73	0.73
... upper quartile bound in the simulation	0.79	0.91	0.91	0.91	0.91	0.91
... rel. freq. of these with price < p*	0.13	0.13	0.13	0.13	0.13	0.13
... rel. freq. of these with price = p*	0.10	0.10	0.10	0.10	0.10	0.10
... rel. freq. of these with price > p*	0.77	0.77	0.77	0.78	0.77	0.77
... avg. buyer proposed contract price - p*	6.17	6.18	6.14	6.19	6.11	6.23
Rel. freq. of contract proposals by sellers	0.27	0.27	0.27	0.27	0.27	0.27
... lower quartile bound in the simulation	0.22	0.11	0.11	0.11	0.11	0.11
... median value in the simulation	0.27	0.28	0.28	0.28	0.28	0.28
... upper quartile bound in the simulation	0.32	0.40	0.40	0.40	0.40	0.40
... rel. freq. of these with price < p*	0.36	0.35	0.36	0.36	0.35	0.35
... rel. freq. of these with price > p*	0.06	0.06	0.06	0.06	0.06	0.06
... rel. freq. of these with price > p*	0.58	0.58	0.58	0.58	0.58	0.59
... avg. seller proposed contract price - p*	1.27	1.32	1.29	1.17	1.30	1.28
Rel. freq. of proposer's curses - all traders	0.67	0.67	0.67	0.67	0.67	0.67
... median value in the simulation	0.68	0.71	0.71	0.71	0.71	0.71
Rel. freq. of proposer's curses - all buyers	0.77	0.77	0.77	0.78	0.77	0.77
Rel. freq. of proposer's curses - all sellers	0.36	0.35	0.36	0.36	0.35	0.35
Rel. freq. of proposer's curses - buyer 1	0.91	0.91	0.91	0.91	0.91	0.91
Rel. freq. of proposer's curses - buyer 2	0.66	0.66	0.66	0.66	0.65	0.66
Rel. freq. of proposer's curses - buyer 3	0.81	0.82	0.82	0.82	0.81	0.82
Rel. freq. of proposer's curses - buyer 4	0.88	0.88	0.88	0.88	0.88	0.88
Rel. freq. of proposer's curses - seller 5	0.47	0.48	0.47	0.47	0.47	0.47
Rel. freq. of proposer's curses - seller 6	0.32	0.32	0.32	0.33	0.32	0.32
Rel. freq. of proposer's curses - seller 7	0.34	0.33	0.36	0.34	0.35	0.33
Rel. freq. of proposer's curses - seller 8	0.33	0.33	0.33	0.33	0.32	0.33

Panel VII.C.1.11 (continued)

	Average of Periods	Period 1	Period 2	Period 3	Period 4	Period 5
Number of market ask and market bid matches	9.84	9.85	9.83	9.84	9.85	9.85
Number of contracts per market ask-bid match	1.05	1.05	1.05	1.05	1.05	1.05
Avg. rank of traded red.value at trading time	1.40	1.40	1.40	1.40	1.40	1.40
... median value in the simulation	1.40	1.39	1.39	1.38	1.39	1.38
Avg. rank of traded unit cost at trading time	2.00	2.00	2.00	2.00	2.00	2.00
... median value in the simulation	2.01	2.01	2.01	2.01	2.01	2.01
Number of submitted bids per contract	14.06	14.06	14.05	14.08	14.04	14.06
... median value in the simulation	13.01	13.01	13.01	13.01	13.01	13.01
... rel. freq. of bids < p* - 200	0.24	0.24	0.24	0.24	0.23	0.24
... rel. freq. of bids < p*	0.77	0.77	0.77	0.77	0.77	0.77
... rel. freq. of bids = p*	0.02	0.02	0.02	0.02	0.02	0.02
... rel. freq. of bids > p*	0.21	0.21	0.21	0.21	0.21	0.21
Number of submitted asks per contract	17.16	17.21	17.16	17.17	17.10	17.18
... median value in the simulation	13.01	13.01	13.01	13.01	13.01	13.01
... rel. freq. of asks < p*	0.03	0.03	0.03	0.03	0.03	0.03
... rel. freq. of asks = p*	0.01	0.01	0.01	0.01	0.01	0.01
... rel. freq. of asks > p*	0.96	0.96	0.96	0.96	0.96	0.96
... rel. freq. of asks > p* + 200	0.43	0.43	0.43	0.43	0.43	0.43
Number of offers for all units in periods > 1	322	-	322	322	321	322
(a) rel. freq. of bids > p-max & asks < p-min	0.01	-	0.01	0.01	0.01	0.01
(b) rel. freq. of bids > p-avg & asks < p-avg	0.10	-	0.10	0.10	0.10	0.10
(c) rel. freq. of bids > p-own & asks < p-own	0.07	-	0.07	0.07	0.07	0.07
(d) rel. freq. of bids < p-own & asks > p-own	0.88	-	0.89	0.88	0.88	0.88
Number of offers for units bought above p-avg or sold below p-avg in the previous period	120	-	121	120	120	121
(C) rel. freq. of bids > p-own & asks < p-own	0.04	-	0.04	0.04	0.04	0.04
Number of offers for units that were not traded in the previous period	52	-	52	52	52	52
(D) rel. freq. of bids < p-own & asks > p-own	0.91	-	0.91	0.91	0.91	0.91
Aggressive offer frequency ratio (c)/(C)	0.57	-	0.57	0.57	0.57	0.57
... median value in the simulation	0.56	-	0.52	0.52	0.52	0.52
Moderate offer frequency ratio (d)/(D)	1.03	-	1.02	1.03	1.03	1.03
... median value in the simulation	1.03	-	1.03	1.03	1.03	1.03

Relative frequencies f(x) and cumulative distributions F(x) of the simulation rounds (5 subsequent periods) in which any n periods have price paths of the same type.

	falling		rising		hanging		standing		unknown	
n	f(x)	F(x)	f(x)	F(x)	f(x)	F(x)	f(x)	F(x)	f(x)	F(x)
5	0.18	0.18	0.00	0.00	0.00	0.00	0.00	0.00	0.00	0.00
4	0.38	0.56	0.00	0.00	0.00	0.00	0.01	0.01	0:00	0.00
3	0.30	0.86	0.00	0.00	0.00	0.00	0.06	0.06	0.00	0.00
2	0.12	0.98	0.02	0.02	0.01	0.01	0.22	0.28	0.00	0.00
1	0.02	1.00	0.19	0.20	0.14	0.15	0.40	0.68	0.00	0.00
0	0.00	1.00	0.80	1.00	0.85	1.00	0.32	1.00	1.00	1.00

Panel VII.C.1.12

Double Auction Market Simulation Results   File: DAMM1821.SIM

Number of simulation rounds	: 10000		Periods per round :			5
Market type	: M1		Number of traders :			8
Market rules	: AltDAM		Technical price range : [1 ... 9999]			
No-Crossing rule	: YES		Ask-Bid-Spread-Reduction rule :			YES
Trader type	: ZI					

Market clearing quantity q*   :   10
Market clearing price   p*   :   740

	Average of Periods	Period 1	Period 2	Period 3	Period 4	Period 5
Deviation of traded quantity from q*	0.33	0.32	0.33	0.33	0.32	0.34
Efficiency	0.99	0.99	0.99	0.99	0.99	0.99
... median value in the simulation	1.00	0.99	0.99	0.99	0.99	0.99
Average contract price (deviation from p*)	5.06	5.07	5.06	5.09	5.06	5.04
... median value in the simulation	5.55	5.60	5.60	5.60	5.60	5.55
Minimum contract price (deviation from p*)	-10.25	-10.20	-10.30	-10.21	-10.20	-10.36
... median value in the simulation	-8.99	-7.99	-7.99	-7.99	-7.99	-8.99
Maximum contract price (deviation from p*)	16.81	16.79	16.84	16.80	16.81	16.81
... median value in the simulation	18.01	18.01	18.01	18.01	18.01	18.01
Market clearing price concordance coefficient	0.53	0.53	0.53	0.53	0.53	0.53
... median value in the simulation	0.54	0.53	0.54	0.53	0.53	0.53
Rel. freq. of falling price paths	0.74	0.74	0.73	0.74	0.74	0.73
Rel. freq. of rising price paths	0.04	0.04	0.04	0.04	0.04	0.04
Rel. freq. of hanging price paths	0.03	0.03	0.03	0.02	0.03	0.03
Rel. freq. of standing price paths	0.20	0.20	0.21	0.20	0.19	0.21
Rel. freq. of unknown price paths	0.00	0.00	0.00	0.00	0.00	0.00
Rel. freq. of contract proposals by buyers	0.74	0.74	0.74	0.74	0.74	0.74
... lower quartile bound in the simulation	0.70	0.61	0.64	0.64	0.64	0.61
... median value in the simulation	0.75	0.81	0.81	0.81	0.81	0.81
... upper quartile bound in the simulation	0.80	0.91	0.91	0.91	0.91	0.91
... rel. freq. of these with price < p*	0.12	0.12	0.12	0.12	0.12	0.12
... rel. freq. of these with price = p*	0.10	0.10	0.10	0.10	0.10	0.10
... rel. freq. of these with price > p*	0.78	0.78	0.78	0.78	0.78	0.78
... avg. buyer proposed contract price - p*	6.35	6.38	6.31	6.36	6.33	6.36
Rel. freq. of contract proposals by sellers	0.26	0.26	0.26	0.26	0.26	0.26
... lower quartile bound in the simulation	0.21	0.11	0.11	0.11	0.11	0.11
... median value in the simulation	0.26	0.20	0.20	0.20	0.20	0.20
... upper quartile bound in the simulation	0.31	0.40	0.37	0.37	0.37	0.40
... rel. freq. of these with price < p*	0.34	0.34	0.34	0.34	0.34	0.34
... rel. freq. of these with price = p*	0.06	0.06	0.06	0.06	0.06	0.06
... rel. freq. of these with price > p*	0.60	0.60	0.60	0.61	0.60	0.60
... avg. seller proposed contract price - p*	1.64	1.64	1.65	1.78	1.59	1.55
Rel. freq. of proposer's curses - all traders	0.68	0.68	0.67	0.68	0.68	0.67
... median value in the simulation	0.68	0.71	0.71	0.71	0.71	0.71
Rel. freq. of proposer's curses - all buyers	0.78	0.78	0.78	0.78	0.78	0.78
Rel. freq. of proposer's curses - all sellers	0.34	0.34	0.34	0.34	0.34	0.34
Rel. freq. of proposer's curses - buyer 1	0.92	0.92	0.91	0.92	0.92	0.92
Rel. freq. of proposer's curses - buyer 2	0.66	0.67	0.66	0.66	0.66	0.66
Rel. freq. of proposer's curses - buyer 3	0.82	0.82	0.82	0.83	0.82	0.82
Rel. freq. of proposer's curses - buyer 4	0.89	0.89	0.89	0.89	0.89	0.89
Rel. freq. of proposer's curses - seller 5	0.46	0.45	0.46	0.45	0.45	0.47
Rel. freq. of proposer's curses - seller 6	0.31	0.32	0.31	0.31	0.32	0.31
Rel. freq. of proposer's curses - seller 7	0.33	0.34	0.32	0.33	0.34	0.33
Rel. freq. of proposer's curses - seller 8	0.31	0.32	0.30	0.31	0.32	0.31

Panel VII.C.1.12 (continued)

	Average of Periods	Period 1	Period 2	Period 3	Period 4	Period 5
Number of market ask and market bid matches	9.87	9.87	9.86	9.87	9.86	9.87
Number of contracts per market ask-bid match	1.05	1.05	1.05	1.05	1.05	1.05
Avg. rank of traded red.value at trading time	1.38	1.38	1.38	1.38	1.38	1.38
... median value in the simulation	1.38	1.37	1.37	1.37	1.37	1.37
Avg. rank of traded unit cost at trading time	2.00	2.00	2.00	2.01	2.00	2.00
... median value in the simulation	2.01	2.01	2.01	2.01	2.01	2.01
Number of submitted bids per contract	13.95	13.97	13.93	13.94	13.95	13.93
... median value in the simulation	13.01	13.01	13.01	13.01	13.01	13.01
... rel. freq. of bids < p* - 200	0.22	0.22	0.22	0.22	0.22	0.22
... rel. freq. of bids < p*	0.76	0.76	0.76	0.76	0.76	0.76
... rel. freq. of bids = p*	0.02	0.02	0.02	0.02	0.02	0.02
... rel. freq. of bids > p*	0.21	0.21	0.21	0.22	0.21	0.21
Number of submitted asks per contract	17.25	17.27	17.24	17.26	17.24	17.24
... median value in the simulation	13.01	13.01	13.01	13.01	13.01	13.01
... rel. freq. of asks < p*	0.03	0.03	0.03	0.03	0.03	0.03
... rel. freq. of asks = p*	0.01	0.01	0.01	0.01	0.01	0.01
... rel. freq. of asks > p*	0.96	0.96	0.96	0.96	0.96	0.96
... rel. freq. of asks > p* + 200	0.43	0.43	0.43	0.43	0.43	0.43
Number of offers for all units in periods > 1	322	·	321	322	322	322
(a) rel. freq. of bids > p-max & asks < p-min	0.01	·	0.01	0.01	0.01	0.01
(b) rel. freq. of bids > p-avg & asks < p-avg	0.11	·	0.11	0.10	0.11	0.11
(c) rel. freq. of bids > p-own & asks < p-own	0.07	·	0.07	0.07	0.07	0.07
(d) rel. freq. of bids < p-own & asks > p-own	0.88	·	0.88	0.88	0.88	0.88
Number of offers for units bought above p-avg or sold below p-avg in the previous period	120	·	120	120	121	120
(C) rel. freq. of bids > p-own & asks < p-own	0.04	·	0.04	0.04	0.04	0.04
Number of offers for units that were not traded in the previous period	52	·	52	52	51	52
(D) rel. freq. of bids < p-own & asks > p-own	0.91	·	0.91	0.91	0.91	0.91
Aggressive offer frequency ratio (c)/(C)	0.58	·	0.58	0.58	0.58	0.58
... median value in the simulation	0.57	·	0.52	0.53	0.53	0.53
Moderate offer frequency ratio (d)/(D)	1.03	·	1.03	1.02	1.03	1.03
... median value in the simulation	1.03	·	1.03	1.03	1.03	1.03

Relative frequencies f(x) and cumulative distributions F(x) of the simulation rounds (5 subsequent periods) in which any n periods have price paths of the same type.

	falling		rising		hanging		standing		unknown	
n	f(x)	F(x)	f(x)	F(x)	f(x)	F(x)	f(x)	F(x)	f(x)	F(x)
5	0.22	0.22	0.00	0.00	0.00	0.00	0.00	0.00	0.00	0.00
4	0.39	0.61	0.00	0.00	0.00	0.00	0.01	0.01	0.00	0.00
3	0.27	0.88	0.00	0.00	0.00	0.00	0.05	0.06	0.00	0.00
2	0.10	0.98	0.01	0.01	0.01	0.01	0.20	0.26	0.00	0.00
1	0.02	1.00	0.16	0.18	0.11	0.12	0.41	0.67	0.00	0.00
0	0.00	1.00	0.82	1.00	0.88	1.00	0.33	1.00	1.00	1.00

Panel VII.C.1.13

Double Auction Market Simulation Results   File: DAMM1831.SIM

Number of simulation rounds	: 10000	Periods per round :	5
Market type	: M1	Number of traders :	8
Market rules	: AltDAM	Technical price range : [1 ... 9999]	
No-Crossing rule	: YES	Ask-Bid-Spread-Reduction rule :	YES
Trader type	: ZI		

Market clearing quantity q*	:	10
Market clearing price p*	:	436

	Average of Periods	Period 1	Period 2	Period 3	Period 4	Period 5
Deviation of traded quantity from q*	0.34	0.34	0.34	0.33	0.33	0.34
Efficiency	0.99	0.99	0.99	0.99	0.99	0.99
... median value in the simulation	1.00	0.99	0.99	0.99	0.99	0.99
Average contract price (deviation from p*)	6.18	6.18	6.19	6.18	6.16	6.17
... median value in the simulation	6.80	6.80	6.80	6.80	6.80	6.80
Minimum contract price (deviation from p*)	-7.26	-7.26	-7.25	-7.32	-7.29	-7.20
... median value in the simulation	-2.99	-2.99	-2.99	-2.99	-2.99	-2.99
Maximum contract price (deviation from p*)	17.09	17.11	17.11	17.09	17.08	17.06
... median value in the simulation	18.01	18.01	18.01	18.01	18.01	18.01
Market clearing price concordance coefficient	0.53	0.53	0.53	0.53	0.53	0.53
... median value in the simulation	0.54	0.53	0.53	0.53	0.53	0.53
Rel. freq. of falling price paths	0.81	0.81	0.81	0.81	0.81	0.81
Rel. freq. of rising price paths	0.01	0.01	0.01	0.01	0.02	0.01
Rel. freq. of hanging price paths	0.01	0.01	0.01	0.01	0.01	0.01
Rel. freq. of standing price paths	0.17	0.16	0.17	0.17	0.17	0.17
Rel. freq. of unknown price paths	0.00	0.00	0.00	0.00	0.00	0.00
Rel. freq. of contract proposals by buyers	0.79	0.79	0.79	0.79	0.79	0.79
... lower quartile bound in the simulation	0.75	0.71	0.71	0.71	0.71	0.71
... median value in the simulation	0.80	0.81	0.81	0.81	0.81	0.81
... upper quartile bound in the simulation	0.85	0.91	0.91	0.91	0.91	0.91
... rel. freq. of these with price < p*	0.09	0.09	0.09	0.09	0.09	0.09
... rel. freq. of these with price = p*	0.10	0.10	0.10	0.10	0.10	0.10
... rel. freq. of these with price > p*	0.81	0.81	0.81	0.81	0.81	0.81
... avg. buyer proposed contract price - p*	6.99	7.00	7.00	6.99	6.98	6.97
Rel. freq. of contract proposals by sellers	0.21	0.21	0.21	0.21	0.21	0.21
... lower quartile bound in the simulation	0.16	0.11	0.11	0.11	0.11	0.11
... median value in the simulation	0.21	0.20	0.20	0.20	0.20	0.20
... upper quartile bound in the simulation	0.26	0.31	0.31	0.31	0.31	0.31
... rel. freq. of these with price < p*	0.27	0.27	0.27	0.27	0.27	0.27
... rel. freq. of these with price = p*	0.06	0.05	0.06	0.06	0.06	0.06
... rel. freq. of these with price > p*	0.67	0.67	0.67	0.67	0.67	0.67
... avg. seller proposed contract price - p*	3.38	3.39	3.42	3.36	3.33	3.40
Rel. freq. of proposer's curses - all traders	0.70	0.70	0.70	0.70	0.70	0.70
... median value in the simulation	0.71	0.71	0.71	0.71	0.71	0.71
Rel. freq. of proposer's curses - all buyers	0.81	0.81	0.81	0.81	0.81	0.81
Rel. freq. of proposer's curses - all sellers	0.27	0.27	0.27	0.27	0.27	0.27
Rel. freq. of proposer's curses - buyer 1	0.95	0.95	0.95	0.95	0.95	0.95
Rel. freq. of proposer's curses - buyer 2	0.69	0.68	0.69	0.68	0.69	0.69
Rel. freq. of proposer's curses - buyer 3	0.86	0.86	0.85	0.86	0.86	0.85
Rel. freq. of proposer's curses - buyer 4	0.93	0.94	0.93	0.93	0.93	0.93
Rel. freq. of proposer's curses - seller 5	0.38	0.38	0.38	0.38	0.37	0.37
Rel. freq. of proposer's curses - seller 6	0.25	0.26	0.24	0.25	0.25	0.25
Rel. freq. of proposer's curses - seller 7	0.26	0.25	0.25	0.26	0.26	0.26
Rel. freq. of proposer's curses - seller 8	0.25	0.25	0.25	0.25	0.24	0.25

Panel VII.C.1.13 (continued)

	Average of Periods	Period 1	Period 2	Period 3	Period 4	Period 5
Number of market ask and market bid matches	9.94	9.95	9.95	9.94	9.93	9.94
Number of contracts per market ask-bid match	1.04	1.04	1.04	1.04	1.04	1.04
Avg. rank of traded red.value at trading time	1.29	1.29	1.29	1.29	1.29	1.29
... median value in the simulation	1.29	1.28	1.28	1.28	1.28	1.28
Avg. rank of traded unit cost at trading time	2.03	2.03	2.03	2.03	2.03	2.03
... median value in the simulation	2.04	2.06	2.06	2.06	2.06	2.06
Number of submitted bids per contract	13.33	13.34	13.35	13.33	13.34	13.31
... median value in the simulation	13.01	13.01	13.01	13.01	13.01	13.01
... rel. freq. of bids < p* - 200	0.16	0.16	0.16	0.16	0.16	0.16
... rel. freq. of bids < p*	0.72	0.72	0.72	0.72	0.72	0.72
... rel. freq. of bids = p*	0.03	0.03	0.03	0.03	0.03	0.03
... rel. freq. of bids > p*	0.25	0.25	0.25	0.25	0.25	0.25
Number of submitted asks per contract	17.55	17.56	17.56	17.54	17.55	17.55
... median value in the simulation	13.01	13.01	13.01	13.01	13.01	13.01
... rel. freq. of asks < p*	0.02	0.02	0.02	0.02	0.02	0.02
... rel. freq. of asks = p*	0.01	0.01	0.01	0.01	0.01	0.01
... rel. freq. of asks > p*	0.97	0.97	0.97	0.97	0.97	0.97
... rel. freq. of asks > p* + 200	0.43	0.43	0.43	0.43	0.43	0.43
Number of offers for all units in periods > 1	319	·	319	319	319	319
(a) rel. freq. of bids > p-max & asks < p-min	0.01	·	0.01	0.01	0.01	0.01
(b) rel. freq. of bids > p-avg & asks < p-avg	0.11	·	0.11	0.11	0.11	0.11
(c) rel. freq. of bids > p-own & asks < p-own	0.06	·	0.06	0.06	0.06	0.06
(d) rel. freq. of bids < p-own & asks > p-own	0.88	·	0.88	0.88	0.88	0.88
Number of offers for units bought above p-avg or sold below p-avg in the previous period	118	·	118	118	118	118
(C) rel. freq. of bids > p-own & asks < p-own	0.04	·	0.04	0.04	0.04	0.04
Number of offers for units that were not traded in the previous period	50	·	50	49	50	50
(D) rel. freq. of bids < p-own & asks > p-own	0.90	·	0.90	0.90	0.90	0.90
Aggressive offer frequency ratio (c)/(C)	0.59	·	0.59	0.59	0.59	0.59
... median value in the simulation	0.58	·	0.53	0.53	0.54	0.54
Moderate offer frequency ratio (d)/(D)	1.03	·	1.03	1.02	1.02	1.03
... median value in the simulation	1.03	·	1.03	1.03	1.03	1.03

Relative frequencies f(x) and cumulative distributions F(x) of the simulation rounds (5 subsequent periods) in which any n periods have price paths of the same type.

n	falling f(x)	falling F(x)	rising f(x)	rising F(x)	hanging f(x)	hanging F(x)	standing f(x)	standing F(x)	unknown f(x)	unknown F(x)
5	0.34	0.34	0.00	0.00	0.00	0.00	0.00	0.00	0.00	0.00
4	0.41	0.75	0.00	0.00	0.00	0.00	0.00	0.00	0.00	0.00
3	0.19	0.95	0.00	0.00	0.00	0.00	0.03	0.04	0.00	0.00
2	0.05	0.99	0.00	0.00	0.00	0.00	0.17	0.20	0.00	0.00
1	0.01	1.00	0.07	0.07	0.04	0.04	0.40	0.60	0.00	0.00
0	0.00	1.00	0.93	1.00	0.96	1.00	0.40	1.00	1.00	1.00

Panel VII.C.1.14

Double Auction Market Simulation Results   File: DAMM1841.SIM

Number of simulation rounds	: 10000	Periods per round :	5
Market type	: M1	Number of traders :	8
Market rules	: AltDAM	Technical price range : [1 ... 9999]	
No-Crossing rule	: YES	Ask-Bid-Spread-Reduction rule :	YES
Trader type	: ZI		

Market clearing quantity q* : 10
Market clearing price   p* : 628

	Average of Periods	Period 1	Period 2	Period 3	Period 4	Period 5
Deviation of traded quantity from q*	0.33	0.34	0.33	0.33	0.33	0.33
Efficiency	0.99	0.99	0.99	0.99	0.99	0.99
... median value in the simulation	1.00	0.99	0.99	0.99	0.99	0.99
Average contract price (deviation from p*)	5.43	5.43	5.44	5.41	5.42	5.46
... median value in the simulation	6.01	6.01	6.10	5.91	6.01	6.01
Minimum contract price (deviation from p*)	-9.30	-9.36	-9.26	-9.41	-9.20	-9.29
... median value in the simulation	-6.99	-6.99	-5.99	-6.99	-5.99	-6.99
Maximum contract price (deviation from p*)	16.91	16.90	16.90	16.91	16.89	16.93
... median value in the simulation	18.01	18.01	18.01	18.01	18.01	18.01
Market clearing price concordance coefficient	0.53	0.53	0.53	0.53	0.53	0.53
... median value in the simulation	0.54	0.53	0.53	0.53	0.53	0.53
Rel. freq. of falling price paths	0.76	0.75	0.76	0.76	0.76	0.77
Rel. freq. of rising price paths	0.03	0.03	0.03	0.03	0.03	0.02
Rel. freq. of hanging price paths	0.02	0.02	0.02	0.02	0.02	0.02
Rel. freq. of standing price paths	0.19	0.20	0.19	0.19	0.19	0.18
Rel. freq. of unknown price paths	0.00	0.00	0.00	0.00	0.00	0.00
Rel. freq. of contract proposals by buyers	0.76	0.76	0.76	0.76	0.76	0.76
... lower quartile bound in the simulation	0.71	0.64	0.64	0.64	0.64	0.64
... median value in the simulation	0.77	0.81	0.81	0.81	0.81	0.81
... upper quartile bound in the simulation	0.81	0.91	0.91	0.91	0.91	0.91
... rel. freq. of these with price < p*	0.11	0.11	0.11	0.11	0.11	0.11
... rel. freq. of these with price = p*	0.10	0.10	0.10	0.10	0.10	0.10
... rel. freq. of these with price > p*	0.79	0.79	0.79	0.79	0.79	0.79
... avg. buyer proposed contract price - p*	6.55	6.54	6.54	6.57	6.55	6.56
Rel. freq. of contract proposals by sellers	0.24	0.24	0.24	0.24	0.24	0.24
... lower quartile bound in the simulation	0.20	0.11	0.11	0.11	0.11	0.11
... median value in the simulation	0.24	0.20	0.20	0.20	0.20	0.20
... upper quartile bound in the simulation	0.30	0.37	0.37	0.37	0.37	0.37
... rel. freq. of these with price < p*	0.32	0.32	0.32	0.32	0.32	0.32
... rel. freq. of these with price = p*	0.06	0.06	0.06	0.06	0.06	0.06
... rel. freq. of these with price > p*	0.62	0.62	0.63	0.62	0.62	0.62
... avg. seller proposed contract price - p*	2.22	2.21	2.23	2.18	2.23	2.28
Rel. freq. of proposer's curses - all traders	0.68	0.68	0.68	0.68	0.69	0.68
... median value in the simulation	0.69	0.71	0.71	0.71	0.71	0.71
Rel. freq. of proposer's curses - all buyers	0.79	0.79	0.79	0.79	0.79	0.79
Rel. freq. of proposer's curses - all sellers	0.32	0.32	0.32	0.32	0.32	0.32
Rel. freq. of proposer's curses - buyer 1	0.93	0.93	0.93	0.93	0.93	0.93
Rel. freq. of proposer's curses - buyer 2	0.67	0.67	0.67	0.67	0.67	0.67
Rel. freq. of proposer's curses - buyer 3	0.83	0.83	0.83	0.83	0.83	0.84
Rel. freq. of proposer's curses - buyer 4	0.91	0.91	0.91	0.91	0.91	0.91
Rel. freq. of proposer's curses - seller 5	0.43	0.43	0.43	0.44	0.43	0.44
Rel. freq. of proposer's curses - seller 6	0.30	0.30	0.29	0.30	0.29	0.30
Rel. freq. of proposer's curses - seller 7	0.31	0.30	0.30	0.32	0.31	0.30
Rel. freq. of proposer's curses - seller 8	0.29	0.30	0.30	0.28	0.29	0.29

Panel VII.C.1.14 (continued)

	Average of Periods	Period 1	Period 2	Period 3	Period 4	Period 5
Number of market ask and market bid matches	9.89	9.90	9.89	9.90	9.88	9.88
Number of contracts per market ask-bid match	1.04	1.04	1.04	1.04	1.05	1.05
Avg. rank of traded red.value at trading time	1.35	1.36	1.35	1.35	1.35	1.35
... median value in the simulation	1.35	1.31	1.31	1.31	1.31	1.31
Avg. rank of traded unit cost at trading time	2.01	2.01	2.02	2.01	2.01	2.02
... median value in the simulation	2.02	2.01	2.01	2.01	2.01	2.01
Number of submitted bids per contract	13.76	13.76	13.76	13.77	13.77	13.76
... median value in the simulation	13.01	13.01	13.01	13.01	13.01	13.01
... rel. freq. of bids < p^* - 200	0.20	0.20	0.20	0.20	0.20	0.20
... rel. freq. of bids < p^*	0.75	0.75	0.75	0.75	0.75	0.75
... rel. freq. of bids = p^*	0.03	0.03	0.03	0.03	0.03	0.03
... rel. freq. of bids > p^*	0.22	0.22	0.23	0.22	0.22	0.23
Number of submitted asks per contract	17.33	17.32	17.34	17.34	17.34	17.33
... median value in the simulation	13.01	13.01	13.01	13.01	13.01	13.01
... rel. freq. of asks < p^*	0.02	0.02	0.02	0.03	0.02	0.02
... rel. freq. of asks = p^*	0.01	0.01	0.01	0.01	0.01	0.01
... rel. freq. of asks > p^*	0.97	0.97	0.97	0.97	0.97	0.97
... rel. freq. of asks > p^* + 200	0.43	0.43	0.43	0.43	0.43	0.43
Number of offers for all units in periods > 1	321	-	321	321	321	321
(a) rel. freq. of bids > p-max & asks < p-min	0.01	-	0.01	0.01	0.01	0.01
(b) rel. freq. of bids > p-avg & asks < p-avg	0.11	-	0.11	0.11	0.11	0.11
(c) rel. freq. of bids > p-own & asks < p-own	0.06	-	0.06	0.06	0.06	0.06
(d) rel. freq. of bids < p-own & asks > p-own	0.88	-	0.88	0.88	0.88	0.88
Number of offers for units bought above p-avg or sold below p-avg in the previous period	120	-	120	119	119	120
(C) rel. freq. of bids > p-own & asks < p-own	0.04	-	0.04	0.04	0.04	0.04
Number of offers for units that were not traded in the previous period	51	-	51	51	51	51
(D) rel. freq. of bids < p-own & asks > p-own	0.90	-	0.90	0.90	0.90	0.91
Aggressive offer frequency ratio (c)/(C)	0.58	-	0.58	0.58	0.58	0.58
... median value in the simulation	0.57	-	0.54	0.53	0.53	0.52
Moderate offer frequency ratio (d)/(D)	1.02	-	1.02	1.03	1.02	1.03
... median value in the simulation	1.03	-	1.03	1.03	1.03	1.03

Relative frequencies f(x) and cumulative distributions F(x) of the simulation rounds (5 subsequent periods) in which any n periods have price paths of the same type.

	falling		rising		hanging		standing		unknown	
n	f(x)	F(x)	f(x)	F(x)	f(x)	F(x)	f(x)	F(x)	f(x)	F(x)
5	0.26	0.26	0.00	0.00	0.00	0.00	0.00	0.00	0.00	0.00
4	0.40	0.66	0.00	0.00	0.00	0.00	0.01	0.01	0.00	0.00
3	0.25	0.91	0.00	0.00	0.00	0.00	0.05	0.05	0.00	0.00
2	0.08	0.98	0.01	0.01	0.00	0.00	0.19	0.24	0.00	0.00
1	0.02	1.00	0.12	0.13	0.09	0.10	0.41	0.66	0.00	0.00
0	0.00	1.00	0.87	1.00	0.90	1.00	0.34	1.00	1.00	1.00

Panel VII.C.1.15

Double Auction Market Simulation Results   File: DAMM2811.SIM

Number of simulation rounds	: 10000	Periods per round :		5
Market type	: M2	Number of traders :		8
Market rules	: AltDAM	Technical price range :	[1 ...	9999]
No-Crossing rule	: YES	Ask-Bid-Spread-Reduction rule :		YES
Trader type	: ZI			

Market clearing quantity q*   :   7
Market clearing price   p*   :   765

	Average of Periods	Period 1	Period 2	Period 3	Period 4	Period 5
Deviation of traded quantity from q*	0.25	0.25	0.26	0.25	0.25	0.25
Efficiency	0.97	0.97	0.97	0.97	0.97	0.97
... median value in the simulation	0.98	0.99	0.99	0.99	0.99	0.99
Average contract price (deviation from p*)	11.22	11.19	11.25	11.21	11.23	11.22
... median value in the simulation	11.88	12.01	12.01	12.01	12.01	11.88
Minimum contract price (deviation from p*)	-8.25	-8.28	-8.18	-8.28	-8.21	-8.32
... median value in the simulation	-5.99	-5.99	-5.99	-5.99	-5.99	-5.99
Maximum contract price (deviation from p*)	32.01	31.99	32.03	32.03	31.99	32.00
... median value in the simulation	34.01	34.01	34.01	34.01	34.01	34.01
Market clearing price concordance coefficient	0.39	0.39	0.39	0.39	0.39	0.39
... median value in the simulation	0.39	0.38	0.38	0.38	0.38	0.37
Rel. freq. of falling price paths	0.78	0.78	0.78	0.77	0.78	0.77
Rel. freq. of rising price paths	0.02	0.02	0.02	0.02	0.02	0.02
Rel. freq. of hanging price paths	0.01	0.01	0.01	0.02	0.02	0.01
Rel. freq. of standing price paths	0.19	0.19	0.19	0.19	0.19	0.20
Rel. freq. of unknown price paths	0.00	0.00	0.00	0.00	0.00	0.00
Rel. freq. of contract proposals by buyers	0.68	0.68	0.68	0.68	0.68	0.68
... lower quartile bound in the simulation	0.63	0.58	0.58	0.58	0.58	0.58
... median value in the simulation	0.69	0.72	0.72	0.72	0.72	0.72
... upper quartile bound in the simulation	0.75	0.86	0.86	0.86	0.86	0.86
... rel. freq. of these with price < p*	0.14	0.14	0.14	0.14	0.14	0.14
... rel. freq. of these with price = p*	0.11	0.11	0.11	0.11	0.11	0.11
... rel. freq. of these with price > p*	0.75	0.75	0.75	0.75	0.75	0.75
... avg. buyer proposed contract price - p*	13.10	13.03	13.19	13.13	13.07	13.09
Rel. freq. of contract proposals by sellers	0.32	0.32	0.32	0.32	0.32	0.32
... lower quartile bound in the simulation	0.26	0.15	0.15	0.15	0.15	0.15
... median value in the simulation	0.32	0.29	0.29	0.29	0.29	0.29
... upper quartile bound in the simulation	0.38	0.43	0.43	0.43	0.43	0.43
... rel. freq. of these with price < p*	0.31	0.31	0.31	0.31	0.30	0.31
... rel. freq. of these with price = p*	0.06	0.06	0.06	0.05	0.06	0.06
... rel. freq. of these with price > p*	0.64	0.64	0.64	0.63	0.64	0.63
... avg. seller proposed contract price - p*	7.50	7.62	7.46	7.34	7.62	7.44
Rel. freq. of proposer's curses - all traders	0.62	0.62	0.62	0.62	0.62	0.62
... median value in the simulation	0.63	0.63	0.63	0.63	0.63	0.63
Rel. freq. of proposer's curses - all buyers	0.75	0.75	0.75	0.75	0.75	0.75
Rel. freq. of proposer's curses - all sellers	0.31	0.31	0.31	0.31	0.30	0.31
Rel. freq. of proposer's curses - buyer 1	0.93	0.93	0.93	0.93	0.93	0.93
Rel. freq. of proposer's curses - buyer 2	0.90	0.90	0.90	0.90	0.90	0.90
Rel. freq. of proposer's curses - buyer 3	0.51	0.51	0.51	0.51	0.51	0.51
Rel. freq. of proposer's curses - buyer 4	0.78	0.77	0.78	0.78	0.78	0.78
Rel. freq. of proposer's curses - seller 5	0.44	0.44	0.44	0.45	0.44	0.45
Rel. freq. of proposer's curses - seller 6	0.37	0.36	0.36	0.37	0.37	0.37
Rel. freq. of proposer's curses - seller 7	0.21	0.21	0.21	0.22	0.20	0.20
Rel. freq. of proposer's curses - seller 8	0.22	0.23	0.22	0.22	0.22	0.22

Panel VII.C.1.15 (continued)

	Average of Periods	Period 1	Period 2	Period 3	Period 4	Period 5
Number of market ask and market bid matches	6.97	6.97	6.97	6.97	6.96	6.96
Number of contracts per market ask-bid match	1.04	1.04	1.04	1.04	1.04	1.04
Avg. rank of traded red.value at trading time	1.33	1.33	1.33	1.33	1.33	1.33
... median value in the simulation	1.33	1.29	1.29	1.29	1.29	1.29
Avg. rank of traded unit cost at trading time	2.00	2.00	2.00	2.00	2.00	2.00
... median value in the simulation	2.01	2.01	2.01	2.01	2.01	2.01
Number of submitted bids per contract	15.05	15.05	15.05	15.05	15.02	15.05
... median value in the simulation	13.01	13.01	13.01	13.01	13.01	13.01
... rel. freq. of bids < p^* - 200	0.24	0.24	0.24	0.24	0.24	0.24
... rel. freq. of bids < p^*	0.78	0.78	0.78	0.78	0.78	0.78
... rel. freq. of bids = p^*	0.02	0.02	0.02	0.02	0.02	0.02
... rel. freq. of bids > p^*	0.20	0.20	0.20	0.20	0.20	0.20
Number of submitted asks per contract	19.66	19.64	19.67	19.67	19.63	19.67
... median value in the simulation	13.01	13.01	13.01	13.01	13.01	13.01
... rel. freq. of asks < p^*	0.03	0.03	0.03	0.03	0.03	0.03
... rel. freq. of asks = p^*	0.01	0.01	0.01	0.01	0.01	0.01
... rel. freq. of asks > p^*	0.96	0.96	0.96	0.96	0.96	0.96
... rel. freq. of asks > p^* + 200	0.45	0.45	0.44	0.45	0.45	0.45
Number of offers for all units in periods > 1	252	-	252	252	251	252
(a) rel. freq. of bids > p-max & asks < p-min	0.01	-	0.01	0.01	0.01	0.01
(b) rel. freq. of bids > p-avg & asks < p-avg	0.12	-	0.12	0.12	0.11	0.11
(c) rel. freq. of bids > p-own & asks < p-own	0.09	-	0.09	0.09	0.09	0.09
(d) rel. freq. of bids < p-own & asks > p-own	0.87	-	0.87	0.87	0.87	0.87
Number of offers for units bought above p-avg or sold below p-avg in the previous period	79	-	79	79	79	79
(C) rel. freq. of bids > p-own & asks < p-own	0.06	-	0.06	0.06	0.06	0.06
Number of offers for units that were not traded in the previous period	79	-	79	79	79	79
(D) rel. freq. of bids < p-own & asks > p-own	0.89	-	0.89	0.89	0.89	0.89
Aggressive offer frequency ratio (c)/(C)	0.68	-	0.68	0.69	0.68	0.69
... median value in the simulation	0.68	-	0.63	0.63	0.63	0.63
Moderate offer frequency ratio (d)/(D)	1.02	-	1.02	1.02	1.02	1.02
... median value in the simulation	1.03	-	1.03	1.03	1.03	1.03

Relative frequencies f(x) and cumulative distributions F(x) of the simulation rounds (5 subsequent periods) in which any n periods have price paths of the same type.

	falling		rising		hanging		standing		unknown	
n	f(x)	F(x)	f(x)	F(x)	f(x)	F(x)	f(x)	F(x)	f(x)	F(x)
5	0.28	0.28	0.00	0.00	0.00	0.00	0.00	0.00	0.00	0.00
4	0.41	0.69	0.00	0.00	0.00	0.00	0.00	0.01	0.00	0.00
3	0.23	0.92	0.00	0.00	0.00	0.00	0.05	0.05	0.00	0.00
2	0.07	0.99	0.00	0.00	0.00	0.00	0.19	0.24	0.00	0.00
1	0.01	1.00	0.08	0.09	0.07	0.07	0.41	0.65	0.00	0.00
0	0.00	1.00	0.91	1.00	0.93	1.00	0.35	1.00	1.00	1.00

Panel VII.C.1.16

Double Auction Market Simulation Results   File: DAMM2821.SIM

Number of simulation rounds	: 10000	Periods per round :		5
Market type	: M2	Number of traders :		8
Market rules	: AltDAM	Technical price range :	[1 ...	9999]
No-Crossing rule	: YES	Ask-Bid-Spread-Reduction rule :		YES
Trader type	: ZI			

Market clearing quantity q*   :   7
Market clearing price   p*   :   416

	Average of Periods	Period 1	Period 2	Period 3	Period 4	Period 5
Deviation of traded quantity from q*	0.26	0.25	0.26	0.26	0.26	0.26
Efficiency	0.97	0.97	0.97	0.97	0.97	0.97
... median value in the simulation	0.98	0.98	0.98	0.98	0.98	0.99
Average contract price (deviation from p*)	12.88	12.85	12.90	12.87	12.88	12.90
... median value in the simulation	13.58	13.58	13.63	13.58	13.58	13.58
Minimum contract price (deviation from p*)	-5.95	-6.01	-5.90	-5.93	-5.91	-6.00
... median value in the simulation	-4.99	-4.99	-4.99	-4.99	-4.99	-4.99
Maximum contract price (deviation from p*)	32.84	32.86	32.80	32.86	32.87	32.82
... median value in the simulation	35.01	35.01	35.01	35.01	35.01	35.01
Market clearing price concordance coefficient	0.37	0.37	0.37	0.37	0.36	0.36
... median value in the simulation	0.37	0.35	0.35	0.35	0.35	0.35
Rel. freq. of falling price paths	0.84	0.84	0.85	0.84	0.84	0.85
Rel. freq. of rising price paths	0.01	0.01	0.01	0.01	0.01	0.01
Rel. freq. of hanging price paths	0.01	0.01	0.01	0.01	0.01	0.01
Rel. freq. of standing price paths	0.14	0.14	0.14	0.15	0.15	0.14
Rel. freq. of unknown price paths	0.00	0.00	0.00	0.00	0.00	0.00
Rel. freq. of contract proposals by buyers	0.73	0.73	0.73	0.73	0.73	0.73
... lower quartile bound in the simulation	0.67	0.58	0.58	0.58	0.58	0.58
... median value in the simulation	0.73	0.72	0.72	0.72	0.72	0.72
... upper quartile bound in the simulation	0.79	0.86	0.86	0.86	0.86	0.86
... rel. freq. of these with price < p*	0.11	0.11	0.11	0.11	0.11	0.11
... rel. freq. of these with price = p*	0.12	0.12	0.12	0.12	0.12	0.12
... rel. freq. of these with price > p*	0.77	0.77	0.78	0.77	0.77	0.78
... avg. buyer proposed contract price - p*	14.06	14.03	14.07	14.08	14.05	14.09
Rel. freq. of contract proposals by sellers	0.27	0.27	0.27	0.27	0.27	0.27
... lower quartile bound in the simulation	0.22	0.15	0.15	0.15	0.15	0.15
... median value in the simulation	0.28	0.29	0.29	0.29	0.29	0.29
... upper quartile bound in the simulation	0.34	0.43	0.43	0.43	0.43	0.43
... rel. freq. of these with price < p*	0.24	0.24	0.24	0.24	0.23	0.24
... rel. freq. of these with price = p*	0.05	0.06	0.05	0.05	0.06	0.05
... rel. freq. of these with price > p*	0.71	0.71	0.70	0.70	0.71	0.71
... avg. seller proposed contract price - p*	10.10	10.09	10.12	9.96	10.21	10.13
Rel. freq. of proposer's curses - all traders	0.64	0.63	0.64	0.64	0.63	0.64
... median value in the simulation	0.64	0.63	0.63	0.63	0.63	0.63
Rel. freq. of proposer's curses - all buyers	0.77	0.77	0.78	0.77	0.77	0.78
Rel. freq. of proposer's curses - all sellers	0.24	0.24	0.24	0.24	0.23	0.24
Rel. freq. of proposer's curses - buyer 1	0.96	0.96	0.96	0.96	0.95	0.96
Rel. freq. of proposer's curses - buyer 2	0.93	0.93	0.93	0.93	0.93	0.93
Rel. freq. of proposer's curses - buyer 3	0.53	0.53	0.53	0.53	0.53	0.53
Rel. freq. of proposer's curses - buyer 4	0.81	0.81	0.81	0.81	0.81	0.81
Rel. freq. of proposer's curses - seller 5	0.36	0.36	0.37	0.37	0.36	0.36
Rel. freq. of proposer's curses - seller 6	0.29	0.30	0.30	0.29	0.29	0.30
Rel. freq. of proposer's curses - seller 7	0.16	0.16	0.16	0.16	0.15	0.15
Rel. freq. of proposer's curses - seller 8	0.17	0.16	0.17	0.17	0.16	0.17

Panel VII.C.1.16 (continued)

	Average of Periods	Period 1	Period 2	Period 3	Period 4	Period 5
Number of market ask and market bid matches	7.02	7.01	7.03	7.02	7.03	7.01
Number of contracts per market ask-bid match	1.03	1.03	1.03	1.03	1.03	1.03
Avg. rank of traded red.value at trading time	1.25	1.25	1.24	1.25	1.25	1.25
... median value in the simulation	1.24	1.17	1.15	1.17	1.17	1.17
Avg. rank of traded unit cost at trading time	2.03	2.03	2.03	2.03	2.03	2.03
... median value in the simulation	2.04	2.01	2.01	2.01	2.01	2.01
Number of submitted bids per contract	14.12	14.14	14.13	14.11	14.13	14.11
... median value in the simulation	13.01	13.01	13.01	13.01	13.01	13.01
... rel. freq. of bids < p* - 200	0.16	0.16	0.16	0.16	0.16	0.16
... rel. freq. of bids < p*	0.75	0.75	0.74	0.75	0.75	0.75
... rel. freq. of bids = p*	0.02	0.02	0.02	0.02	0.02	0.02
... rel. freq. of bids > p*	0.24	0.23	0.24	0.24	0.24	0.24
Number of submitted asks per contract	20.03	19.99	20.07	20.04	20.02	20.01
... median value in the simulation	13.01	13.01	13.01	13.01	13.01	13.01
... rel. freq. of asks < p*	0.02	0.02	0.02	0.02	0.02	0.02
... rel. freq. of asks = p*	0.01	0.01	0.01	0.01	0.01	0.01
... rel. freq. of asks > p*	0.97	0.97	0.97	0.97	0.97	0.97
... rel. freq. of asks > p* + 200	0.44	0.44	0.44	0.44	0.44	0.44
Number of offers for all units in periods > 1	248	.	248	248	248	248
(a) rel. freq. of bids > p-max & asks < p-min	0.01	.	0.01	0.01	0.01	0.01
(b) rel. freq. of bids > p-avg & asks < p-avg	0.12	.	0.12	0.12	0.12	0.12
(c) rel. freq. of bids > p-own & asks > p-own	0.09	.	0.09	0.09	0.09	0.09
(d) rel. freq. of bids < p-own & asks > p-own	0.87	.	0.87	0.87	0.87	0.87
Number of offers for units bought above p-avg or sold below p-avg in the previous period	78	.	78	78	78	77
(C) rel. freq. of bids > p-own & asks < p-own	0.06	.	0.06	0.06	0.06	0.06
Number of offers for units that were not traded in the previous period	77	.	77	76	77	77
(D) rel. freq. of bids < p-own & asks > p-own	0.89	.	0.89	0.89	0.89	0.89
Aggressive offer frequency ratio (c)/(C)	0.70	.	0.70	0.70	0.70	0.70
... median value in the simulation	0.69	.	0.65	0.66	0.65	0.65
Moderate offer frequency ratio (d)/(D)	1.02	.	1.02	1.02	1.02	1.02
... median value in the simulation	1.03	.	1.03	1.03	1.03	1.03

Relative frequencies f(x) and cumulative distributions F(x) of the simulation rounds (5 subsequent periods) in which any n periods have price paths of the same type.

	falling		rising		hanging		standing		unknown	
n	f(x)	F(x)	f(x)	F(x)	f(x)	F(x)	f(x)	F(x)	f(x)	F(x)
5	0.43	0.43	0.00	0.00	0.00	0.00	0.00	0.00	0.00	0.00
4	0.39	0.82	0.00	0.00	0.00	0.00	0.00	0.00	0.00	0.00
3	0.15	0.97	0.00	0.00	0.00	0.00	0.02	0.02	0.00	0.00
2	0.03	1.00	0.00	0.00	0.00	0.00	0.13	0.15	0.00	0.00
1	0.00	1.00	0.04	0.04	0.03	0.03	0.38	0.54	0.00	0.00
0	0.00	1.00	0.96	1.00	0.97	1.00	0.46	1.00	1.00	1.00

Panel VII.C.1.17

Double Auction Market Simulation Results   File: DAMM2841.SIM

Number of simulation rounds	: 10000		Periods per round :		5
Market type	: M2		Number of traders :		8
Market rules	: AltDAM		Technical price range : [	1 ...	9999]
No-Crossing rule	: YES		Ask-Bid-Spread-Reduction rule :		YES
Trader type	: ZI				

Market clearing quantity q*	: 7				
Market clearing price p*	: 559				

	Average of Periods	Period 1	Period 2	Period 3	Period 4	Period 5
Deviation of traded quantity from q*	0.26	0.25	0.26	0.26	0.26	0.26
Efficiency	0.97	0.97	0.97	0.97	0.97	0.97
... median value in the simulation	0.98	0.99	0.99	0.98	0.98	0.99
Average contract price (deviation from p*)	12.14	12.13	12.08	12.13	12.21	12.15
... median value in the simulation	12.86	12.86	12.86	12.86	13.01	12.86
Minimum contract price (deviation from p*)	-6.99	-7.06	-7.01	-6.99	-6.94	-6.94
... median value in the simulation	-5.99	-5.99	-5.99	-5.99	-5.99	-5.99
Maximum contract price (deviation from p*)	32.48	32.42	32.45	32.44	32.58	32.51
... median value in the simulation	35.01	34.01	35.01	34.01	35.01	35.01
Market clearing price concordance coefficient	0.38	0.38	0.38	0.38	0.37	0.38
... median value in the simulation	0.38	0.36	0.37	0.36	0.36	0.36
Rel. freq. of falling price paths	0.82	0.82	0.81	0.82	0.82	0.82
Rel. freq. of rising price paths	0.01	0.01	0.01	0.01	0.01	0.01
Rel. freq. of hanging price paths	0.01	0.01	0.01	0.01	0.01	0.01
Rel. freq. of standing price paths	0.16	0.17	0.16	0.16	0.16	0.16
Rel. freq. of unknown price paths	0.00	0.00	0.00	0.00	0.00	0.00
Rel. freq. of contract proposals by buyers	0.70	0.70	0.70	0.70	0.71	0.71
... lower quartile bound in the simulation	0.65	0.58	0.58	0.58	0.58	0.58
... median value in the simulation	0.72	0.72	0.72	0.72	0.72	0.72
... upper quartile bound in the simulation	0.78	0.86	0.86	0.86	0.86	0.86
... rel. freq. of these with price < p*	0.12	0.12	0.12	0.12	0.12	0.12
... rel. freq. of these with price = p*	0.11	0.11	0.11	0.12	0.11	0.12
... rel. freq. of these with price > p*	0.76	0.77	0.76	0.76	0.77	0.76
... avg. buyer proposed contract price · p*	13.65	13.72	13.61	13.58	13.76	13.57
Rel. freq. of contract proposals by sellers	0.30	0.30	0.30	0.30	0.29	0.29
... lower quartile bound in the simulation	0.23	0.15	0.15	0.15	0.15	0.15
... median value in the simulation	0.29	0.29	0.29	0.29	0.29	0.29
... upper quartile bound in the simulation	0.36	0.43	0.43	0.43	0.43	0.43
... rel. freq. of these with price < p*	0.27	0.27	0.27	0.27	0.27	0.26
... rel. freq. of these with price = p*	0.06	0.06	0.06	0.05	0.06	0.06
... rel. freq. of these with price > p*	0.68	0.67	0.68	0.68	0.67	0.68
... avg. seller proposed contract price · p*	8.94	8.83	8.92	8.98	8.93	9.06
Rel. freq. of proposer's curses · all traders	0.63	0.63	0.62	0.63	0.63	0.62
... median value in the simulation	0.63	0.63	0.63	0.63	0.63	0.63
Rel. freq. of proposer's curses · all buyers	0.76	0.77	0.76	0.76	0.77	0.76
Rel. freq. of proposer's curses · all sellers	0.27	0.27	0.27	0.27	0.27	0.26
Rel. freq. of proposer's curses · buyer 1	0.94	0.94	0.94	0.94	0.94	0.95
Rel. freq. of proposer's curses · buyer 2	0.92	0.92	0.91	0.92	0.92	0.92
Rel. freq. of proposer's curses · buyer 3	0.52	0.53	0.51	0.52	0.53	0.51
Rel. freq. of proposer's curses · buyer 4	0.80	0.80	0.80	0.79	0.79	0.80
Rel. freq. of proposer's curses · seller 5	0.40	0.40	0.39	0.40	0.40	0.40
Rel. freq. of proposer's curses · seller 6	0.33	0.33	0.33	0.33	0.33	0.31
Rel. freq. of proposer's curses · seller 7	0.18	0.19	0.18	0.18	0.18	0.18
Rel. freq. of proposer's curses · seller 8	0.19	0.20	0.19	0.20	0.19	0.19

Panel VII.C.1.17 (continued)

	Average of Periods	Period 1	Period 2	Period 3	Period 4	Period 5
Number of market ask and market bid matches	7.00	6.99	6.99	7.00	7.00	7.00
Number of contracts per market ask-bid match	1.04	1.04	1.04	1.04	1.04	1.04
Avg. rank of traded red.value at trading time	1.28	1.28	1.29	1.28	1.28	1.28
... median value in the simulation	1.28	1.26	1.26	1.26	1.26	1.26
Avg. rank of traded unit cost at trading time	2.02	2.02	2.01	2.02	2.02	2.01
... median value in the simulation	2.02	2.01	2.01	2.01	2.01	2.01
Number of submitted bids per contract	14.58	14.59	14.57	14.58	14.59	14.58
... median value in the simulation	13.01	13.01	13.01	13.01	13.01	13.01
... rel. freq. of bids < p* - 200	0.20	0.20	0.20	0.20	0.20	0.20
... rel. freq. of bids < p*	0.76	0.76	0.77	0.76	0.76	0.76
... rel. freq. of bids = p*	0.02	0.02	0.02	0.02	0.02	0.02
... rel. freq. of bids > p*	0.22	0.22	0.22	0.22	0.22	0.22
Number of submitted asks per contract	19.88	19.85	19.85	19.89	19.91	19.87
... median value in the simulation	13.01	13.01	13.01	13.01	13.01	13.01
... rel. freq. of asks < p*	0.02	0.02	0.02	0.02	0.02	0.02
... rel. freq. of asks = p*	0.01	0.01	0.01	0.01	0.01	0.01
... rel. freq. of asks > p*	0.97	0.97	0.97	0.97	0.97	0.97
... rel. freq. of asks > p* + 200	0.45	0.44	0.45	0.44	0.45	0.45
Number of offers for all units in periods > 1	250	-	250	250	250	250
(a) rel. freq. of bids > p-max & asks < p-min	0.01	-	0.01	0.01	0.01	0.01
(b) rel. freq. of bids > p-avg & asks < p-avg	0.12	-	0.12	0.12	0.12	0.12
(c) rel. freq. of bids > p-own & asks < p-own	0.09	-	0.09	0.09	0.09	0.09
(d) rel. freq. of bids < p-own & asks > p-own	0.87	-	0.87	0.87	0.87	0.87
Number of offers for units bought above p-avg or sold below p-avg in the previous period	78	-	78	79	79	78
(C) rel. freq. of bids > p-own & asks < p-own	0.06	-	0.06	0.06	0.06	0.06
Number of offers for units that were not traded in the previous period	78	-	78	78	78	78
(D) rel. freq. of bids < p-own & asks > p-own	0.89	-	0.89	0.89	0.89	0.89
Aggressive offer frequency ratio (c)/(C)	0.69	-	0.70	0.70	0.69	0.69
... median value in the simulation	0.68	-	0.65	0.64	0.65	0.64
Moderate offer frequency ratio (d)/(D)	1.02	-	1.02	1.02	1.02	1.02
... median value in the simulation	1.03	-	1.03	1.03	1.03	1.03

Relative frequencies f(x) and cumulative distributions F(x) of the simulation rounds
(5 subsequent periods) in which any n periods have price paths of the same type.

	falling		rising		hanging		standing		unknown	
n	f(x)	F(x)	f(x)	F(x)	f(x)	F(x)	f(x)	F(x)	f(x)	F(x)
5	0.37	0.37	0.00	0.00	0.00	0.00	0.00	0.00	0.00	0.00
4	0.41	0.78	0.00	0.00	0.00	0.00	0.00	0.00	0.00	0.00
3	0.17	0.95	0.00	0.00	0.00	0.00	0.03	0.04	0.00	0.00
2	0.04	1.00	0.00	0.00	0.00	0.00	0.15	0.18	0.00	0.00
1	0.00	1.00	0.05	0.05	0.04	0.04	0.41	0.59	0.00	0.00
0	0.00	1.00	0.95	1.00	0.96	1.00	0.41	1.00	1.00	1.00

Panel VII.C.1.18

Double Auction Market Simulation Results   File: DAMM2A11.SIM

Number of simulation rounds	:	10000
Market type	:	M2
Market rules	:	AltDAM
No-Crossing rule	:	YES
Trader type	:	ZI
Market clearing quantity q*	:	9
Market clearing price p*	:	797

Periods per round :	5
Number of traders :	10
Technical price range : [	1 ... 9999]
Ask-Bid-Spread-Reduction rule :	YES

	Average of Periods	Period 1	Period 2	Period 3	Period 4	Period 5
Deviation of traded quantity from q*	0.51	0.51	0.51	0.50	0.50	0.52
Efficiency	0.97	0.97	0.97	0.97	0.97	0.97
... median value in the simulation	0.98	0.98	0.98	0.98	0.98	0.98
Average contract price (deviation from p*)	7.35	7.34	7.34	7.36	7.37	7.35
... median value in the simulation	7.78	7.78	7.71	7.78	7.78	7.78
Minimum contract price (deviation from p*)	-6.58	-6.57	-6.61	-6.63	-6.54	-6.58
... median value in the simulation	-5.99	-5.99	-5.99	-5.99	-4.99	-5.99
Maximum contract price (deviation from p*)	21.92	21.86	21.92	21.92	21.95	21.96
... median value in the simulation	23.01	23.01	23.01	23.01	23.01	23.01
Market clearing price concordance coefficient	0.38	0.38	0.38	0.38	0.38	0.38
... median value in the simulation	0.38	0.38	0.37	0.37	0.37	0.37
Rel. freq. of falling price paths	0.77	0.77	0.77	0.78	0.77	0.77
Rel. freq. of rising price paths	0.01	0.01	0.01	0.01	0.01	0.01
Rel. freq. of hanging price paths	0.01	0.01	0.01	0.01	0.01	0.01
Rel. freq. of standing price paths	0.21	0.21	0.21	0.20	0.21	0.21
Rel. freq. of unknown price paths	0.00	0.00	0.00	0.00	0.00	0.00
Rel. freq. of contract proposals by buyers	0.69	0.69	0.69	0.69	0.69	0.69
... lower quartile bound in the simulation	0.64	0.56	0.56	0.56	0.56	0.56
... median value in the simulation	0.70	0.71	0.71	0.71	0.71	0.71
... upper quartile bound in the simulation	0.75	0.81	0.81	0.81	0.81	0.81
... rel. freq. of these with price < p*	0.15	0.15	0.15	0.15	0.14	0.15
... rel. freq. of these with price = p*	0.09	0.09	0.09	0.09	0.09	0.09
... rel. freq. of these with price > p*	0.76	0.76	0.76	0.76	0.76	0.76
... avg. buyer proposed contract price - p*	8.62	8.59	8.57	8.62	8.65	8.67
Rel. freq. of contract proposals by sellers	0.31	0.31	0.31	0.31	0.31	0.31
... lower quartile bound in the simulation	0.26	0.20	0.20	0.20	0.20	0.20
... median value in the simulation	0.31	0.31	0.31	0.31	0.31	0.31
... upper quartile bound in the simulation	0.37	0.45	0.45	0.45	0.45	0.45
... rel. freq. of these with price < p*	0.31	0.31	0.31	0.31	0.31	0.31
... rel. freq. of these with price = p*	0.06	0.06	0.06	0.06	0.06	0.06
... rel. freq. of these with price > p*	0.63	0.64	0.63	0.63	0.64	0.63
... avg. seller proposed contract price - p*	4.83	4.87	4.84	4.83	4.90	4.71
Rel. freq. of proposer's curses - all traders	0.63	0.63	0.63	0.63	0.63	0.63
... median value in the simulation	0.64	0.67	0.67	0.67	0.67	0.67
Rel. freq. of proposer's curses - all buyers	0.76	0.76	0.76	0.76	0.76	0.76
Rel. freq. of proposer's curses - all sellers	0.31	0.31	0.31	0.31	0.31	0.31
Rel. freq. of proposer's curses - buyer 1	0.92	0.92	0.92	0.92	0.92	0.92
Rel. freq. of proposer's curses - buyer 2	0.92	0.92	0.92	0.92	0.92	0.92
Rel. freq. of proposer's curses - buyer 3	0.88	0.88	0.89	0.88	0.88	0.88
Rel. freq. of proposer's curses - buyer 4	0.52	0.52	0.52	0.52	0.52	0.53
Rel. freq. of proposer's curses - buyer 5	0.69	0.68	0.69	0.69	0.69	0.68
Rel. freq. of proposer's curses - seller 6	0.44	0.44	0.43	0.44	0.45	0.44
Rel. freq. of proposer's curses - seller 7	0.40	0.40	0.40	0.41	0.40	0.40
Rel. freq. of proposer's curses - seller 8	0.32	0.31	0.33	0.32	0.31	0.33
Rel. freq. of proposer's curses - seller 9	0.20	0.20	0.21	0.19	0.20	0.20
Rel. freq. of proposer's curses - seller 10	0.22	0.20	0.22	0.21	0.22	0.22

Panel VII.C.1.18 (continued)

	Average of Periods	Period 1	Period 2	Period 3	Period 4	Period 5
Number of market ask and market bid matches	9.00	8.99	9.02	8.99	9.00	9.02
Number of contracts per market ask-bid match	1.06	1.06	1.05	1.06	1.06	1.06
Avg. rank of traded red.value at trading time	1.47	1.47	1.47	1.47	1.47	1.47
... median value in the simulation	1.47	1.45	1.45	1.45	1.45	1.45
Avg. rank of traded unit cost at trading time	2.35	2.35	2.35	2.35	2.35	2.35
... median value in the simulation	2.35	2.38	2.38	2.34	2.36	2.34
Number of submitted bids per contract	17.07	17.06	17.09	17.05	17.08	17.10
... median value in the simulation	13.01	13.01	13.01	13.01	13.01	13.01
... rel. freq. of bids < p* - 200	0.25	0.25	0.25	0.25	0.25	0.25
... rel. freq. of bids < p*	0.80	0.80	0.80	0.80	0.80	0.80
... rel. freq. of bids = p*	0.02	0.02	0.02	0.02	0.02	0.02
... rel. freq. of bids > p*	0.18	0.18	0.18	0.18	0.18	0.18
Number of submitted asks per contract	21.66	21.65	21.70	21.62	21.69	21.65
... median value in the simulation	13.01	13.01	13.01	13.01	13.01	13.01
... rel. freq. of asks < p*	0.02	0.02	0.02	0.02	0.02	0.02
... rel. freq. of asks = p*	0.01	0.01	0.01	0.01	0.01	0.01
... rel. freq. of asks > p*	0.97	0.97	0.97	0.97	0.97	0.97
... rel. freq. of asks > p* + 200	0.44	0.44	0.44	0.44	0.44	0.44
Number of offers for all units in periods > 1	368	-	369	367	368	369
(a) rel. freq. of bids > p-max & asks < p-min	0.01	-	0.01	0.01	0.01	0.01
(b) rel. freq. of bids > p-avg & asks < p-avg	0.10	-	0.10	0.10	0.10	0.10
(c) rel. freq. of bids > p-own & asks < p-own	0.07	-	0.07	0.07	0.07	0.07
(d) rel. freq. of bids < p-own & asks > p-own	0.89	-	0.88	0.89	0.89	0.89
Number of offers for units bought above p-avg or sold below p-avg in the previous period	116	-	116	116	116	116
(C) rel. freq. of bids > p-own & asks < p-own	0.05	-	0.05	0.04	0.05	0.05
Number of offers for units that were not traded in the previous period	110	-	110	109	110	111
(D) rel. freq. of bids < p-own & asks > p-own	0.89	-	0.89	0.89	0.89	0.89
Aggressive offer frequency ratio (c)/(C)	0.63	-	0.63	0.62	0.62	0.63
... median value in the simulation	0.62	-	0.58	0.57	0.58	0.59
Moderate offer frequency ratio (d)/(D)	1.01	-	1.01	1.01	1.01	1.01
... median value in the simulation	1.02	-	1.02	1.02	1.02	1.02

Relative frequencies f(x) and cumulative distributions F(x) of the simulation rounds (5 subsequent periods) in which any n periods have price paths of the same type.

	falling		rising		hanging		standing		unknown	
n	f(x)	F(x)	f(x)	F(x)	f(x)	F(x)	f(x)	F(x)	f(x)	F(x)
5	0.27	0.27	0.00	0.00	0.00	0.00	0.00	0.00	0.00	0.00
4	0.40	0.68	0.00	0.00	0.00	0.00	0.01	0.01	0.00	0.00
3	0.24	0.92	0.00	0.00	0.00	0.00	0.06	0.06	0.00	0.00
2	0.07	0.99	0.00	0.00	0.00	0.00	0.22	0.28	0.00	0.00
1	0.01	1.00	0.06	0.06	0.04	0.04	0.40	0.69	0.00	0.00
0	0.00	1.00	0.94	1.00	0.96	1.00	0.31	1.00	1.00	1.00

Vol. 367: M. Grauer, D. B. Pressmar (Eds.), Parallel Computing and Mathematical Optimization. Proceedings. V, 208 pages. 1991.

Vol. 368: M. Fedrizzi, J. Kacprzyk, M. Roubens (Eds.), Interactive Fuzzy Optimization. VII, 216 pages. 1991.

Vol. 369: R. Koblo, The Visible Hand. VIII, 131 pages. 1991.

Vol. 370: M. J. Beckmann, M. N. Gopalan, R. Subramanian (Eds.), Stochastic Processes and their Applications. Proceedings, 1990. XLI, 292 pages. 1991.

Vol. 371: A. Schmutzler, Flexibility and Adjustment to Information in Sequential Decision Problems. VIII, 198 pages. 1991.

Vol. 372: J. Esteban, The Social Viability of Money. X, 202 pages. 1991.

Vol. 373: A. Billot, Economic Theory of Fuzzy Equilibria. XIII, 164 pages. 1992.

Vol. 374: G. Pflug, U. Dieter (Eds.), Simulation and Optimization. Proceedings, 1990. X, 162 pages. 1992.

Vol. 375: S.-J. Chen, Ch.-L. Hwang, Fuzzy Multiple Attribute Decision Making. XII, 536 pages. 1992.

Vol. 376: K.-H. Jöckel, G. Rothe, W. Sendler (Eds.), Bootstrapping and Related Techniques. Proceedings, 1990. VIII, 247 pages. 1992.

Vol. 377: A. Villar, Operator Theorems with Applications to Distributive Problems and Equilibrium Models. XVI, 160 pages. 1992.

Vol. 378: W. Krabs, J. Zowe (Eds.), Modern Methods of Optimization. Proceedings, 1990. VIII, 348 pages. 1992.

Vol. 379: K. Marti (Ed.), Stochastic Optimization. Proceedings, 1990. VII, 182 pages. 1992.

Vol. 380: J. Odelstad, Invariance and Structural Dependence. XII, 245 pages. 1992.

Vol. 381: C. Giannini, Topics in Structural VAR Econometrics. XI, 131 pages. 1992.

Vol. 382: W. Oettli, D. Pallaschke (Eds.), Advances in Optimization. Proceedings, 1991. X, 527 pages. 1992.

Vol. 383: J. Vartiainen, Capital Accumulation in a Corporatist Economy. VII, 177 pages. 1992.

Vol. 384: A. Martina, Lectures on the Economic Theory of Taxation. XII, 313 pages. 1992.

Vol. 385: J. Gardeazabal, M. Regúlez, The Monetary Model of Exchange Rates and Cointegration. X, 194 pages. 1992.

Vol. 386: M. Desrochers, J.-M. Rousseau (Eds.), Computer-Aided Transit Scheduling. Proceedings, 1990. XIII, 432 pages. 1992.

Vol. 387: W. Gaertner, M. Klemisch-Ahlert, Social Choice and Bargaining Perspectives on Distributive Justice. VIII, 131 pages. 1992.

Vol. 388: D. Bartmann, M. J. Beckmann, Inventory Control. XV, 252 pages. 1992.

Vol. 389: B. Dutta, D. Mookherjee, T. Parthasarathy, T. Raghavan, D. Ray, S. Tijs (Eds.), Game Theory and Economic Applications. Proceedings, 1990. IX, 454 pages. 1992.

Vol. 390: G. Sorger, Minimum Impatience Theorem for Recursive Economic Models. X, 162 pages. 1992.

Vol. 391: C. Keser, Experimental Duopoly Markets with Demand Inertia. X, 150 pages. 1992.

Vol. 392: K. Frauendorfer, Stochastic Two-Stage Programming. VIII, 228 pages. 1992.

Vol. 393: B. Lucke, Price Stabilization on World Agricultural Markets. XI, 274 pages. 1992.

Vol. 394: Y.-J. Lai, C.-L. Hwang, Fuzzy Mathematical Programming. XIII, 301 pages. 1992.

Vol. 395: G. Haag, U. Mueller, K. G. Troitzsch (Eds.), Economic Evolution and Demographic Change. XVI, 409 pages. 1992.

Vol. 396: R. V. V. Vidal (Ed.), Applied Simulated Annealing. VIII, 358 pages. 1992.

Vol. 397: J. Wessels, A. P. Wierzbicki (Eds.), User-Oriented Methodology and Techniques of Decision Analysis and Support. Proceedings, 1991. XII, 295 pages. 1993.

Vol. 398: J.-P. Urbain, Exogeneity in Error Correction Models. XI, 189 pages. 1993.

Vol. 399: F. Gori, L. Geronazzo, M. Galeotti (Eds.), Nonlinear Dynamics in Economics and Social Sciences. Proceedings, 1991. VIII, 367 pages. 1993.

Vol. 400: H. Tanizaki, Nonlinear Filters. XII, 203 pages. 1993.

Vol. 401: K. Mosler, M. Scarsini, Stochastic Orders and Applications. V, 379 pages. 1993.

Vol. 402: A. van den Elzen, Adjustment Processes for Exchange Economies and Noncooperative Games. VII, 146 pages. 1993.

Vol. 403: G. Brennscheidt, Predictive Behavior. VI, 227 pages. 1993.

Vol. 404: Y.-J. Lai, Ch.-L. Hwang, Fuzzy Multiple Objective Decision Making. XIV, 475 pages. 1994.

Vol. 405: S. Komlósi, T. Rapcsák, S. Schaible (Eds.), Generalized Convexity. Proceedings, 1992. VIII, 404 pages. 1994.

Vol. 406: N. M. Hung, N. V. Quyen, Dynamic Timing Decisions Under Uncertainty. X, 194 pages. 1994.

Vol. 407: M. Ooms, Empirical Vector Autoregressive Modeling. XIII, 380 pages. 1994.

Vol. 408: K. Haase, Lotsizing and Scheduling for Production Planning. VIII, 118 pages. 1994.

Vol. 409: A. Sprecher, Resource-Constrained Project Scheduling. XII, 142 pages. 1994.

Vol. 410: R. Winkelmann, Count Data Models. XI, 213 pages. 1994.

Vol. 411: S. Dauzère-Péres, J.-B. Lasserre, An Integrated Approach in Production Planning and Scheduling. XVI, 137 pages. 1994.

Vol. 412: B. Kuon, Two-Person Bargaining Experiments with Incomplete Information. IX, 293 pages. 1994.

Vol. 413: R. Fiorito (Ed.), Inventory, Business Cycles and Monetary Transmission. VI, 287 pages. 1994.

Vol. 414: Y. Crama, A. Oerlemans, F. Spieksma, Production Planning in Automated Manufacturing. X, 210 pages. 1994.

Vol. 415: P. C. Nicola, Imperfect General Equilibrium. XI, 167 pages. 1994.

Vol. 416: H. S. J. Cesar, Control and Game Models of the Greenhouse Effect. XI, 225 pages. 1994.

Vol. 417: B. Ran, D. E. Boyce, Dynamic Urban Transportation Network Models. XV, 391 pages. 1994.

Vol. 418: P. Bogetoft, Non-Cooperative Planning Theory. XI, 309 pages. 1994.

Vol. 419: T. Maruyama, W. Takahashi (Eds.), Nonlinear and Convex Analysis in Economic Theory. VIII, 306 pages. 1995.

Vol. 420: M. Peeters, Time-To-Build. Interrelated Investment and Labour Demand Modelling. With Applications to Six OECD Countries. IX, 204 pages. 1995.

Vol. 421: C. Dang, Triangulations and Simplicial Methods. IX, 196 pages. 1995.

Vol. 422: D. S. Bridges, G. B. Mehta, Representations of Preference Orderings. X, 165 pages. 1995.

Vol. 423: K. Marti, P. Kall (Eds.), Stochastic Programming. Numerical Techniques and Engineering Applications. VIII, 351 pages. 1995.

Vol. 424: G. A. Heuer, U. Leopold-Wildburger, Silverman's Game. X, 283 pages. 1995.

Vol. 425: J. Kohlas, P.-A. Monney, A Mathematical Theory of Hints. XIII, 419 pages, 1995.

Vol. 426: B. Finkenstädt, Nonlinear Dynamics in Economics. IX, 156 pages. 1995.

Vol. 427: F. W. van Tongeren, Microsimulation Modelling of the Corporate Firm. XVII, 275 pages. 1995.

Vol. 428: A. A. Powell, Ch. W. Murphy, Inside a Modern Macroeconometric Model. XVIII, 424 pages. 1995.

Vol. 429: R. Durier, C. Michelot, Recent Developments in Optimization. VIII, 356 pages. 1995.

Vol. 430: J. R. Daduna, I. Branco, J. M. Pinto Paixão (Eds.), Computer-Aided Transit Scheduling. XIV, 374 pages. 1995.

Vol. 431: A. Aulin, Causal and Stochastic Elements in Business Cycles. XI, 116 pages. 1996.

Vol. 432: M. Tamiz (Ed.), Multi-Objective Programming and Goal Programming. VI, 359 pages. 1996.

Vol. 433: J. Menon, Exchange Rates and Prices. XIV, 313 pages. 1996.

Vol. 434: M. W. J. Blok, Dynamic Models of the Firm. VII, 193 pages. 1996.

Vol. 435: L. Chen, Interest Rate Dynamics, Derivatives Pricing, and Risk Management. XII, 149 pages. 1996.

Vol. 436: M. Klemisch-Ahlert, Bargaining in Economic and Ethical Environments. IX, 155 pages. 1996.

Vol. 437: C. Jordan, Batching and Scheduling. IX, 178 pages. 1996.

Vol. 438: A. Villar, General Equilibrium with Increasing Returns. XIII, 164 pages. 1996.

Vol. 439: M. Zenner, Learning to Become Rational. VII, 201 pages. 1996.

Vol. 440: W. Ryll, Litigation and Settlement in a Game with Incomplete Information. VIII, 174 pages. 1996.

Vol. 441: H. Dawid, Adaptive Learning by Genetic Algorithms. IX, 166 pages.1996.

Vol. 442: L. Corchón, Theories of Imperfectly Competitive Markets. XIII, 163 pages. 1996.

Vol. 443: G. Lang, On Overlapping Generations Models with Productive Capital. X, 98 pages. 1996.

Vol. 444: S. Jørgensen, G. Zaccour (Eds.), Dynamic Competitive Analysis in Marketing. X, 285 pages. 1996.

Vol. 445: A. H. Christer, S. Osaki, L. C. Thomas (Eds.), Stochastic Modelling in Innovative Manufactoring. X, 361 pages. 1997.

Vol. 446: G. Dhaene, Encompassing. X, 160 pages. 1997.

Vol. 447: A. Artale, Rings in Auctions. X, 172 pages. 1997.

Vol. 448: G. Fandel, T. Gal (Eds.), Multiple Criteria Decision Making. XII, 678 pages. 1997.

Vol. 449: F. Fang, M. Sanglier (Eds.), Complexity and Self-Organization in Social and Economic Systems. IX, 317 pages, 1997.

Vol. 450: P. M. Pardalos, D. W. Hearn, W. W. Hager, (Eds.), Network Optimization. VIII, 485 pages, 1997.

Vol. 451: M. Salge, Rational Bubbles. Theoretical Basis, Economic Relevance, and Empirical Evidence with a Special Emphasis on the German Stock Market.IX, 265 pages. 1997.

Vol. 452: P. Gritzmann, R. Horst, E. Sachs, R. Tichatschke (Eds.), Recent Advances in Optimization. VIII, 379 pages. 1997.

Vol. 453: A. S. Tangian, J. Gruber (Eds.), Constructing Scalar-Valued Objective Functions. VIII, 298 pages. 1997.

Vol. 454: H.-M. Krolzig, Markov-Switching Vector Autoregressions. XIV, 358 pages. 1997.

Vol. 455: R. Caballero, F. Ruiz, R. E. Steuer (Eds.), Advances in Multiple Objective and Goal Programming. VIII, 391 pages. 1997.

Vol. 456: R. Conte, R. Hegselmann, P. Terna (Eds.), Simulating Social Phenomena. VIII, 536 pages. 1997.

Vol. 457: C. Hsu, Volume and the Nonlinear Dynamics of Stock Returns. VIII, 133 pages. 1998.

Vol. 458: K. Marti, P. Kall (Eds.), Stochastic Programming Methods and Technical Applications. X, 437 pages. 1998.

Vol. 459: H. K. Ryu, D. J. Slottje, Measuring Trends in U.S. Income Inequality. XI, 195 pages. 1998.

Vol. 460: B. Fleischmann, J. A. E. E. van Nunen, M. G. Speranza, P. Stähly, Advances in Distribution Logistic. XI, 535 pages. 1998.

Vol. 461: U. Schmidt, Axiomatic Utility Theory under Risk. XV, 201 pages. 1998.

Vol. 462: L. von Auer, Dynamic Preferences, Choise Mechanisms, and Welfare. XII, 226 pages. 1998.

Vol. 463: G. Abraham-Frois (Ed.), Non-Linear Dynamics and Endogenous Cycles. VI, 204 pages. 1998.

Vol. 464: A. Aulin, The Impact of Science on Economic Growth and its Cycles. IX, 204 pages. 1998.

Vol. 465: T. J. Stewart, R. C. van den Honert (Eds.), Trends in Multicriteria Decision Making. X, 448 pages. 1998.

Vol. 466: A. Sadrieh, The Alternating Double Auction Market. VII, 350 pages. 1998.